Praise for this book

"Scrum is described as taking 10 minutes to learn and a lifetime to master. In this book, Richard provides tips and tricks to mastering Scrum. He marries the practical with the abstract, providing a foundation of learning that helps Developers deliver high-value products and solve complex problems. If you are using Azure DevOps and want to get better at doing it, then this is the book for you."

—Dave West, Scrum.org Product Owner and CEO

"Like it or not, many teams need tooling to help them with their Scrum implementation. That's where Richard comes in. His knowledge and passion shine through in all that he touches—especially in this essential guide for how to use Azure DevOps for Scrum Teams. If you know anything about Richard, and you are using Azure DevOps with Scrum, then you'll know this book is a must-read."

—Daniel Vacanti, Co-founder, ActionableAgile

"In this book, Richard Hundhausen does a great job explaining and connecting the domains of Professional Scrum with professional development using Microsoft Azure DevOps. Richard introduces the history and current state in both domains and makes the book even richer with personal tips and illustrations through case studies."

—Gunther Verheyen, independent Scrum Caretaker,
Professional Scrum Trainer

"Scrum is a framework that is easy to understand but difficult to master. Richard takes the difficult out of the equation for you. What sets him apart from all others is his ability to help others not only understand Scrum, but become masters at it."

—Chris Roan, Wells Fargo Agile Transformation Leader

"If you're working on a Scrum team, do yourself a favor and read this book. In it, Richard distills his many years of practical experience leading Scrum teams in order to help you and your team accelerate your DevOps transformation. If you want to deliver more customer value at higher velocity, there's no better place to start."

—Jeff Beehler, Senior Director, Product Operations, GitHub, Inc.

"During my time on the Azure DevOps team, I became aware of Rich   Professional Scrum and his desire for us to build the tool in a way that   would love it. The essence of DevOps is to get a right blend of proces   people working seamlessly to deliver customer value. Combine that

D1211164

you have a winner. Richard does a great job of taking the theory of Scrum and converting it into specific sets of actions that everyone in a team (Product Owners, Developers, Testers, stakeholders, etc.) can follow. If you want to be an expert at Scrum while putting it into day-to-day practice using Azure DevOps, this is the book for you!"

—*Ravi Shanker, Principal Group Program Manager and former Product Owner for Azure Test Plans*

"Richard successfully weaves three important concepts: Azure Devops, Scrum, and creating quality code. This book is a must-read for anyone interested in end-to-end solutioning within the Microsoft development environment."

—*Donis Marshall, Microsoft MVP, Professional Scrum Developer, President of Innovation in Software*

"Richard has been at the forefront of agile and Scrum since the beginning and was the first ALM/DevOps MVP. The book shows his vast knowledge and understanding of Professional Scrum and Azure DevOps. It's a must-have for teams to continue on their improvement journey."

—*Philip Japikse, CTO Pintas & Mullins, Microsoft MVP, Professional Scrum Trainer*

"Scrum is simple—or it seems that way until you actually try to implement it. The great thing about Richard's book is that it gives readers practical implementation advice to translate the simple words in the Scrum Guide into valuable actions by their teams."

—*Steve Porter, Scrum.org Professional Series Manager, Professional Scrum Trainer*

"Azure DevOps is a suite of tools and Scrum is a framework used to deliver a product in an iterative and incremental way. Both have a lot in common but are totally different beasts. Richard blends them together in a surprisingly delightful and easy-to-digest way that clearly explains how and where to apply both to help teams deliver better and more valuable software together."

—*Jesse Houwing, Lead Consultant at Xpirit, Professional Scrum Trainer, Microsoft MVP*

"If you are working with Azure DevOps, then this book should be required reading for everyone who touches that system. With 80 percent of all development teams using Scrum, this is the book to help you do Professional Scrum within Azure DevOps and improve the likelihood of success of your team."

—Martin Hinshelwod, naked Agility, Professional Scrum Trainer, Azure DevOps MVP

"Scrum is simple to understand but extremely hard to implement well. In his book Richard offers battle-hardened experience and practices with which mastering Scrum becomes achievable using Azure DevOps."

—Ognjen Bajic, Professional Scrum Trainer, Azure DevOps MVP

"Richard draws on his expertise as a Professional Scrum Trainer and DevOps guru to write a fantastic book that describes the "why" and the "how" of doing Scrum with Azure DevOps. Clear and well written, this should be on your bookshelf if you are using Azure DevOps with Scrum."

—Simon Reindl, Professional Scrum Trainer

"Just by following Richard's actionable recommendations, even first-time users will get a properly configured and productive implementation of Professional Scrum with Azure DevOps. Experienced practitioners will be able to confront their ways of work and experiences with real-world advice from the book."

—Ana Roje Ivancic, Professional Scrum Trainer, Azure DevOps MVP

"I use Azure DevOps every day. I did not realize how much I had to learn until I read *Professional Scrum Development with Azure DevOps*. It is filled with expert guidance for maximizing value for your team!"

—Cory Isakson, Microsoft Senior Consultant

"If you have to read one book about Scrum, make it the one you are holding right now. This book teaches you everything you need to know about Professional Scrum Development—clear and concise, without the fluff."

—Martin Kulov, Microsoft DevOps MVP, Microsoft Regional Director

Professional Scrum Development with Azure DevOps

Richard Hundhausen

PROFESSIONAL SCRUM DEVELOPMENT WITH AZURE DEVOPS

Published with the authorization of Microsoft Corporation by:

Pearson Education, Inc.

Hoboken, New Jersey

ISBN-13: 978-0-13-678923-9

ISBN-10: 0-13-678923-4

Library of Congress Control Number: 2021930493

1 2021

TRADEMARKS

Microsoft and the trademarks listed at http://www.microsoft.com on the "Trademarks" webpage are trademarks of the Microsoft group of companies. "Planning Poker" in Chapter 5 is a registered trademark of Mountain Goat Software, LLC. Figure 5-21 provided by SpecFlow (https://specflow.org) the #1 BDD framework for AzureDevops. All other marks are property of their respective owners.

WARNING AND DISCLAIMER

SPECIAL SALES

For information about buying this title in bulk quantities, or for special sales opportunities (which may include electronic versions; custom cover designs; and content particular to your business, training goals, marketing focus, or branding interests), please contact our corporate sales department at corpsales@pearsoned.com or (800) 382-3419.

For government sales inquiries, please contact governmentsales@pearsoned.com.

For questions about sales outside the U.S., please contact intlcs@pearson.com.

CREDITS

EDITOR-IN-CHIEF
Brett Bartow

EXECUTIVE EDITOR
Loretta Yates

SPONSORING EDITOR
Charvi Arora

DEVELOPMENT EDITOR
Songlin Qiu

MANAGING EDITOR
Sandra Schroeder

SENIOR PROJECT EDITOR
Tracey Croom

COPY EDITOR
Liz Welch

INDEXER
Tim Wright

PROOFREADER
Donna Mulder

TECHNICAL EDITOR
Donis Marshall

EDITORIAL ASSISTANT
Cindy Teeters

COVER DESIGNER
Twist Creative, Seattle

COMPOSITOR
codeMantra

This book is dedicated to my Scrum Team:
Esmay, Isla, Berlin, Blaize, Sawyer, and Kristen.
—Richard Hundhausen

Contents at a Glance

Contents

PART I SCRUMDAMENTALS

Chapter 1 **Professional Scrum** **3**

Chapter 2 **Azure DevOps** **49**

PART III   IMPROVING

About the Author

RICHARD HUNDHAUSEN is the president of Accentient, a company that helps software organizations and teams deliver better products by understanding and leveraging Azure DevOps and Scrum. He is a Professional Scrum Trainer and co-creator of the Nexus Scaled Scrum framework.

As a software developer, consultant, and trainer with nearly 40 years of experience, he understands that software is built and delivered by people and not by processes or tools. You can reach Richard at *richard@accentient.com*.

Foreword

By 2001, the software industry was in trouble—more projects were failing than succeeding. Customers began demanding contracts with penalties and sending work offshore. Some software developers, though, had increasing success with a development process known as "lightweight." Almost uniformly, these processes were based on the well-known iterative, incremental process.

In February 2001, these developers issued a manifesto—the Agile Manifesto. The Manifesto called for Agile software development based on four principle values and twelve underlying principles. Two of the principles were 1) to satisfy customers through early and continuous delivery of working software, and 2) to deliver working software frequently, from a couple of weeks to a couple of months, with a preference for the shorter timescale.

By 2009, the Scrum Agile process was used predominantly. A simple framework, it provided an easily adopted iterative incremental framework for software development. It also incorporated the Agile Manifesto's values and principles. The two authors of Scrum, Jeff Sutherland and myself, also were among the authors of the Agile Manifesto.

I had anticipated some of the difficulties organizations (and even teams) would face when they adopted Scrum. However, I believed that developers would bloom in a Scrum environment. Stifled and choked by waterfall, developers would stand tall, employing development practices, collaboration, and tooling that nobody had time to use in waterfall projects.

Much to my surprise, this was only true for perhaps 20 percent of all software developers.

In 2009, Martin Fowler characterized most Agile software development as "flaccid":

> There's a mess I've heard about with quite a few projects recently.
> It works out like this:

- They want to use an Agile process, and pick Scrum.

- They adopt the Scrum practices, and maybe even the principles.

- After a while, progress is slow because the codebase is a mess.

> What's happened is that they haven't paid enough attention to the internal quality of their software. If you make that mistake you'll soon find your productivity dragged down because it's much harder to add new features than you'd like. You've taken on a crippling Technical Debt

and your Scrum has gone weak at the knees. (And if you've been in a real scrum, you'll know that's a Bad Thing.)" http://martinfowler.com/bliki/FlaccidScrum.html

Martin's description of flaccid Scrum resonated with our experience. Most developers were skilled, but not adequately skilled in the three dimensions required to rapidly build complete increments of usable functionality. These dimensions are:

- **People**   The ability to work in a small, cross-functional, self-managing team.

- **Practices**   The knowledge of and ability to apply modern engineering practices that short cycle development mandates.

- **Tooling**   Tools that integrate and automate these practices so that successive increments can be rapidly integrated without the drag of exponentially accruing artifacts that must be handled manually.

We put our business on hold while we worked through 2009 to create what has become known as the Professional Scrum Developer program. Offered in a three-day format, we formulated a workshop. The input was developers whose knowledge and capabilities produced flaccid increments. The output were teams of developers who had developed solid increments of software called for by the Agile Manifesto and demanded by the modern, competitive organization.

Richard has been there since the beginning. His book, *Professional Scrum Development with Azure DevOps*, continues his participation in the movement started by us few in 2009.

When you read Richard's book, you can learn the three dimensions needed for Agile software development: people, practices, and tools. Just like in the course, Richard intertwines them into something you can absorb. If you are on a Scrum team, read Richard's book. List the called-for practices. Identify which practices pose challenges to your team. Order them by their greatest impact. Then remediate them, one by one.

Many people spend money going to Agile conferences. Save the money and more by buying this book, discussing it with others, and going to meetups and code camps—the "un-conferences" for the serious.

Richard and I look forward to your increased skill. Our industry and our society need it. Software is the last great scalable resource needed by our increasingly complex society. The effective, productive teamwork of Agile teams is the basis of problem solving that our society also needs.

Scrum on!

Ken Schwaber

Co-creator of Scrum

Introduction

Scrum is a framework for developing and sustaining complex products such as software. Scrum is just a set of rules, as defined in the *Scrum Guide* (*https://scrumguides.org*), and it describes the roles, events, and artifacts, as well as the rules that bind them together. When used correctly, this framework enables a team to address complex problems while productively and creatively delivering products of the highest possible value. Scrum is an Agile method. In fact, it is the most popular Agile method in use today.

Scrum employs an iterative and incremental approach to optimizing predictability and controlling risk. This is due to the empirical process control nature of Scrum. Through proper use of inspection, adaptation, and transparency, a Scrum Team can try a new way of doing something (an experiment) and gauge its usefulness after a short iteration. They can then collectively decide to embrace, extend, or drop the practice. This includes the tools a team uses and how they use them.

Combining Scrum with the tools found in Microsoft Azure DevOps is a powerful marriage. It is the purpose of this book to establish a baseline understanding of Scrum and how Scrum is supported in Azure DevOps. I will also illustrate which practices provide more value when executed without the use of tools. In addition, I will point out the tools that have been erroneously marketed as agile and contrast them with more preferred practices.

In software development, anything and everything can change in a moment's notice. Healthy teams know this. They also know that continuously inspecting and adapting the way things are done is a way of life. High-performance Scrum Teams take this a step further. They know that within every impediment or dysfunction is an opportunity to learn and improve. Reading this book is a great first step.

Who Should Read This Book

This book will be of value to any member of a software development team that uses Scrum or is considering using Scrum. I primarily focus on the responsibilities and tasks of the developer (which in Scrum includes designers, architects, coders, testers, technical writers, etc.). Product Owners and Scrum Masters will also derive value from this book, as they will be using many of the same Azure DevOps tools to plan and manage their work and assess progress. Stakeholders, including customers, users, sponsors, and managers, will also gain value from this book, especially when they learn what they should and should not do according to the rules of Scrum and which tools in Azure DevOps support these rules.

This book primarily focuses on using Scrum for software products, mostly because that's the target domain for Azure DevOps. Much of this book, however, is applicable beyond software development and IT projects. Since Scrum is a lightweight framework for developing adaptive solutions for all types of complex problems, the guidance in this book can apply to developing any kind of product, such as a service, a physical product, or something more abstract.

This Book Might Not Be for You If . . .

This book is intended for teams using Scrum and Azure DevOps together as they develop complex products, such as software. It won't provide as much value for non-Scrum teams or Scrum teams developing products that are not complex. It won't provide any value for teams running formal waterfall or sequential software development projects, except to hopefully change the minds of such proponents. Likewise, if a team is using Scrum but not yet using Azure DevOps, the bulk of the book won't be very interesting, except to define and highlight Professional Scrum and point out what goodness those teams might be missing out on. This is also the case for teams using older versions of Team Foundation Server, which won't contain the latest, high-value, team-based tools for planning and managing work and enabling team collaboration.

If you are looking for "best practices," then you have the wrong book and the wrong author. I refuse to use that term because it implies a couple of wrong assumptions: (1) that this practice truly is "best" for all teams working on all products in all organizations, and (2) that a team can stop looking and experimenting once they've found that best practice. I prefer the term "proven practice" instead. Regardless of what you or I call it, this book is full of many practices for you and your team to consider on its improvement journey.

Organization of This Book

This book is divided into three sections, each of which focuses on a different aspect of the marriage of Professional Scrum and Azure DevOps. Part I, "Scrumdamentals," sets a baseline understanding of the Scrum framework, Professional Scrum, Azure DevOps, and specifically the Azure Boards service. Part II, "Practicing Professional Scrum," consists of several chapters detailing the practical application of how a Professional Scrum Team would use the relevant features of Azure DevOps to create and manage a Product Backlog, plan a Sprint, create a Sprint Backlog, and effectively collaborate during the Sprint. Part III, "Improving," includes a chapter on defining and improving a Scrum Team's flow, identifying common challenges and dysfunctions in order to remove them, and using techniques to continually improve your game of Scrum. There is also a chapter on how to improve at scale by adopting Scaled Professional Scrum using the Nexus scaled Scrum framework. By reading all sections sequentially, you will see how Azure DevOps and Scrum can be used together in an effective way and how a Scrum Team can evolve into a Professional Scrum Team and, further, into a high-performance Professional Scrum Team.

Throughout each chapter, I suggest and recommend many practices and patterns of working. I use terms like *Professional* Scrum Team and *high-performance* Scrum Team to differentiate from garden-variety Scrum Teams—those practicing mechanical Scrum without attention to inspection, adaptation, and improving. At times you may dismiss my guidance as "magical thinking" and assume that I don't live in the real world. You may think that the ideas I propose won't work for your team, with your people, in your organization. Although it's true that I don't know the specifics of your organization, I'm confident that improvement can be made regardless of the amount of friction you might face. I've seen it and hundreds of my Professional Scrum Trainer colleagues have, too. Keep in mind that my descriptions of these high-performance behaviors should be considered a vision or "perfection goal" of what your team can achieve. It will be hard. It will take time. It will take help. Ultimately it will be people like you who lead the improvement journey.

Finding Your Best Starting Point in This Book

The different sections of *Professional Scrum Development with Azure DevOps* cover a range of topics. Depending on your needs and your existing understanding of Scrum, Azure DevOps, and the related practices, you may wish to focus on specific areas of the book. Use the following table to determine how best to proceed through the book.

If you are	Follow these steps
New to Scrum or have never heard of it	Read the *Scrum Guide* and then read Chapter 1
New to Professional Scrum or have never heard of it	Read Chapter 1
New to Azure DevOps or its suite of tools	Read Chapter 2
New to the Azure Boards service or want to know how to create a custom, Professional Scrum process	Read Chapter 3
Familiar with Scrum and Azure DevOps and only want to learn how to set up Azure DevOps for a Scrum Team	Read Chapter 4

If you are	Follow these steps
Familiar with Scrum and Azure DevOps and only want to learn how to plan a Sprint and create a Sprint Backlog	Read Chapter 6
New to the concept of acceptance test-driven development and how to plan and track a Sprint using Azure Test Plans	Read Chapter 7
New to the concept of flow or how a Scrum Team can use the Kanban board to visualize work and manage its flow	Read Chapter 9
Facing common Scrum challenges and are interested in overcoming dysfunctional behavior	Read Chapter 10
Facing a scaling situation where several Scrum Teams are collaborating to build a common product	Read Chapter 11

Conventions and Features in This Book

This book presents information using conventions designed to make the information readable and easy to follow.

- Screenshots from relevant Azure DevOps features are provided for your reference.

- Boxed elements with labels such as "Note" or "Tip" provide additional information and guidance related to the subject.

- Some notes and tips are practical guidance provided by fellow Professional Scrum Developers and Professional Scrum Trainers who have helped review this book.

In addition, I have included two additional boxed elements, one labeled "Smells" and the other labeled "Fabrikam Fiber Case Study."

Smell Throughout this book, I point out specific situations and traps that a Scrum Team or its members should avoid. I refer to these as *smells*. These smells typically—but not always—indicate an underlying dysfunction or other unhealthy behavior. For teams new to Scrum, these smells may be hard to identify. Once they are brought to light, however, they should be mitigated and used as learning opportunities. As a team improves, it should be able to recognize dysfunction on its own, as well as remove it. Professional Scrum Teams have the ability to identify potential waste or dysfunction, evaluate the risks, and even decide to opt in to specific behaviors, including those that may be a smell to the uneducated.

Fabrikam Fiber Case Study

As you flip through the pages, you will read about Fabrikam Fiber as our case study. Fabrikam Fiber is a fictional broadband communications provider (think: Cox, Sparklight, Charter/Spectrum, Comcast, etc.). Fabrikam Fiber is a large corporation that provides services for multiple U.S. states. They also use an on-premises web application for their customer service representatives to create and manage tickets for customer support issues. The team has been using Scrum for some time and has recently moved to Azure DevOps. My opinions on healthy and unhealthy behaviors are made evident through the choices made by the Fabrikam Fiber Scrum Team.

System Requirements

Although this book does not contain any hands-on exercises, I encourage you to sign up for Azure DevOps Services in order to experiment and learn as you read. It takes only a few minutes to create an organization, and the first five users are free on the Basic Plan—which is more than adequate for you and some colleagues to use all the features mentioned in this book. Azure DevOps Services is a cloud-based SaaS offering delivering new features every three weeks, which means that the screenshots in this book may not match what you see in your browser.

In addition, you may want to download Visual Studio Community Edition or Visual Studio Code to explore how they connect to Azure DevOps and how they can be used for collaboration using Azure Boards and Azure Repos. Both of these products are free.

Downloads: Code Samples

This book contains no code samples.

Acknowledgments

There are several people who helped me write this book. Thanks to: Loretta Yates for giving me another opportunity to write for Microsoft Press; Charvi Arora, Tracey Croom, Elizabeth Welch, Songlin Qiu, Vaishnavi Venkatesan and Donna Mulder for patiently reviewing my content and helping me get the styles right; Donis Marshall for inspiring me to write another book and giving me such direct (and valuable) feedback; Dan Hellem for answering scores of Azure Boards questions and reviewing chapters; Phil Japikse, Simon Reindl, Brian Randell, Ognjen Bajić, Ana Roje Ivančić, Martin Kulov, Cory Isakson, David Corbin, Charles Revell, Daniel Vacanti, and Christian Hassa for providing some great ideas and helping me sharpen the message; and Ken Schwaber and Jeff Sutherland for updating the *Scrum Guide* after I was almost done writing this book. ☺

Errata, Updates, and Book Support

We've made every effort to ensure the accuracy of this book and its companion content. You can access updates to this book—in the form of a list of submitted errata and their related corrections—at:

MicrosoftPressStore.com/ProfScrumDevelopment/errata

If you discover an error that is not already listed, please submit it to us at the same page.

For additional book support and information, please visit *www.MicrosoftPressStore.com/Support*.

Please note that product support for Microsoft software and hardware is not offered through the previous addresses. For help with Microsoft software or hardware, go to *http://support.microsoft.com*.

Stay in Touch

Let's keep the conversation going! We're on Twitter: *http://twitter.com/MicrosoftPress*

Scrumdamentals

The chapters in this section establish a baseline understanding of the three areas that every Professional Scrum practitioner using Microsoft's DevOps tools must know:

- Scrum, and more specifically, Professional Scrum

- Microsoft Azure DevOps (broadly)

- Microsoft Azure Boards (specifically)

I begin by looking at Scrum and the rules of Scrum. The focus is on how and when a Developer interacts with the Product Owner and Scrum Master, participates in the various Scrum events, and interacts with the various Scrum artifacts. It's important for all Developers to understand the rules of Scrum and what's expected of them and their team, as well as when and how they should interact with the Product Owner, Scrum Master, stakeholders, and various artifacts.

Note In the 2020 *Scrum Guide*, the *Development Team* role was replaced with the Developer role. The goal was to eliminate the concept of a separate team within a team that led to "proxy" or "us and them" behavior between the Product Owner and Development Team. There is now just one team—the Scrum Team—and it is focused on the same objective, with three different sets of accountabilities: Product Owner, Scrum Master, and Developers. Remember, Scrum recognizes a tester, coder, designer, architect, analyst, database professional, technical writer as simply . . . a Developer.

The remaining chapters in this section are more technical in nature and cover the tools found in Azure DevOps, specifically Azure Boards. I focus on the cloud-hosted Azure DevOps *Services*, rather than the on-premises Azure DevOps *Server*. Although many DevOps tools are available to a Scrum Team, I endeavor to list and discuss only those relevant to practicing Professional Scrum. I also point out which shiny tools are better left in the toolbox, allowing the team to exercise higher-valued collaborative practices instead. After all, we value individuals and interactions over process and tools, right?

CHAPTER 1

Professional Scrum

Scrum is a lightweight framework that helps people, teams, and organizations generate value through adaptive solutions for complex problems. Software is a complex problem. Therefore, Scrum is ideal for managing the development of software in order to find solutions in an adaptive way. Software development doesn't generate the same output every time, given a certain input. Scrum embraces this fact, and because of its empirical nature, it promotes the use of experimentation in order to inspect and adapt.

Scrum is not a methodology or a process. It is only a framework. In other words, if you take Scrum and add your own complementary practices, such as acceptance test-driven development, what you will end up with is *your* process. Teams can leverage Scrum's empirical attributes to regularly see the effectiveness of its practices and make changes accordingly. I will introduce you to many complementary practices in this book for your consideration.

> **Note** Scrum is founded on empirical process control theory and lean thinking. Empiricism asserts that knowledge comes from experience and making decisions based on what is known. Lean thinking leads to reduction in waste while focusing on the essentials.

Even today, more than 60 years into the evolution of software development, the chances are still good that a medium-sized to large-sized software project will fail. Fortunately, our industry has finally noticed, understands, and has started to respond to this problem. Some organizations have improved their odds. Evidence shows that agile practices, such as Scrum, are leading these successes.

> **Tip** Using a software development analogy, you can think of agile as being an *interface*. Agile defines four abstract values and 12 abstract principles (*http://agilemanifesto.org*). Although there are many ways to implement these values and principles, the Agile Manifesto does not describe them. Scrum does. You can think of Scrum as a *concrete class* that *implements* the agile values and principles through its roles, events, artifacts, and rules.

Agile teams know that they must continuously inspect and adapt—not just their product, but their process and practices as well. Being book-smart on Scrum, DevOps, and Microsoft's tools is a good start. Having experience using them together in practice is better. Being able to identify and act on opportunities for improvement as you use them is awesome! It's being a professional. That should be your goal. Don't just settle for a project that doesn't fail. Strive for completing the project better, with more value, and with more learning than the spectators thought possible.

The Scrum Guide

Scrum has been around since the early 1990s. During that time, Scrum's definition and related practices have come from books, presentations, and professionals doing their best to explain it. Unfortunately, those messages were not always accurate and almost never consistent. Even the two creators, Ken Schwaber and Jeff Sutherland, were inconsistent at times. Scrum, as it has emerged today, doesn't look like it did 25 years ago.

In 2009, Scrum.org codified Scrum by creating and publishing the *Scrum Guide*. This free guide represents the official rules of Scrum and is maintained by Scrum's creators, Schwaber and Sutherland. It is very concise. In fact, the 2020 version PDF is only 14 pages. It is available in 30 languages and downloadable at *https://scrumguides.org*.

The *Scrum Guide* is a great reference that you can use even as you are reading this book. As you read the guide, you will see that Scrum is lightweight and quite easy to understand. Unfortunately, it is extremely difficult to master. The *Scrum Guide* will continue to be updated and may supersede the guidance you read in this chapter and the rest of the book.

> **Tip** You can think of Scrum as being like the game of chess. Both have rules. For example, Scrum doesn't allow two Product Owners just as chess doesn't allow a player to have two kings. When you play chess, it is expected that you play by the rules. If you don't, then you're not playing chess. This is the same with Scrum. Another way to think about it is that both Scrum and chess do not fail or succeed. Only the players fail or succeed. Those who keep playing by the rules will eventually improve, though it will take time to master the game.

The Scrum framework consists of the Scrum Team and the associated roles, events, artifacts, and rules. Each of these elements serves a specific purpose, as you will see in this chapter. The rules of Scrum, as defined in the *Scrum Guide*, bind together these roles, events, and artifacts. Following these rules is essential to the success of a team's ability to use Scrum and, more importantly, to the successful development and delivery of a high-value, high-quality product. Changing the core design or ideas of Scrum, leaving out elements, or not following its rules covers up problems and limits the benefits Scrum provides, potentially even rendering it useless.

Scrum is free and offered in the *Scrum Guide*. Scrum's roles, events, artifacts, and rules are immutable, and although implementing only parts of Scrum is possible, the result is not Scrum.

Scrum exists only in its entirety and functions well as a container for other practices, techniques, and methodologies. I often describe Scrum as "a framework within which a team can experiment with various complementary practices."

The Pillars of Scrum

Scrum is founded on empiricism and lean thinking. Empiricism asserts that knowledge comes from experience and making decisions based on what is observed and known. Lean thinking reduces waste and focuses on the essentials. Scrum employs an iterative, incremental approach to optimize predictability and to control risk. Scrum engages groups of people who collectively have all the skills and expertise to do the work and share or acquire such skills as needed.

Scrum combines four formal events for inspection and adaptation within a containing event—the Sprint. These events work because they implement the empirical Scrum pillars of inspection, adaptation, and transparency. I will cover events later in this chapter, but I want to spend a moment defining these three pillars of Scrum:

- **Inspection**   The Scrum artifacts and the progress toward agreed goals must be inspected frequently and diligently to detect potentially undesirable variances or problems. To help with inspection, Scrum provides cadence in the form of its events. Inspection enables adaptation. Inspection without adaptation is considered pointless. Scrum's events are designed to provoke change and improvement.

- **Adaptation**   If any aspects of a process deviate outside acceptable limits or if the resulting product is unacceptable, the process being applied or the item being produced must be adjusted. The adjustment must be made as soon as possible to minimize further deviation. Adaptation becomes more difficult when the people involved are not empowered or self-managing. A Scrum Team is expected to adapt the moment it learns anything new through inspection.

- **Transparency**   The emergent process and work must be visible to those performing the work as well as those receiving the work. With Scrum, important decisions are based on the perceived state of its three formal artifacts. Artifacts that have low transparency can lead to decisions that diminish value and increase risk. Transparency enables inspection. Inspection and adaptation without transparency is misleading and wasteful.

Scrum in Action

If you study the *Scrum Guide*, you can understand the components and related rules, but you won't necessarily see how they flow together. Doing so requires you to actually experience Scrum while developing a product on a team. As a substitute for that experience, Figure 1-1 was created by Scrum. org to help illustrate the Scrum framework in action.

In Scrum, the product is the vehicle that delivers value. The product could be a service, a physical product, or something more abstract. The product has a clear boundary, known stakeholders, and well-defined users or customers. Software fits this definition nicely, although Scrum can be used for so much more than software. That said, and because this is also a book about Azure DevOps, I will be describing Scrum and Professional Scrum in the context of *software* being the product.

The Product Goal describes a future state of the product which can serve as a target for the Scrum Team to plan against. The Product Goal is the long-term objective for the Scrum Team. They must fulfill (or abandon) one objective before taking on the next. An example of a Product Goal for software might be "to achieve 100k downloads from the app store."

The Product Backlog is an ordered list of everything that is known to be needed to achieve Product Goals. It is the single source for any changes to be made to the product and it is emergent – meaning that it will evolve continuously due to variation in business conditions, the domain, and technology. Each item in this list is a Product Backlog item (PBI). For software, the Product Backlog includes features to be implemented, bugs to be fixed, and experiments to be conducted. The Product Owner is accountable for managing the product's expectations, risks, and outcomes as well as ensuring that the Product Backlog is ordered (prioritized), made transparent (available), and understood.

The Developers collaborate with the Product Owner, and others as needed, during Sprint Planning and Product Backlog refinement to understand, estimate, and forecast PBIs. The Product Owner orders the Product Backlog according to one or more factors, such as ROI (value/size) of those PBIs, risk, business priority, dependency, and learning opportunity.

The Sprint is a fixed-length period of time that contains the other Scrum events. A Sprint should be one month or less in duration in order to lower risk and create consistency. A new Sprint starts immediately after the conclusion of the previous Sprint.

The first event within a Sprint is Sprint Planning. In this timeboxed event, the Scrum Team collaborates to forecast and plan the work of the Sprint. The PBIs toward the top of the Product Backlog that can best achieve the Product Goal will be considered. The Scrum Team collaborates to forecast those items that it believes it can complete by the end of the Sprint. A Sprint Goal is crafted, and the Sprint Backlog emerges. The Sprint Backlog contains those forecasted items plus a plan for delivering them. The Sprint Backlog shows the work remaining in the Sprint at all times.

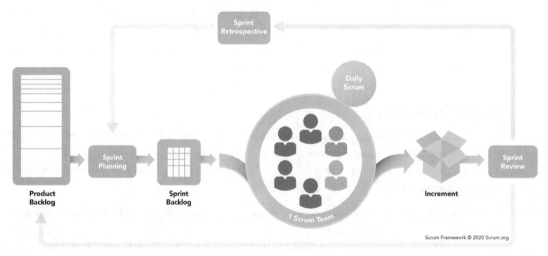

FIGURE 1-1 The Scrum Framework.

The bulk of the Sprint will be spent working to achieve the Sprint Goal through the development of the items in the Sprint Backlog. The rules of Scrum are fairly quiet on what occurs each day during development. The Developers must meet regularly for the Daily Scrum to synchronize on the plan for the next 24 hours.

If necessary, the Developers should also meet with the Product Owner to refine the Product Backlog. During refinement, items in the Product Backlog are given additional detail and estimates. This keeps the Product Backlog healthy so that the Product Owner can have more meaningful conversations with stakeholders, plan releases, and make better decisions on what to do next.

During the Sprint, the Scrum Team completes items in their Sprint Backlog according to the Definition of Done. This definition lists the practices and standards that must be met for every item before it can be considered complete. If a Definition of Done doesn't already exist, then it is created by the Scrum Team. It's important for everyone to understand the definition because it will be used to facilitate the inspection of progress and quality. Work that does not meet the Definition of Done is . . . not done, and cannot be released. It should also not be inspected at the Sprint Review.

Ideally, the Developers collaborate with the Product Owner throughout the Sprint to ensure that a great product emerges. If the Developers complete their forecasted work early, they should collaborate with the Product Owner to find additional PBIs to work on. Conversely, at the first inclination that the Developers suspect that they won't be able to complete their forecasted work, they should collaborate with the Product Owner to identify and discuss trade-offs and modify the Sprint Backlog in a way that does not sacrifice quality or alter the Sprint Goal.

An Increment is a useful and valuable body of inspectable work that meets the Definition of Done. An Increment is made up of one or more of the done PBIs that were forecasted for the Sprint. The Increment is inspected during Sprint Review. The Product Owner may invite various stakeholders to the Sprint Review for their feedback. This feedback is captured and might end up as new items in the Product Backlog. Existing PBIs may also need to be updated or removed. Stakeholder feedback may influence the Product Owner's decision on releasing, continuing development, or stopping it altogether. These latter decisions should be based on business reasons, not quality reasons. Regardless of when the Increment is released, the Scrum Team should always develop the Increment as though it were going to be released. This may include being released multiple times during the Sprint—which Scrum totally supports.

The last event in the Sprint is the Sprint Retrospective. This is an opportunity for the Scrum Team to inspect themselves and how they work in order to adapt their practices and process so that they can improve. If improvement experiments are identified, the team should create an actionable plan for the next Sprint. To ensure continuous improvement, the Scrum Team should identify at least one high-priority process improvement and enact it during the next Sprint. Nothing is out of scope during a Sprint Retrospective—people, relationships, process, practices, and tools can all be discussed. The Scrum Team may also decide to adjust its Definition of Done in order to increase quality. After the Sprint Retrospective, the next Sprint begins, and the cycle repeats.

Scrum Roles

The group of individuals who are responsible for, and committed to, building the product and meeting its goals is known as the Scrum Team. The Scrum Team consists of the following Scrum roles:

- Product Owner
- Developers
- Scrum Master

If you think of the roles in terms of providing service, the Developers serve the Product Owner, whereas the Scrum Master serves everyone. Therefore, the Developers can exert a strong influence to select (that is, hire or fire) the Scrum Master. The Product Owner can exert a strong influence to select the Developers that they want doing the work (HR policies and procedures notwithstanding). Because of this separation of duties, the roles should be played by separate individuals. This mitigates any chance of a conflict of interest. Smaller teams may find it necessary to combine roles.

Developers

Developers are the professionals on the Scrum Team who are capable of designing, building, testing, and delivering a Done, releasable product. There are between three and nine Developers on a Scrum Team—a small enough number to enable them to easily communicate and be nimble, while being large enough to be cross-functional and collaborate in the complex space, such as software development.

A team with only two Developers doesn't need Scrum, since they can simply communicate directly and be productive. Also, there is a greater chance that those two Developers won't have all the skills required to do all the work. On the other hand, teams with more than nine Developers require too much coordination. These larger teams tend to generate too much complexity to derive value from Scrum's empiricism. For situations with more than nine Developers, multiple Scrum Teams may need to be formed, or even a Nexus. I'll talk more about the Nexus and Scaled Professional Scrum in Chapter 11, "Scaled Professional Scrum."

> **Note** The Product Owner and Scrum Master are not considered in this count of 3–9 Developers unless they are *also* a Developer who will be working out of the Sprint Backlog during the Sprint. Regardless, the Scrum Team is typically 10 or fewer people.

It's important to note just because a person is playing the Scrum role of Developer, that does not necessarily mean that they are a developer in the classic sense—someone who develops/writes code. Depending on the task, they may be *developing* architecture, *developing* user interfaces, *developing* test cases, *developing* database schema, *developing* deployment pipelines, *developing* test results, *developing* installers, or *developing* documentation. Everyone *develops* something.

Table 1-1 lists the high-level activities that the Developers on a Scrum Team will perform.

TABLE 1-1 Developer activities within Scrum.

Activity	When
Collaborating with the Product Owner to forecast the Sprint's work and craft a Sprint Goal	Sprint Planning
Collaborating with fellow Developers on a plan to implement the forecasted work	Sprint Planning, Daily Scrum, or as needed
Participating in the Daily Scrum	Daily
Developing the Increment according to the Definition of Done	After Sprint Planning and prior to Sprint Review
Collaborating with the Product Owner to refine the Product Backlog	During the Sprint as determined by the Scrum Team
Collaboratively identifying additional development when forecasted work is completed early	During the Sprint as needed
Collaboratively discussing trade-offs and creating a contingency plan for when forecasted work can't be completed	During the Sprint as needed
Assisting stakeholders with inspecting the Increment and eliciting feedback	Sprint Review or any time during the Sprint as needed
Reflecting on the process and practices in order to identify experiments for improvement	Sprint Retrospective
Inspecting, adapting, learning, and improving—living the Scrum Values	Always

Don't assume that a Developer will execute only those types of tasks they are good at or familiar with. For example, just because Dieter has a background in database programming, that doesn't mean he'll be the one executing those types of tasks. If, during the Sprint, it is decided that the next logical task to execute requires database programming and Dieter is not available, another Developer should jump in and take on that work if at all possible. During development, the person who is best suited to perform a given task will emerge based on many factors, including expertise and availability. It is for this reason that any Sprint Backlog estimates are made by the Developers as a collective, not individuals—even if those individuals are specialists or even experts within those domains. It's also why a Scrum Team should have more than one Developer with a necessary skillset. I will dive deeper into this in Chapter 8, "Effective Collaboration."

> **Tip** I find very few Scrum Teams whose members refer to each other as "Developers." There is still a reflex to equate "developer" to programmer or coder. Our industry reinforces this. For these teams, and for the time being, using the term "team member" may be a suitable substitute—and be more inclusive for testers, designers, and other nonprogrammers.

As a collective, the Developers must be cross-functional. This means that there must be at least one Developer on the Scrum Team who has the necessary skill to execute every type of work required. In other words, the Developers—as a whole—must have all the skills needed to complete their work. Being cross-functional doesn't mean that each Developer is cross-functional—although that would be awesome. Ideally, there will be more than one Developer who has a required skill. If not, then the team should strive to correct that by pairing and sharing, or by leveraging some other instructional techniques during development. Having only one Developer on a team with a key skill is a risk.

The composition of the Scrum Team does not change during the Sprint. If it must change, it should only change "in-between" Sprints. This is typically the result of a decision made collaboratively during the Sprint Retrospective. Changes may include adding a new team member, swapping a member with another team, removing a team member, or changing a team member's role or capacity. Keep in mind that any change to the team composition is a disruption. Productivity may initially decrease for a time and then should (hopefully) increase again. If the Scrum Team tracks velocity or Throughput, the disruption will be apparent.

> **Note** Velocity is a historical measure of PBIs that a Scrum Team has been able to deliver. Velocity can be measured in the number, size/effort (such as points), or business value of those items. Velocity of a single Sprint is not useful, but tracking it over several Sprints shows the general direction of productivity of a team. Once velocity has normalized, it can be useful in planning Sprints and releases. For example, if a team has an average velocity of 20 points per Sprint and the Product Backlog shows 12 PBIs totaling 96 points yet to be developed to reach a Product Goal, you can expect the release to be available in roughly five Sprints, or 2 1/2 months given a 2-week Sprint cadence. Velocity is not mentioned in the *Scrum Guide* and is considered a complementary practice. Teams may also consider tracking and using flow metrics such as Throughput as a measure of past performance. I'll cover flow and flow metrics in Chapter 9, "Improving Flow."

Fabrikam Fiber Case Study

The Developers on the Fabrikam Fiber Scrum Team consist of five cross-functional individuals with varying backgrounds, skill sets, and skill levels. They are Andy, Dave, Dieter, Toni, and Richard (yours truly). Andy and Toni have architecture, design, and some C# experience. Dave, Dieter, and I have solid C# experience. Dieter and I also have SQL and Azure development experience, including Windows PowerShell. As a team, we have all taken Scrum.org's Professional Scrum Developer training and achieved passing assessment scores.

The Product Owner

The Product Owner represents the voice of the user. This means the Product Owner knows not only the product, its domain, its vision, and its goals, but also its users. Just knowing how the product works and what to fix is not enough to be a competent Product Owner. Good Product Owners are in touch with the needs of their users. Great Product Owners will actually share in their passion and feel empathy for their struggles and expectations. The Product Owner also represents the Developers to the organization—at least until the Developers are introduced and begin working with others in the organization directly.

Note Over the years I've heard that the Product Owner is the voice of the stakeholder or customer. Although true, I prefer thinking of the Product Owner primarily as the voice of the *user*. What's the difference? The stakeholder is anyone who is interested in the product or its development. The customer is typically the one who is sponsoring or paying for the product. The user is the one who actually uses it. A Professional Scrum Product Owner strives to satisfy all stakeholders.

The Product Owner must represent the needs of the user and drive value in their direction, rather than just trying to satisfy the person writing the check. There is only one Product Owner on a Scrum Team, an arrangement that helps avoid confusion. When the Developers have a question about the product or need to be introduced to a stakeholder, their first instinct should be to talk with the Product Owner. The Product Owner may have to consult others for answers, especially for large and overly complex products. The Product Owner should be considered the go-to person for all questions about the product's vision, value, goals, and functionality.

The Product Owner is responsible for maximizing the value of the product through the work of the Developers. The Product Owner's primary communication tool for doing this is a refined and ordered Product Backlog. The Product Owner collaborates with the Developers on what and when to develop. Table 1-2 lists the common Developer interactions with the Product Owner.

A common misconception—one that was corrected in the 2020 *Scrum Guide*—is that the Developers develop the product. In fact, it's the Scrum Team that "develops" the product—through the cooperation and collaboration of everyone on the team.

Tip A Professional Scrum Product Owner should know the product, know the product's domain, know the product's customers, know the product's users, know Scrum, have authority to make decisions related to the direction of the product, be highly available to the rest of the Scrum Team, and have good people skills. I've not yet met a Product Owner who ticked all of these boxes. I have met many Product Owners who wanted to improve in all these areas and work toward that goal.

TABLE 1-2 Developer interactions with the Product Owner.

Interaction	When
Collaboratively plan the Sprint and forecast PBIs.	Sprint Planning
Get product/domain questions answered and get introduced to stakeholders.	During the Sprint as needed
Refine the Product Backlog.	During the Sprint as determined by the Scrum Team
Collaborate to take on additional work.	During the Sprint as needed
Collaborate to plan contingency work.	During the Sprint as needed
Assist the Product Owner with inspecting the Increment and other emerging work.	During the Sprint as needed

Interaction	When
Assist the Product Owner with stakeholder inspection and to elicit feedback.	At least during Sprint Review, but also during the Sprint as needed
Collaborate to inspect the Scrum Team's practices and plan for improvement.	Sprint Retrospective
Collaborate to create the Definition of Done.	Sprint Retrospective

Professional Scrum Teams understand the separation of duties between the Product Owner and the Developers and have come to rely on each role doing their part. Although the *Scrum Guide* doesn't explicitly state that the Product Owner cannot be the Scrum Master or a Developer, I think that's a good separation to maintain. Keeping the Product Owner focused on *what* to develop, the Developers focused on *how* to develop it, and the Scrum Master focused on ensuring that everyone understands and follows the rules of Scrum is a recipe for success.

Since the organization may hold the Product Owner accountable for the profit or loss of the product, the Product Owner should maintain a constant vigil for optimizing the product's value. Professional Scrum Product Owners are engaging Product Owners. They continuously want what is best for their product and, more importantly, what is best for its users.

Fabrikam Fiber Case Study

Paula is the Product Owner of the Fabrikam Fiber web application. Having started as a technician, she knows the pain points of the product and the struggles of its users. This awareness inspires her to constantly improve and evolve the capabilities of the product. She even likes to brag that she's the app's most prolific user. Her vision is to improve Fabrikam Fiber one ticket at a time until most customers can get their issues resolved the same day. Paula is an informed and engaging Product Owner who is available when necessary and has the authority to make the necessary decisions. Paula has been using Scrum for about three years. She has been through Scrum.org's Professional Scrum Foundations and Professional Scrum Product Owner training.

The Scrum Master

The Scrum Master fosters the Scrum Values, practices, and rules throughout the Scrum Team and the organization. The Scrum Master ensures that the Product Owner and the Developers are functional and productive by providing necessary guidance and support. The Scrum Master is also responsible for ensuring that Scrum is understood by all involved parties and that everyone plays by the rules.

Note .The Scrum Master is not the same thing as a project manager—not even close. They *are* considered a manager—but of Scrum and its implementation, not of a project, or people, or the product. Therefore, they should be given authority to make changes and remove impediments.

The Scrum Master must be vigilant while giving the organization time to acclimate and realize the benefits of Scrum. This means keeping any dysfunctional (such as old "waterfall") habits at bay. It also means keeping any unenlightened managers at bay, while continually quashing the illusion that command and control and opaqueness equate to better and faster value delivery. Sometimes the Scrum Master may become the de facto change agent, leading the effort of organizational adoption of Scrum. If this is the case, then the Scrum Master's steadfastness must be able to scale. This also illustrates the need for a Scrum Master to have great interpersonal skills.

A Scrum Master can be called on to act as a coach, ensuring that the team self-manages and is functional and productive. This role may include shielding them from focus-busting interruptions and other external conflicts while also removing any impediments to their progress. The ability of the Scrum Master to serve the team by removing impediments to their success is a vital piece of Scrum.

As a servant leader, the Scrum Master achieves results by giving priority to the needs of the team. Scrum Masters may also be of service to stakeholders and others in the organization, helping them understand the Scrum framework and expectations from the various players. Servant leaders are often seen as humble stewards of the people and processes in which they are involved. By having a "What can I do for you today?" attitude, the Scrum Master fosters an environment of collaboration and respect, providing fertile soil for a high-performance Scrum Team. Lao Tzu, the ancient Chinese philosopher, said it best:

> When the master governs, the people are hardly aware that they exist. Next best is a leader who is loved. Next, one who is feared. The worst is one who is despised. If you don't trust people, you make them untrustworthy. The master doesn't talk, they act. When their work is done, the people say, "Amazing: we did it, all by ourselves!"

The Scrum Master is not a technical role. Having a strong background in the pertinent domain (such as software development) is not necessary, though it can be helpful at times. Scrum Masters must really know Scrum—that's their domain. That's not negotiable. A good Scrum Master will also have good communication and interpersonal skills. They may have to facilitate interactions with other team members or enable cooperation across roles, events, or others in the organization. It's important to have those soft skills. Keep this in mind when considering who might make a good Scrum Master. Table 1-3 lists the ways in which the Scrum Master serves the Developers.

> **Tip** In my opinion, traditional project managers *don't* make good Scrum Masters. Unfortunately, this is a common reflex for an organization adopting Scrum. For example, the decision makers decide to send "Roger," their PMI-certified, Gantt chart–loving, Microsoft Project expert to Professional Scrum Master training. The expectation is that Roger will lead the change. What I've seen happen is that either Roger's project management "muscle memory" adversely affects the adoption of Scrum, or his old colleagues and managers do.

TABLE 1-3 Ways in which the Scrum Master serves the Developers.

Service	When
Help facilitate Scrum events, when the Developers can't/won't facilitate for themselves.	During the Sprint as needed
Identify, document, and remove impediments.	During the Sprint as needed
Provide training, coaching, mentoring, and motivation.	During the Sprint as needed
Coach the Developers on self-management.	During the Sprint as needed
Be the Developers' emissary to the organization.	During the Sprint as needed
Attend nonproductive but "required" meetings on the Developers' behalf.	During the Sprint as needed
Shield the Developers from interruption and noise in order to protect their focus.	During the Sprint as needed
Be relied on less and less.	Over time as the Developers improve

The duties of the Scrum Master may not require a full-time commitment. Professional Scrum Teams recognize this and may select a Developer to play the part-time role of Scrum Master. This role may rotate between other Developers over time. Full-time Scrum Masters may morph into a Developer or move on to assist other Scrum Teams as they emerge in the organization. The Scrum Master role is more flexible than the other roles in this regard. As long as a Scrum Team understands and follows the rules of Scrum and has access to someone who can perform the duties of a Scrum Master when needed, party on.

Tip The skills of a Scrum Master are unique and important. The best Scrum Master I ever met was a man named Brian. He didn't have a business background or a technical background. He was formerly a drug and alcohol counselor, which meant that he could listen, encourage, motivate, and also tell when someone was slacking, not doing all they were capable of doing, or not telling the truth. Being a Scrum Master is a career choice for some. In my experience, they tend to be outgoing, motivated, and dedicated to continuously improving their skills as they serve the team. These Scrum Masters should remain just that and shouldn't be dismissed or converted to another role. They will bring more value to the team and the organization as a full-time Scrum Master. They are worth their weight in sticky notes.

Fabrikam Fiber Case Study

Scott was brought on last year to serve as Scrum Master. Initially, he served another team, providing the necessary coaching in order to transform them into a high-performance, Professional Scrum Team. Management agrees that he'll be a fine Scrum Master for Paula's team. They also plan to use Scott to help other teams within the organization learn and adopt Scrum. Scott has many years of practical, hands-on Scrum experience with various companies and teams. He has been through several Professional Scrum training classes and is active in the Scrum.org community.

Stakeholders

Although not an officially defined role in the *Scrum Guide*, stakeholders include everyone else involved or interested in the development of the product. Stakeholders can be managers, directors, executives, board members, analysts, domain experts, attorneys, sponsors, members from other teams, customers, and users of the software. Stakeholders are very important. They represent the product's necessity from various perspectives. They also drive the vision, goals, and usability of the product by influencing the Product Owner. Without stakeholders, who would use the product, pay for its development, or derive benefit from it?

In my experience, Developers can tend to discount nontechnical individuals. This is unfortunate—stakeholders should not be ignored. They should be engaged. That said, some stakeholders can take *too much* interest in the development effort and its status, becoming a distraction and killing focus. A lot of misunderstanding exists on when and why stakeholders and Developers interact. Reading the *Scrum Guide*, you might think interaction only occurs during Sprint Review. As you can see in Table 1-4, stakeholders and Developers can interact at multiple points throughout the Sprint.

TABLE 1-4 Developer interactions with stakeholders.

Interaction	When
Collaborate to refine the Product Backlog.	During the Sprint as determined by the Scrum Team
Collaborate to answer questions that Developers might have about a PBI (estimating, planning, designing, building, testing, etc.).	During the Sprint as needed
Collaborate to inspect the Increment and to elicit feedback.	At least during Sprint Review, but also during the Sprint as needed

Inspecting and providing feedback on the product, such as requesting a capability/feature, should involve the Product Owner. Inspecting and providing feedback on the development process, such as inquiring about status, should be handled by the Scrum Master. In other words, stakeholders should almost always be kept out of the development process—unless they are invited in by the Developers. Following this guideline will help protect the Developers' focus.

> **Tip** Burndown, burnup, or other analytics posted in a common area or on a dashboard are a great way to keep stakeholders informed. This keeps the interruptions of the Scrum Team to a minimum. If anyone has questions, the Scrum Master can educate them.

The Scrum Master should strive to keep stakeholders out of the various Scrum events, with the exception of the Sprint Review and maybe Sprint Planning. Stakeholders should not be involved in any planning, development, or refinement activities unless their input is required. Attendance to any event is by invitation of the Scrum Team only. Stakeholders should not attend the Daily Scrum, as its purpose is to allow the Developers to synchronize with one another on the upcoming work. Even the Product Owner's presence at the Daily Scrum may be considered a distraction from its purpose.

Fabrikam Fiber Case Study

The Fabrikam Fiber company has been around a few years. It has changed its name once and has gone through a few reorgs. The primary stakeholders are its customers, who are users of the company's primary value stream: internet connectivity. Along with the service technicians, other business units, including sales and marketing, are all stakeholders of the Fabrikam Fiber web app and its support of operations. Of the various groups of stakeholders, most are not technical when it comes to software. Some internal stakeholders, however, have deep expertise in the domain of hardware and networking. All stakeholders have been pretty open about providing feedback on the web application. Paula understands the importance of capturing customer feedback. To that end, she insisted on setting up a *wish@fabrikam.com* email address to receive email-based feedback. These emails are routed to a support person who triages the requests and then works with Paula to add PBIs to the Product Backlog.

Scrum Events

The Scrum framework uses events to structure the various workflows of incremental development. Each event is timeboxed, which means that there is a fixed period of time to execute the activities within that event. Timeboxing minimizes waste by ensuring that an appropriate amount of time is spent focusing on a particular activity. These Scrum events are meant to establish regularity and a cadence. They are also meant to minimize the need for wasteful or impromptu meetings that are not part of Scrum.

All Scrum events are a formal opportunity to inspect and adapt something. Inspecting allows the team to assess progress toward a goal, as well as identify any variance in their current plan. If an inspection identifies any unacceptable deviation, an adjustment (adaptation) must be made. These adjustments should be made as soon as possible to minimize further deviation. Failure to include or participate in any of the Scrum events results in reduced transparency and is a lost opportunity to inspect and adapt.

Referring back to Figure 1-1, you can see that there are five events in Scrum:

- **Sprint** A container for the other four events. All the work necessary to achieve the Product Goal, including Sprint Planning, Daily Scrums, Sprint Review, and Sprint Retrospective, happens within a Sprint. Sprints are fixed-length events of one month or less. A new Sprint starts immediately after the conclusion of the previous Sprint.

- **Sprint Planning** Initiates the Sprint by laying out the work to be performed for the Sprint. This planning is performed through the collaborative work of the entire Scrum Team.

- **Daily Scrum** Used to inspect progress toward the Sprint Goal and adapt the Sprint Backlog as necessary, adjusting the upcoming planned work. Only Developers participate in the Daily Scrum.

- **Sprint Review** Used to inspect the outcome of the Sprint and determine future adaptations. The Scrum Team presents the results of their work to key stakeholders, the Product Backlog is updated, and progress toward the Product Goal is discussed.

- **Sprint Retrospective** Used to plan ways to increase quality and effectiveness.

> **Note** A notion exists that the Sprint is that time period *after* Sprint Planning and ending before the Sprint Review in which the actual development occurs. This is incorrect. The Sprint is actually an outer "container" for the four other events. This means that the Sprint has already begun when Sprint Planning commences. The time between Sprint Planning and Sprint Review doesn't technically have a name, but most refer to it as "development." The Sprint concludes after the Sprint Retrospective, and then starts again with the next Sprint Planning.

The Sprint

A Sprint is the set period of time in which an Increment of the product is developed. A *Sprint* is Scrum's term for an iteration. Sprints are one month or less in length and run end to end, one after another. The frequency of feedback, experience and technical excellence of the team, and the organization and Product Owner's need for agility are key factors in determining the length of a Sprint. For example, if the product is an enterprise desktop application with fairly well-defined goals and not much deviation in plans, longer Sprints are fine. If the application is a cloud-based software as a service (SaaS) product with several competitors and demanding customers, shorter Sprints are more desirable. The stakeholders and the Scrum Team must collaborate to determine the ideal length of the Sprint.

Sprint Planning, development, Daily Scrums, the Sprint Review, and the Sprint Retrospective all take place within the Sprint. After you start using Scrum, you are *always* in a Sprint—assuming there is still a product, stakeholders, and a desire to invest in new capability. When a Sprint Retrospective ends, the next Sprint begins and you repeat the inner events again. There should never be any breaks in between Sprints.

I asked Ken Schwaber once how long a Sprint should be. His answer was, "as short as possible and no shorter." Sprints of longer than four weeks (one month) have a smell—the smell of water falling. When a Sprint's length is longer than a month, the definition of what is being built may change or complexity and risk may increase. By limiting the maximum length of a Sprint, at most one month of development effort would be wasted, rather than several months in a classic, "waterfallian" project.

Conversely, Sprints with a length of less than one week are possible but should be executed only by high-performance Professional Scrum Teams. Even with very short Sprints, the overhead of the inner events must be factored in, leaving even less time—as a percentage—for actual development. Teams working in "micro Sprints" like these need to be on their A-game every day.

Ideally, the length of the Sprint does not change. If it must, it can only change in between Sprints, as a result of a decision made during a Sprint Retrospective. Any change to the length of a Sprint will cause disruption to the Developers' cadence, thus impacting forecasts and plans. Although this will correct over time, it's not a good idea to constantly introduce such chaos.

Each Sprint is like a mini-project. In fact, when I'm introducing Scrum to a new organization or team, I suggest replacing their use of the term "project" with "Sprint" or "Sprints" in conversation. For example, instead of referring to the "CRM integration project," just refer to that initiative as Sprints 17–19, where PBIs relating to CRM integration will be forecasted and developed.

The Sprint has a definition of "what" is to be developed. It also includes a flexible approach on "how" to develop it. During the Sprint, all aspects of the development work are executed. With a software product, this will typically be more than just designing, coding, and testing. The scope of work may be clarified as more is learned, and the Product Owner may collaborate with the Developers to renegotiate and add new items or swap different items in the Sprint Backlog—as long as they fit with the Sprint Goal. The Developers may not decrease any quality goals in order to finish its work. The resulting product Increment is produced, inspected by stakeholders, and perhaps even released.

The choice of which day of the week to start (and end) a Sprint is entirely up to the Scrum Team. Some practitioners prefer Mondays or Fridays. I prefer midweek so that the chances are highest that the whole team will be present and participating with maximum focus. Events can be rescheduled around holidays, but I recommend the Sprints remained fixed. For example, if Sprint 26 ends during a company holiday, leave its start and end date alone to maintain the two-week cadence, and just schedule the Sprint Review and the Sprint Retrospective at the end of the first week. Minor adjustments like these may need to be done throughout the year.

Fabrikam Fiber Case Study

Originally, the Scrum Team tried four-week Sprints. They felt that the longer timebox would be closer to the quarterly delivery schedule they had been accustomed to. Unfortunately, since the team was new to Scrum, they continued to take a sequential approach to development. They spent a lot of time on analysis and design at the beginning of the Sprint and deferred testing until the end. The resulting crunch in the last days of the Sprint was not sustainable and was really just a backslide into waterfallian habits (also known as "Scrummerfall"). The team did not experience the productivity gains everyone anticipated. When they brought on Scott (the Scrum Master), he recommended moving to two-week Sprints. This caused the developers to experience an increased sense of urgency and change the way they worked, maintaining a comfortable level of intensity throughout the Sprint. Scott also recommended starting the Sprint on a Wednesday. This change increased the chances of the whole team being available and operating at peak focus. It also allowed stakeholders to fly in for an in-person Sprint Review and subsequent Sprint Planning without having to stay over a weekend. The Scrum Team has completed many successful Sprints while on this two-week cadence.

Sprint Planning

Sprint Planning is for selecting and planning the work that will be performed during the Sprint. This is the first event that occurs within the Sprint. The entire Scrum Team attends and participates. A refined and ordered (prioritized) Product Backlog is required as an input for Sprint Planning. Developers collaborate with the Product Owner on the scope of work that can be accomplished. This is called the *forecast*. The forecasted work, along with a Sprint Goal and a plan for doing the work, are the outputs. The Sprint Backlog holds these outputs.

Sprint Planning is timeboxed, so everyone needs to be laser-focused. Distractions, such as off-topic conversations, should be minimized. The maximum length of Sprint Planning is eight hours but, in practice, the length should be a function of the length of the Sprint, as you can see in Table 1-5.

TABLE 1-5 Length of Sprint Planning.

Sprint length	Sprint Planning length
4 weeks	~ 8 hours or less
3 weeks	~ 6 hours or less
2 weeks	~ 4 hours or less
1 week	~ 2 hours or less
Less than a week	In proportion to the above lengths

Sprint Planning consists of three topics. Each topic answers a question: why, what, and how. Each question is answered by an output of Sprint Planning. *Why* is answered with the Sprint Goal. *What* is answered with the forecast. *How* is answered with the plan. The following pages will go into detail on each of these topics.

Topic 1: Why is this Sprint valuable?   During Sprint Planning, the Product Owner proposes how the product could increase its value and utility in the current Sprint. The whole Scrum Team then collaborates to define a Sprint Goal that communicates why the Sprint is valuable to stakeholders. The Sprint Goal is an objective, in narrative format, that guides the Developers as they develop the Increment. The Sprint Goal also provides stakeholders the ability to see a synopsis of what the Scrum Team is working on. The Sprint Goal must be finalized prior to the end of Sprint Planning.

Although the Product Owner may bring the Product Goal and other business objectives into Sprint Planning, it's important that the whole Scrum Team finalize the Sprint Goal together and agree on its verbiage and meaning. Everyone on the Scrum Team should then commit it to memory. Stakeholders should have access to it as well. After development has begun (that is, Sprint Planning is over), the Sprint Goal should not be changed. It is the *theme* that the team commits to achieving. If the Developers aren't able to achieve the Sprint Goal, or the goal becomes obsolete, the Product Owner might decide to cancel the Sprint—another indication of the Sprint Goal's importance. Cancelling a Sprint is discussed in Chapter 6, "The Sprint."

The *Scrum Guide* doesn't say which comes first, the Sprint Goal or the forecast. Some Scrum Teams like to craft the Sprint Goal first, or at least in parallel with the forecasting of work. This way, there is more cohesion with the goal and the PBIs that are developed during the Sprint. This cohesion makes it easier to understand the value of the Increment and how it fits into the goals of the product or release. Other teams may want to forecast the highest-ordered items in the Product Backlog and then craft a narrative Sprint Goal around those items. Both approaches can be difficult for teams who need to deliver disparate features and bug fixes for a given Sprint.

The Sprint Goal gives the Developers some flexibility and guidance about the functionality implemented within the Sprint. Even if they deliver fewer PBIs than were forecasted in Sprint Planning, they can still achieve their Sprint Goal. For example, let's say the Developers forecast the following PBIs during Sprint Planning:

1. Add a Twitter feed to the homepage.

2. Create a Facebook page for the company.

3. Create and host a wiki for product support.

Given this forecast, the Sprint Goal might read, "To increase community awareness of our company and its products" or simply, "Socialize Fabrikam Fiber." As the Developers work, they keep this goal in mind. If the team is unable to finish the third PBI (wiki), they didn't fail because they were still able to achieve the Sprint Goal by completing the first two PBIs.

If it sounds like Sprint Goals give the Developers "wiggle room," you are correct. Remember that what developers do is difficult and full of risk. That's why they should *forecast* the individual items they think they can deliver but *commit* to the Sprint Goal that embodies them.

> **Note** The Sprint Goal should not be too broad, such as "To improve the product." Nor should the Sprint Goal be too specific, such as "To implement PBI #1 and PBI #2 and PBI #3 and PBI #4 and PBI #5 . . ." It may be difficult to achieve vague Sprint Goals, and compound Sprint Goals tend to split the team's focus and not allow flexibility. A Sprint Goal should describe the reason for undertaking the Sprint, not just a review of the forecasted PBIs. A good Sprint Goal helps a team understand the purpose and impact of the work they are doing, which is a motivator.

Topic 2: What can be done in this Sprint?   During Sprint Planning, the Developers consider the highest-ordered PBIs from the Product Backlog one at a time. The order is decided by the Product Owner. Each PBI's details and acceptance criteria are discussed. Clarification is provided by the Product Owner as well as other stakeholders, who may be able to provide domain expertise and clarification.

After obtaining a sufficient understanding of the PBI, the Developers collaborate to determine whether the PBI is small enough to fit into their Sprint capacity. If the Developers honestly believe that they can deliver the item in this Sprint—according to their Definition of Done—the item is added to the forecast. This may require estimation or use of a flow metric such as Service Level Expectation. I'll discuss flow and flow metrics in Chapter 9.

If the Developers are not in agreement on whether an item should be forecasted, this may require the PBI to be further analyzed and discussed. It may end up being split or deferred until a later Sprint, when more is known. The Developers then move to the next item—in order—in the Product Backlog. Since the Product Owner is part of Sprint Planning, the order of the items in the Product Backlog can be updated as needed.

This process of forecasting is repeated until the Developers think that they have a comfortable amount of work for the Sprint, given their capacity/availability, past performance, and other factors. These forecasted PBIs are moved from the Product Backlog to the Sprint Backlog.

New Scrum Teams that don't yet have a handle on their past performance may just use their instinct to decide what *feels* like the right amount of work. High-performance Professional Scrum Teams may do the same. If, during the Sprint, the Developers complete their forecasted work early, they should collaborate with the Product Owner mid-Sprint to identify and develop additional PBIs. Ideally, these items should align with the Sprint Goal. The Developers should never forecast more work than they *know* they can complete.

> **Note** In 2011, the *Scrum Guide* introduced a somewhat controversial change to Sprint Planning. The word "commit" was replaced with "forecast." Scrum practitioners had an issue with the word "commit" for some time. The problem was that "commit" implied that the team was obligated to deliver all PBIs at the end of the Sprint. This was especially true when stakeholders, who tend to not understand the complexities of developing complex products such as software, heard the word. Since complex product development is difficult and full of risk, delivering all PBIs every Sprint is unrealistic. The Developers might have to cut quality in order to make good on their "commitment"—and that is forbidden in Scrum, as well as in ethical product management.
>
> The term "forecast" is more realistic and easier to understand by stakeholders who have probably heard terms like "sales forecast." It suggests that, although the Developers will do their best, given what they know new information may emerge during the Sprint that might impede their best-laid plans. If you and your organization still use the term "commit," it may take some time to get used to the term "forecast." It may sound like a weasel word to some, but overall, its usage is more honest and transparent.

Topic 3: How will the chosen work get done?   Sprint Planning is not complete until the Developers have also devised a plan for how they will develop the forecasted PBIs. The plan must ensure that all PBI acceptance criteria are satisfied while meeting the Definition of Done. The plan might be visualized as a collection of sticky notes in the same row as the associated PBI sticky note. In a software tool, such as Azure DevOps, it might be several child records related to a parent record. Regardless of the tool the team uses, the Sprint Backlog contains the Sprint Goal, the forecasted PBIs that will achieve the Sprint Goal, and the plan (tasks, tests, diagrams, etc.) to develop those PBIs.

> **Tip** Go lightweight during Sprint Planning. Whiteboards are a great medium for sketching ideas and brainstorming tasks. Laptops aren't. Whiteboards can be easily photographed and wiped clean afterward. Files on laptops or work items in Azure Boards tend to linger and yearn to be kept and updated. They also indicate a finality set in stone that is not necessarily the truth. Using sticky notes to brainstorm the plan is also good. They can be moved and removed easily from the board. A Professional Scrum Team will avoid using any tool during Sprint Planning unless its value outweighs its distraction factor. Sticky notes and whiteboard sketches can be translated later, after the Developers agree on a plan.

Because of the Sprint Planning timebox, the Developers probably won't be able to identify every detail required to develop each PBI. For expediency creating the plan, a minimum amount of information should be recorded—perhaps just a title and estimate of effort. Sprint Planning is not the time for detailed design. Instead, the Developers need to focus on the high-level plan. Implementation details will emerge throughout the Sprint.

For example, let's assume that the team will have to create several entity models, controllers, and views. Rather than go down a design "rat hole" during Sprint Planning, Developers should just identify a couple of high-level tasks: *create models*, *create views*, and so forth. If the Developers estimate time for those tasks, those estimates would simply be an aggregate of effort to perform all the related activities per task.

Assuming the Developers are using tasks to formulate the plan, the tasks to be executed earlier in the Sprint should be more decomposed and detailed than those further down the Sprint Backlog. Estimates, if the Developers decide to use them, can be in whatever unit of measure they decide. For tasks, hours are the most common unit. I've also seen teams use days or task points (similar to story points). Personally, I think using points for estimating tasks can lead to confusion. Rarely would you want to relatively compare the estimates of two tasks that could end up being done by different team members. Regardless of the unit of measure, all of these numeric values will power a Sprint burndown chart, should the team choose to employ one. I'll cover estimating (sizing) in Chapter 5, "The Product Backlog."

It's important for the Developers to leave Sprint Planning with the why, what, and how—all of which is documented in the Sprint Backlog. As you will learn in Chapter 6, task ownership is not a required outcome of Sprint Planning. In fact, it's better to leave "to do" tasks unassigned so that team members who have capacity can pick a relevant task to work on next. That said, it is common for each Developer to own a task prior to leaving Sprint Planning. The Developers will then self-manage to undertake the work in the Sprint Backlog as needed throughout the Sprint.

Fabrikam Fiber Case Study

The first Sprint Planning sessions were chaotic. The Developers were introduced to new PBIs for the first time at Sprint Planning. Paula (the Product Owner) wasn't always prepared and the domain experts were sometimes unavailable. Most of Sprint Planning was spent understanding *what* was to be developed, and planning the *how* got deferred until the first few days of the Sprint. The *why* (Sprint Goal) was missing in several early Sprints as well. With coaching from Scott, the Scrum Team improved over time as everyone got used to Scrum and got into a cadence. Sprint Planning also became more efficient when the team started meeting regularly to refine the Product Backlog.

The Daily Scrum

The Daily Scrum is a 15-minute, timeboxed meeting for the Developers to inspect progress toward the Sprint Goal and adapt the Sprint Backlog as necessary, adjusting the upcoming planned work. The Developers can select whatever structure and techniques they want, as long as their Daily Scrum focuses on progress toward the Sprint Goal and produces an actionable plan for the next 24 hours of work. The Daily Scrum should be held on all working days of the Sprint—even on days containing Sprint Planning, Sprint Review, and Sprint Retrospective events.

Daily Scrums improve communications, identify impediments, promote quick decision making, and consequently eliminate the need for other meetings. The Daily Scrum allows Developers to listen to what other Developers have done and are about to do. This leads to increased collaboration, as well as accountability. If one Developer hears that another Developer is about to work in a similar area of the product, they may choose to pair up for the day. On the other hand, if the team hears that a Developer is on day 3 of a two-hour task, it may be time to pair up or inquire about the root cause of the delay. Developers need to understand that commitments are being made at this meeting and that these commitments will be tested 24 hours from now.

> **Note** I hear a lot of teams refer to this event as the "daily standup." It's called the "Daily Scrum." If the team decides to stand, they may do so. They can also sit, squat, or plank.

The Developers can use the dialogue heard during the Daily Scrum to assess their progress. By hearing what is or isn't being accomplished each day, the team can determine whether they are on their way to achieving the Sprint Goal. As teams improve in their collaboration, this vibe will become more noticeable—even outside the Daily Scrum. High-performance Professional Scrum Teams may even outgrow the need for a formal assessment tool such as a Sprint burndown.

The Daily Scrum should be held in the same place and at the same time every day to reduce complexity and to maximize the likelihood of attendance. Ideally, the Daily Scrum is held in the morning so that the Developers are able to synchronize their work for that day. Dislocated teams may need to be more flexible in their start times.

The Daily Scrum is not a status meeting. Problem solving should not be attempted at the Daily Scrum because it can often lead to the team violating the 15-minute timebox. Those conversations should be deferred until after the Daily Scrum. Use a *parking lot* or other practice to track off-topics. A parking lot is a tool, such as a whiteboard or wiki, to capture comments or questions not related to the agenda. In the case of the Daily Scrum, this would be anything not related to inspecting progress or adjusting the Sprint Backlog.

The Daily Scrum is not meant for anyone other than the Developers to participate. This includes the Product Owner. In fact, the Scrum Master is not even required to attend. The Scrum Master needs only to ensure that the Daily Scrum takes place and that the rules are followed. Any impediments can be identified, tracked, and even mitigated by a Developer. If the Product Owner or Scrum Master is actively working on items in the Sprint Backlog, then they are considered a Developer and will participate in the Daily Scrum.

Tip Keep laptops, boards, burndown charts, and other tools or artifacts out of the Daily Scrum. These tend to distract from the purpose of the Daily Scrum—which is to synchronize with one another. Each Developer should know their own information without having to look up anything. Observations and impediments can be recorded on a whiteboard or by using sticky notes. High-performance teams will use a *parking lot* to track anything not relevant to the Daily Scrum, and follow-up conversations can tackle those issues. The Developers are self-managing and can decide to meet formally or informally at any time during the day for any reason. In fact, the Scrum framework has no guidance on what the Developers do the other 7 hours and 45 minutes of the day. I would hope they would interact with other individuals as necessary in order to build an awesome product.

Fabrikam Fiber Case Study

The Developers have their Daily Scrum at 9 a.m. each day. Before the meeting, each Developer updates their work remaining estimates on their tasks on the board. Doing so gives them a fresh perspective on their remaining work and enriches the conversation. A side benefit is that this keeps the burndown reports accurate, which is good if they are consulted throughout the day. The actual Daily Scrum finds each Developer talking briefly about what they've accomplished yesterday and what they plan to do today. Any impediments are also raised. Sticky notes are created and placed in a parking lot as needed. The Daily Scrum usually takes less than 10 minutes. Scott rarely shows up anymore.

Sprint Review

After the Sprint's development timebox has expired, a Sprint Review is held. The entire Scrum Team attends, as well as any stakeholders the Product Owner invites. This informal meeting is for inspecting the Increment developed by the team. Stakeholders get to observe an informal demonstration of—or get their hands on—working product. Their feedback is elicited and captured. This collaboration can yield new, updated, or removed PBIs. Progress toward the Product Goal is also discussed, as everyone collaborates on what to do next. The Sprint Review is a working session and the Scrum Team should avoid limiting it to a simple demonstration or presentation.

The Sprint Review is timeboxed to a maximum length of four hours. For shorter Sprints, the Sprint Review is usually shorter. In practice, the length should be a function of the length of the Sprint—essentially half that of Sprint Planning—as you can see in Table 1-6.

TABLE 1-6 Length of the Sprint Review.

Sprint length	Sprint Review length
4 weeks	~ 4 hours or less
3 weeks	~ 3 hours or less
2 weeks	~ 2 hours or less
1 week	~ 1 hour or less
Less than a week	In proportion to the above lengths

During the Sprint Review, the Sprint Goal and forecasted PBIs should be restated. Keeping their audience in mind, the Scrum Team may give a short summary about what went well, what didn't, and how they overcame any problems. Completed PBIs are inspected through demonstration or by having the stakeholders interact directly with the product. Inspection should not be done by showing secondary artifacts, such as slides, mock-ups, diagrams, or passing tests. Because this is a working session, the Scrum Team may need to explain what the stakeholders are seeing and answer any of their questions.

> **Tip** The Developers should never surprise their Product Owner at a Sprint Review. In other words, this should not be the first time the Product Owner sees the completed work. Professional Scrum Teams know the value of collaboration with the Product Owner. At a minimum, the Developers should ask the Product Owner's opinion on individual PBIs as they approach completion throughout the Sprint. There is no built-in Product Owner "acceptance" in Scrum. Should a Definition of Done include it, however, don't delay until the end of the Sprint—or Sprint Review—to obtain it. You don't want a Sprint Review to become a "sign-off" meeting. Sprint Reviews are more about improving the product through collaboration and feedback.

A Sprint Review can generate one or more outcomes:

- Feedback (new requests, ideas, experiments, or hypotheses) is added to the Product Backlog.
- Unnecessary items are removed from the Product Backlog.
- The Product Backlog is refined.

- The Product Backlog is reordered.

- The Product Owner decides to release the Increment.

- The Product Owner decides to turn on a feature flag.

- The Product Owner decides to cancel development.

As previously mentioned, the Sprint Review is an informal meeting. The Scrum Team should not spend much time preparing for it. Nobody should feel like they are attending a technical presentation at a conference. On the other hand, the team should be organized enough so that it doesn't waste the stakeholders' time. If necessary, the Scrum Master can intervene and make corrections to maximize the meeting's value for everyone. Any process corrections can be discussed at the Sprint Retrospective and implemented in the next Sprint.

There are many ways to run a Sprint Review. Some Scrum Teams like it to have some structure. Others don't. Some like the Scrum Master to kick it off. Others like it to be the Product Owner. Some like to rotate Developers so everyone gets a chance to talk and drive the review. Others like their strongest communicator driving. Still others like stakeholders to put hands on the product directly. Regardless, the Sprint Review should be down to earth and foster an environment of collaboration and discussion. The Scrum Team should be inquisitive, and all feedback should be welcomed and captured. Later, the Product Owner can go through the feedback, adding description and detail. Inane ideas will eventually sink to the depths of the Product Backlog or be deleted altogether.

Being mindful of the timebox, the Scrum Team can discuss unfinished or not started PBIs with the stakeholders. If the stakeholders have blocked time out of their busy day, don't squander an opportunity to get their feedback on PBIs that might be coming up in an approaching Sprint. These refinement discussions can generate valuable input for planning.

Fabrikam Fiber Case Study

Sprint Reviews have always been a big deal for the Scrum Team. They meet every other Tuesday morning, inviting all the pertinent stakeholders and sometimes members from other teams. Paula (the Product Owner) kicks off the meeting with a review of the Product Goal, Sprint Goal, and the forecasted work for the Sprint. Scott (the Scrum Master) then gives a summary of the Sprint, including the team's progress (using analytics), any obstacles that emerged, and how the Scrum Team overcame them. The bulk of the typically hourlong review is spent by the stakeholders inspecting the completed functionality. This is typically done through demonstration—in a storytelling way, with the Developers playing different personas as they take various journeys through the product. This engaging approach makes everyone in the room feel safe and comfortable in sharing their opinions and ideas. Feedback is captured by members of the Scrum Team. Stakeholders sometimes send feedback after the Sprint Review as well. Paula then wraps up by discussing her ideas for the next Sprint and updates everyone on progress toward the Product Goal. The Sprint Reviews are recorded, and those videos are shared with others who couldn't attend.

Sprint Retrospective

The last event in the Sprint is the Sprint Retrospective, where the Scrum Team will inspect and adapt its own behavior and practices, looking for opportunities to improve. The Sprint Retrospective occurs after the Sprint Review but before the next Sprint Planning. The exact time and location are up to the Scrum Team. It's important for the Product Owner, Scrum Master, and all Developers to attend—but not stakeholders. Having stakeholders in the Sprint Retrospective may cause transparency to drop. The maximum length of a Sprint Retrospective is three hours but, in practice, the length should be a function of the length of the Sprint, as you can see in Table 1-7.

TABLE 1-7 Length of the Sprint Retrospective.

Sprint length	Sprint Retrospective length
4 weeks	~ 3 hours or less
3 weeks	~ 2.25 hours or less
2 weeks	~ 1.5 hours or less
1 week	~ 45 minutes or less
Less than a week	In proportion to the above lengths

The purpose of the Sprint Retrospective is for everyone to share their observations, thoughts, and ideas on what went well and what didn't with regard to people, relationships, process, tools, and so forth. These discussions can get deep and sometimes heated, especially when you are talking about social interactions with individuals. The meeting should remain constructive, and it's the Scrum Master's responsibility to keep it that way.

> **Note** Impediments and struggles with the development process and practices can be inspected and adapted at any time, such as during the Daily Scrum or throughout the day. The Sprint Retrospective provides the *formal* opportunity for such inspection, as well as time for planning any adaptations.

The output of a Sprint Retrospective is a plan for implementing improvements. These improvements can target the development process as a whole or individual practices within it. Improvements might include changing the way the Developers work, where, or when. Improvements might also include changing how the Developers use their tools or what tools they use. Adaptations might be more social in nature, such as experimenting with ways to make the team environment more or less stimulating.

Any potential improvement is just an experiment, since the Scrum Team constantly inspects and adapts its practices. Some of these changes can be pretty disruptive, so they should be executed only with the consensus of the full Scrum Team having a complete understanding of the ramifications of making the change. Any change made must still abide by the rules of Scrum. Table 1-8 lists some changes that the Scrum Team might make during a Sprint Retrospective.

TABLE 1-8 Changes that can be made at the Sprint Retrospective.

Change	Example(s)
Increase product quality by updating the Definition of Done.	Adding a rule that code coverage cannot decrease as new code is pushed
Change the person playing the Scrum Master role.	Relieving Scott of his duty while attributing the role to Dave
Change the team composition.	Adding another Developer or dropping Toni's capacity to 50%
Change the Sprint length.	Changing from three weeks to two weeks

> **Tip** Don't just hold a Sprint Retrospective for the sake of checking a box. If problems are identified, make sure improvement experiments are also identified. If improvement experiments are identified, make sure that they are actually performed in the next Sprint. Inspect, adapt, and repeat! I'll go deeper on this in Chapter 10, "Continuous Improvement."

A Scrum Team can choose from among many techniques during a Sprint Retrospective. The most common is to have Scrum Team members answer three questions:

- What did we do well this Sprint?

- What could we have done better?

- What will we do differently next Sprint?

There are other approaches to start the conversation, elicit feedback, and brainstorm ideas. Entire books and websites have been devoted to running successful retrospectives. Table 1-9 lists some of the techniques that my fellow Professional Scrum Trainers have employed successfully. You can search the web for additional information, such as the instructions for using each technique or how they may be combined.

TABLE 1-9 Sprint Retrospective techniques and activities.

Technique	Description
Timeline	A timeline for the Sprint is marked on a wall, and team members add sticky notes to it to indicate good and bad events that occurred at that point in time.
Emotional Seismograph	Similar to the Timeline, but team members mark their emotional level as a high or low point on a y-axis throughout the Sprint.
Happiness Metric	Similar to the Emotional Seismograph, but team members track their happiness levels throughout the Sprint using a scale of 1–5 with comments. A chart is produced for the Sprint Retrospective, and the peaks and valleys are discussed.
Mad, Sad, Glad	Team members brainstorm on the events that made them mad, sad, or glad during the Sprint. Sticky notes are combined, grouped, and discussed.
The 4 L's	Four posters or whiteboards are created for Liked, Learned, Lacked, and Longed For. Team members add sticky notes to the respective board. They are grouped and discussed.

Technique	Description
The 5 Why's	A question-asking technique used to explore cause-and-effect relationships underlying a particular issue or impediment.
Remember the Future	Used to create a vision of what the team wants to achieve by inquiring about a future point in time that follows another future point in time where the hypothetical change was made.
Car Speeding Toward Abyss	A technique in which you draw a picture of a speeding car heading toward an abyss and use this analogy to identify the engine, parachute, abyss, and bridge comparisons to the Scrum Team's current situation. The Speedboat and Sailboat are variations of this technique.
Perfection Game	A technique used to maximize the value of ideas. Team members rate an idea from 1 to 10 and provide positive feedback on how to make it a 10. No feedback means they've given it a 10.
Fishbowl	Arranging chairs in an inner and an outer circle in order to attract team members to an empty chair in the inner circle (the fishbowl) and participate in the conversation.
Starfish	Using a starfish diagram, team members add sticky notes in these categories: do the same (=), do less of (<), stop doing (–), start doing (+), do more of (>). The findings are normalized and discussed.
Problem Tree Diagram, or Ishikawa (Fishbone) Diagram	A technique for visualizing the cause-and-effect relationships pertaining to a particular problem.
Team Radar	The team defines the factors (feedback, communication, collaboration, etc.), and then each team member rates their interpretation of that factor on a scale of 0–10, where 0 means not at all and 10 means as much as possible. The chart is discussed and saved for later comparison.
Circles and Soup	A technique for helping identify what is and what is not the responsibility of the Scrum Team. This is similar to the Circle of Concern and Circle of Influence techniques.

It's also important during the Sprint Retrospective to celebrate the team's victories. The good things that occurred should be encouraged to persist. Likewise, challenges during the Sprint should be seen as opportunities for victory in the next. This continuous improvement mindset is foundational in a Professional Scrum Team, and they live it every day. Since not every team member is wired this way, respect, encouragement, and team building are important and should be part of the Sprint Retrospective, too, if necessary.

Fabrikam Fiber Case Study

In the early Sprints, the Sprint Retrospectives would not generate much return on the time invested. The entire Scrum Team would go through the basic questions. To them, it just felt like a longer version of the Daily Scrum and a waste of time. Retrospective notes were captured and sometimes a plan for improving was executed. When Scott joined as Scrum Master, this changed. He introduced new techniques to get everyone involved. He focused on what went well and on team building. He also worked hard so that any action items were implemented during the following Sprint.

Product Backlog Refinement

Maintaining a well-refined Product Backlog helps the development of a successful product. Product Backlog refinement is the periodic meeting of the Product Owner and the Developers to add detail to upcoming PBIs. This is the time when the requirements and acceptance criteria are explored and revised. When the Developers have sufficient understanding of a PBI, they are then able to estimate its size—or at least determine whether it is small enough to fit into a Sprint. This estimate may change over time as more is learned about the item. In fact, the Developers may continue to refine and estimate the same PBI several times before they deem it "ready."

Although Product Backlog refinement is a necessary and important part of product development, it is not a formal event in Scrum. It is optional. If the Scrum Team decides to refine the Product Backlog, it should take no more than 10 percent of the capacity of the Developers (for example, no more than 8 hours total spent refining for an ideal, two-week Sprint). The exact *where* and *when* of refinement are left up to the Scrum Team to decide.

Some teams try to avoid a refinement near the beginning or end of the Sprint so that it doesn't collide with the Scrum events. It is important to have the entire Scrum Team—including all Developers—involved in refinement because the analysis, conversation, and learning will be more meaningful. Diligently refining the Product Backlog minimizes the risk of developing the wrong product.

Fabrikam Fiber Case Study

With the adoption of two-week Sprints, the Developers now spend every Friday morning with Paula for "story time" —their term for Product Backlog refinement. All Developers attend because each sees work through their own unique lens and has valuable input to the discussion. Because of these regular refinement sessions, Sprint Plannings have become more productive—and shorter. The Scrum Team now spends less time forecasting because the most important PBIs are "ready" and fresh in everyone's minds. As the Product Backlog goes through seasons of increased or decreased churn, Product Backlog refinement can increase or decrease proportionally.

Scrum Artifacts

Scrum's artifacts represent the work to be done in the product and the Sprint, as well as the work and value that have already been done within the product. Each artifact has a clear commitment and clear ownership. Each artifact is structured in a way that maximizes transparency of key information while providing opportunities for inspection and adaptation.

There are three artifacts in Scrum:

- Product Backlog

- Sprint Backlog

- The Increment

The Scrum artifacts are designed to maximize transparency of key information. Thus, everyone inspecting them has the same basis for adaptation. Each artifact also includes a commitment to ensure that it provides information that enhances transparency and focus against which progress can be measured. Table 1-10 lists the commitment for each Scrum artifact. These commitments exist to reinforce empiricism and the Scrum Values for the Scrum Team and their stakeholders.

TABLE 1-10 Scrum artifacts and related commitment.

Artifact	Commitment
Product Backlog	Product Goal
Sprint Backlog	Sprint Goal
The Increment	Definition of Done

Note Burndowns (product, release, Sprint, etc.)—as artifacts—were removed from the *Scrum Guide* in 2011. Their inclusion was considered too prescriptive. Although it's important for the Scrum Team to monitor progress toward a goal, many practices could support this. Burndowns are certainly a popular option and are still acceptable and used by many Scrum Teams. No technique will replace the importance of empiricism. In complex environments, such as software development, what will happen is unknown. The Scrum Team can only use what *has happened* to influence its decision making.

Product Backlog

The Product Backlog is an ordered list of everything requested of the product. It is the single source of requirements for any potential changes to be made. Each item in the Product Backlog is called a Product Backlog item, or simply a PBI. A PBI can be a *happy* thing that doesn't yet exist in the product, like a feature or an enhancement. PBIs can also be *sad* things, like a bug or defect to be fixed. PBIs can range from extremely important and urgent to silly and trivial. Because of this variety, I affectionately refer to the Product Backlog as a list of *desirements*. Fellow Professional Scrum Trainer Phil Japikse calls them *requestaments*. At some point, somebody, somewhere, for some reason *desired* or *requested* each item in the Product Backlog. Whether or not that item gets developed is a separate decision—one made by the Product Owner.

Note The Product Backlog is a dynamic, living document. It is never complete and will constantly change as desires change. The Product Backlog will exist as long as the product exists.

These types of items are considered valid PBIs:

- User story
- Epic

- Feature
- Enhancement
- Behavior
- Journey
- Scenario
- Use case
- Bug/defect

These types of items are generally *not* considered valid PBIs:

- Task ("refactor code," "create acceptance tests," etc.)
- Acceptance criteria ("page content in German and English," "report exportable as PDF," etc.)
- Acceptance tests
- Nonfunctional requirements (such as "ticket search returns in less than 5 seconds")
- Definition of Done ("code is peer-reviewed," "build pipeline exists," "all tests pass," etc.)
- Impediment ("forgotten password," "expired subscription," "crashed hard drive," etc.)
- Anything else that doesn't provide direct value to the users/stakeholders

Each PBI should be clearly identified by a title. This is the minimum amount of information required to add it to the Product Backlog. If the Product Owner is interested and decides it's worth the time to describe it further, then a short description should be added. This description should be written in an easy-to-understand, business-friendly language. The PBI should also be assigned a business value. Once the Developers size it, it can be ordered with the other items in the Product Backlog. These conversations and sizing can be done during Product Backlog refinement or Sprint Planning. Table 1-11 lists the ways in which the Developers interact with the Product Backlog.

TABLE 1-11 Developer interactions with the Product Backlog.

Activity	When
Inspect it.	Any time.
Add a new PBI to it.	Any time (if allowed by the Product Owner).
Refine it (includes changing or removing PBIs).	Product Backlog refinement, Sprint Planning, or Sprint Review (with whole Scrum Team).
Reorder it.	Any time (if allowed/approved by the Product Owner).
Forecast work from it.	Sprint Planning (with whole Scrum Team) or any time during the Sprint after the forecast has been met and capacity allows.

I'm often asked if being *responsible* for the Product Backlog means that the Product Owner has to be the person who actually creates the PBIs (such as writing the user stories). The answer is no. The

Product Owner can have a Developer or stakeholder (business analyst, customer, or user) create and even manage the PBIs. The Product Owner has the authority to update any item, such as making it more understandable, changing acceptance criteria, or removing any item deemed unnecessary.

A PBI represents a request from somebody. It can take any number of shapes or forms. A user story is quite popular for teams practicing agile software development. This is primarily because user stories are lightweight and not technical. User stories describe the ask from the customer's or user's perspective. It is not a requirements document or a technical specification, nor is it a communiqué between the requirements giver and the Scrum Team.

A user story represents a "what" that the product should do. A well-written user story description will explain who wants or who would benefit from the capability, as well as how and why it will be useful. In a single sentence, a user story description provides lots of context, as well as a value proposition. Even bug reports can be written in the form of a user story.

The most popular format of a user story description looks like this: *As a (role), I want (something) so that (benefit)*. An example would be, "As a returning technician, I want the app to remember my credentials so that I don't have to sign in each time." Another example would be, "As a visitor to the Fabrikam Fiber website, I want to see a list of recent tweets so that I know that Fabrikam Fiber and its products are alive and well." Anyone looking at either PBI instantly knows the context and value to the end user.

> **Tip** Don't put the user story description in the PBI's title. I've seen teams do this and the cards and lists get very messy. Keep the description short—and the title even shorter.

Having a title and a short, meaningful description is a good start. To properly complete a user story, communication between the Scrum Team and knowledgeable stakeholders is required. A complete user story includes the *three C's:* card, conversation, and confirmation. I'll elaborate.

The *card* is already done at this point. You have written a title and the description (in user story format) on a sticky note, an index card, or a work item. This allows somebody to reference the user story during conversation, update it, estimate it, rank it, and so forth.

Next, the *conversation* takes place with the customers, users, or domain experts. This conversation is meant to exchange thoughts and opinions. It can take place at any time with the Product Owner, the stakeholders, and the Developers as needed. If the Developers are to be involved, it could take place at Product Backlog refinement, Sprint Planning, or Sprint Review. Conversations that yield examples, especially executable and testable examples, are preferred over formal documents and mock-ups.

Finally, the *confirmation* occurs. Here the user story's acceptance criteria are agreed on and recorded. These criteria help determine when the PBI is done—when all criteria are met and, in accordance with the team's Definition of Done, the PBI is done. If and when the PBI gets forecasted for a Sprint, the Developers will create the appropriate acceptance tests to validate the acceptance criteria.

Tip Don't create tasks, tests, or code for a PBI *before* the Sprint in which it has been forecasted for development. Conditions can change rapidly, forcing a change to the PBI or its acceptance criteria. Time spent creating these kinds of artifacts ahead of time will often be wasted. The plan for developing a PBI, as well as any code or tests, should be created at the latest *responsible* moment—such as during the Sprint. Even though you will always know more tomorrow than today, you should avoid falling into the trap of doing things at the last *possible* moment. Besides, you have plenty of work in the Sprint Backlog without designing or coding ahead.

Whoever creates a PBI should be sure to INVEST in it. The mnemonic *INVEST* is a reminder of the characteristics of a good PBI:

- **I–Independent**   As much as possible, the PBI should stand alone, without any dependency on another PBI. Try to create PBIs that don't have long "dependency chains."

- **N–Negotiable**   The PBI can be changed and rewritten up until it gets forecasted, but significant changes after being forecasted should be avoided and minimized. Minor tweaks are okay as long as they don't greatly affect the size that the Developers estimated.

- **V–Valuable**   The PBI must deliver value to the user or the customer. This value is often delivered through a visible, tangible user interface, but not always.

- **E–Estimable**   The Developers must be able to size the PBI. If too little is known, it will be difficult for them to have a meaningful conversation and come to any kind of shared understanding or consensus on size.

- **S–Small**   The PBI must be small enough that the team can develop it—according to the Definition of Done—in a single Sprint and preferably within a few days. There are many suitable techniques for decomposing/splitting PBIs.

- **T–Testable**   The acceptance criteria are clearly understood and can be tested. This is probably the most important characteristic. It relates to the third "C"—confirmation.

You can think of the Product Backlog as an iceberg, as shown in Figure 1-2. PBIs on the top, above the surface, are what the Developers have forecasted for the current Sprint. These items are very clear (denoted by a lighter color), sized small enough, and ready to be worked. Below the surface, the Product Owner knows what other PBIs they would like in the future, but it won't be clear which ones surface until the next Sprint Planning. These items are generally understood and sized so that a release plan can be devised. These are the items that will be in scope during upcoming Product Backlog refinement sessions.

Below the release, you will find all the other PBIs that may or may not make it into a future release. Some of these may have only a title or a vague description (denoted by darker colors) of the desired functionality. Some PBIs may even remain in these cold, chilly depths for eternity—or until they are deleted.

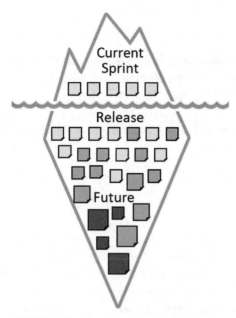

FIGURE 1-2 The Product Backlog iceberg.

The Product Owner is accountable for managing the Product Backlog. This includes the clarity and precision of its contents. They should also ensure that the Product Backlog is visible to all interested parties. The Product Owner will order (prioritize) the PBIs to meet the Product Goal. The PBIs at the top of the ordered Product Backlog will, more than likely, be what the Developers work on next. The Product Goal should be discernible by studying the order and content of the PBIs. If necessary, the Scrum Master should coach the Product Owner on how to manage the Product Backlog more effectively.

Note Starting in 2011, the *Scrum Guide* began using the term "order" instead of "prioritize." This subtle change has led to some confusion, which is why I typically use both terms together. For example, a Product Owner might urgently want to sell products and accept payments, but this capability requires a dependent shopping cart feature to be developed first. Although selling products and accepting payments is a higher *priority*, developing the shopping cart is of higher *order*—because of the dependency.

Creating an effective Product Backlog can be difficult. It can be time-consuming. It can become political. However, once you've gone through the exercise of creating the Product Backlog, you'll wonder how you ever got along without one. Just having everything in one ordered list can be a game changer!

Fabrikam Fiber Case Study

Creating the initial Product Backlog was difficult. Requirements, feature requests, and bugs were tracked by different people in different formats. Giving up control of those lists started a turf war—but in the end, it was best for the product. When possible, PBIs were converted to a user story format. Today, the Scrum Team maintains its Product Backlog in Azure DevOps. The administrator gave permission to anyone on the Scrum Team to manage the Product Backlog. Everyone else can only view it. Paula (the Product Owner) is considering granting access to some additional stakeholders to help her create and manage PBIs.

Sprint Backlog

The Sprint Backlog contains the Sprint Goal, the PBIs forecasted to be developed during the Sprint, and the plan (for example, tasks or tests) for developing them, and the Sprint Goal. The Sprint Goal and forecasted PBIs were agreed on and selected through collaboration of the Scrum Team during Sprint Planning. The plan for developing them was agreed on through collaboration of the Developers. The Sprint Backlog is the output of the Sprint Planning and represents the Developers' forecast of *what* functionality will be in the next software product Increment, the plan for *how* it will happen, and the Sprint Goal explaining *why* the Scrum Team is doing it.

The Developers own the Sprint Backlog—they are wholly responsible for how to implement the PBIs, as long as they do so in accordance with the Definition of Done. Nobody can tell the Developers how to develop the Increment. Nobody except the Developers themselves can add, edit, or remove items from the Sprint Backlog. The Sprint Backlog should be kept up to date and visible to the Developers. It provides a real-time picture of the work that they plan to accomplish during the Sprint. The Product Owner, the Scrum Master, and stakeholders don't necessarily need access to the Sprint Backlog.

Tip Increasing the Sprint Backlog's visibility beyond the Developers is an invitation for the three "M's": misunderstanding, meddling, and micromanaging. Allowing stakeholders, or any interested parties, to view the Product Backlog or burndown charts (if used) is preferable to seeing all of the technical and tactical details in the Sprint Backlog. In fact, even the Product Owner might want to stay out of the Sprint Backlog to avoid misunderstanding, meddling, and micromanaging.

Table 1-12 lists the ways in which the Developers interact with the Sprint Backlog.

TABLE 1-12 Developer interactions with the Sprint Backlog.

Activity	When
Inspect it.	Any time
Move a PBI from the Product Backlog into it.	Sprint Planning or any time afterward (with Product Owner collaboration)

Activity	When
Add, update, split, or remove a task or test in it.	Sprint Planning or any time afterward until Sprint Review
Take ownership of a new task or test in it.	Any time (as the plan demands)
Update status of a PBI, task, or test in it.	Any time (as status changes)
Estimate remaining work.	At least daily, if not more often

All Developers should collaborate on the plan and create the tasks or tests. Scrum Teams must be cross-functional for just this reason. Everyone can and should contribute. This will create a richer and more honest Sprint Backlog than if only one or two "specialists" create the plan. A good approach is to start with a conversation in order to understand the PBI and discuss any potential plan. The plan could have begun emerging weeks earlier at Product Backlog refinement. The plan can emerge onto sticky notes or a whiteboard, and then be transferred to a tool like Azure DevOps. There's zero technical debt and waste in a discussion and close to zero in a set of sticky notes.

The Developers must do their best to identify *all* work in the Sprint Backlog, not just the design, coding, and testing items. There may be learning, installing, data entry, meetings, documentation, deployment, training, and other tasks. The Definition of Done might drive the creation of additional tasks for each PBI as well.

> **Tip** Have the Definition of Done posted nearby during Sprint Planning. It will help the Developers as they brainstorm, forecast their work, and create the plan. Also, depending on how the last Sprint went, there may be additional work in the Sprint Backlog related to improvements identified at the Sprint Retrospective.

The Developers should estimate their Sprint Backlog items at least daily. This can be done before or after the Daily Scrum but preferably not during. Most teams I coach prefer to estimate work remaining before the Daily Scrum so that they will have accurate analytics to reference. Some high-performance Scrum Teams won't bother tracking hours or estimating remaining work on tasks. They focus on the Sprint Goal and delivering the PBIs, not the tasks. It is more difficult to assess progress without this information, but it's safe to assume that they've built a ton of trust with the Product Owner and stakeholders at that point.

> **Note** Professional Scrum Teams don't consider the time spent working on a task. Tracking actual hours is counterproductive to obtaining the Sprint Goal. I would even call it wasteful. If, however, an organization requires its employees to track their time to get paid or bill a client, that's a separate discussion. The worry is that once such a metric is created, it would be used in a *command and control* way. For example, a manager might see that a set of UX design tasks took 28 hours and then use that as an estimate for future work, or as a stick to beat the Developer with if their next set of tasks goes beyond that number—which it could, because development in a complex space is very difficult and full of risk.

The Sprint Backlog will be empty at the start of a Sprint. It will begin to emerge during Sprint Planning, and (ideally) be populated with tasks by the first few days of the Sprint. Tasks can be identified all the way through the Sprint, but it is difficult to assess progress if you don't know what the plan is. Even Professional Scrum Teams must change their plan sometimes. Each PBI introduces new complexities that can derail an execution plan. New tasks may have to be created mid-Sprint. This is not uncommon.

Tip In Scrum, work should never be directed or assigned. When creating a new Sprint Backlog task, don't assign it to anyone. For example, you should resist the urge to assign the *testing* tasks to Toni (even though she has a background in testing). Doing so will decrease opportunity for other team members to collaborate and learn. When the time is right, the team should decide who will take on a task. The team will take many factors into account, including the background, experience, availability, and capacity of the Developer. I will spend more time on this topic in Chapter 8.

As the Scrum Team improves, it will learn to manage risk better by taking on riskier work early. The team will also become better at creating a plan and identifying the full spectrum of tasks, at least at a high level, during Sprint Planning. It's okay for the more distant tasks to be coarsely defined. As the time nears for that piece of work to begin, the team can decompose and reestimate.

If Sprint burndown charts are being used, their trend lines will help predict when the Developers will be done with their work. Observers of the burndown charts need to understand that the Developers will know more tomorrow than they did today—so they should expect change. This means that the burndown may occasionally go flat or go up. The Scrum Master should be able to provide this education to anyone with questions.

Fabrikam Fiber Case Study

During Sprint Planning, the Developers brainstorm the plan for developing the Increment and achieving the Sprint Goal. When they were just starting out with Scrum, they would only get one or two PBIs planned out. They delayed the planning of the rest of the PBIs until later in the Sprint. Over time, they've improved in the way they decompose and plan their work, now being able to create most of the Sprint plan during Sprint Planning. Regular Product Backlog refinement has made Sprint Planning more efficient as well. The Developers still estimate tasks in hours but have made improvements in that process. Originally, they would have the "experts" in the various task areas do the estimates. That made estimation go quicker, but during development, they would usually blow their estimates because the expert wasn't always the one who performed the work. They now estimate the tasks collaboratively. They also find that they are under as many times as they are over on their estimates. They can live with that.

The Increment

Scrum is an iterative and incremental product development framework. The word *incremental* means "occurring in especially small increments." Each Sprint is an especially small period of time during which the team develops one of these small increments. These small periods of time (Sprints) reduce risk by maximizing collaboration and feedback. Incremental delivery of a Done product ensures that a useful version of the working product is always available.

An Increment is a concrete stepping-stone toward the Product Goal. Each Increment is additive to all prior Increments and thoroughly verified, ensuring that all Increments work together. The entire Scrum Team is accountable for creating a valuable, useful Increment every Sprint. In order to provide value, the Increment must be released.

Multiple Increments may, in fact, be created within a Sprint. The sum of the Increments is presented at the Sprint Review to support empiricism. An Increment may be delivered to stakeholders *prior* to the end of the Sprint. The Sprint Review should never be considered a gate to releasing value. Work cannot be considered part of an Increment unless it meets the Definition of Done.

> **Tip** If possible, make the Increment available to the Product Owner and stakeholders throughout the Sprint. Think of it as a hands-on demo. As the Developers finish a PBI, a demo environment is provisioned for people to inspect the product. This doesn't have to be any kind of a formal testing, such as user acceptance testing (UAT) with formal testing agendas such as manual acceptance tests—just something that supports exploratory testing and encourages feedback during the Sprint. This is better than waiting until Sprint Review for feedback, especially from the Product Owner, and especially if they don't like it!

> **Note** The notion of the Increment being releasable means that it *could* be released (to the customer, to production, to the app store, etc.) if the Product Owner chooses to do so. This is possible because the Increment contains only Done PBIs. PBIs aren't done until they meet the level of quality defined in the Definition of Done. The Product Owner may decide to wait until several related PBIs are completed (release by feature/scope), until a certain point in time (release by date), or as each PBI is done (continuously). Older *Scrum Guides* referred to the increment being *potentially* releasable—to stress the fact that it's the Product Owner's decision and doesn't necessarily happen automatically.

Definition of Done

The Definition of Done is a formal description of the state of the Increment when it meets the quality measures required for the product. The moment a PBI meets the Definition of Done, an Increment is born. In other words, the Definition of Done creates transparency by providing everyone with a shared understanding of what work was completed as part of the Increment.

If a Product Backlog item does not meet the Definition of Done, it cannot be released or even inspected at the Sprint Review. Instead, it returns to the Product Backlog for future consideration. If the Definition of Done for an Increment is part of the standards of the organization, all Scrum Teams must follow it as a minimum. If no such standards exist, the Scrum Team must create a Definition of Done appropriate for the product. Once created, the Developers are required to conform to the Definition of Done.

> **Note** If there are multiple Scrum Teams working together on a product, they must mutually define and adhere to a common Definition of Done. Individual Scrum Teams may choose to apply a more stringent Definition of Done within their own teams, but they cannot apply less rigorous criteria. I cover this topic in more detail in Chapter 11.

The Definition of Done is not a formal artifact in Scrum, but it might as well be. It is very important. It serves as the commitment for the Increment and, as a corollary, as a commitment to each PBI going into the Increment. In other words, Done is the state when a PBI has been developed according to team's Definition of Done. Scaling that up, Done is also the state when the Increment containing all the done PBIs becomes Done and, thus, becomes releasable.

The Definition of Done is a simple, auditable checklist created by the Scrum Team. It may be based on organizational standards, product constraints, and Developer practices. The Definition of Done must be made transparent to the stakeholders, and they must understand its content and purpose. This is why it must be simple and explainable to stakeholders.

Here is a simple Definition of Done:

- All acceptance criteria have been met.
- A build pipeline exists.
- No code analysis errors or warnings exist.
- All new code is covered by unit tests.
- All automated unit and acceptance tests pass.
- A release note exists.
- The Product Owner likes the work.

Definitions of Done can be quite long and complex. Everything in the definition should be achievable, although some items may not be applicable. For example, if the Developers are working on a PBI that is mostly graphic design–centric, there may not be any code to unit-test. For all PBIs that have code, however, the team must create unit tests.

The Definition of Done is there for a reason. The Developers should never cut corners by ignoring any part of it in order to complete the forecast. The team has already unanimously decided that quality, as defined in the Definition of Done, is more important than getting done more quickly.

> **Note** The Definition of Done is a *minimum* standard. There may be times when the Scrum Team may want to do more than the minimum. This is acceptable as long as the extra effort is justified and not considered "gold plating." Gold plating is when one or more Developers continue to work on a PBI beyond what is necessary. This extra work is typically not worth the value that it adds to the product. When in doubt, check with the Product Owner before adding new functionality.

The Definition of Done should be inspected during the Sprint Retrospective. Over time it should increase, as the Scrum Team wishes to improve its quality practices. It's a smell—and possibly a dysfunction—if a Definition of Done remains unchanged Sprint after Sprint.

> **Note** In this book I will differentiate between done and Done. I will use lowercase *done* to simply mean finished with a task or piece of work. I will use uppercase *Done* to mean that the PBI or Increment is completed according to the Definition of Done.

Undone Work

An explicit and concrete Definition of Done may seem like a small piece of the development process, but it can be the most critical checkpoint during a Sprint. Without a consistent meaning of what Done means, Developers cannot have meaningful conversations with stakeholders, progress cannot be accurately assessed, planning will be impacted, and even the Increment's value becomes questionable. Having a transparent Definition of Done—that the team adheres to—ensures that everyone knows the Increment produced at the end of Sprint is of high quality.

Professional Scrum Developers should not generate undone work. They should also make sure that "done" means Done. In the long run, it will be cheaper to hold fast to the Definition of Done by improving their practices than to keep sprinting with an unknown amount of work still to be finished at the end of the release. For example, I have worked with teams that will complete the design and coding Sprint after Sprint but accrue the testing work until the Sprints just before release. Don't do this.

Note I will often ask organizational managers and decision makers which they value more: output or outcome. They always answer outcome. However, when I test them, their reflex is to value output instead. For example, let's assume a Scrum Team has five Developers: three coding specialists, one testing specialist, and one operations specialist. If each works to capacity doing their specialty, they are maximizing output, but not necessarily getting anything Done. Lots of code gets written, but only some testing. If they collaborate and work outside of their specialties while they swarm or mob, they can complete more PBIs per Sprint, maximizing outcome. To the managers, however, it appears as though the team is slacking because they may not be "keeping busy to their capacity" or "doing the (specialty) job they were hired for." This is an opportunity for the Scrum Master to coach the organization on this new, weird thing called Scrum and let them know that sometimes a team has to slow down to get more things Done! I cover this weird way of working in more detail in Chapter 8.

If the Product Owner looks at an Increment and doesn't know how much work still needs to be done, they won't really know when it will be ready to release. Any projections might be off . . . wildly. There may be a need for one or more "stabilization" Sprints—which are not a thing in Scrum—at the end of the release just to tackle all the accumulated undone work. If this occurs, conversations with stakeholders can become contentious.

What's even worse is that the undone work from the Sprints accumulates exponentially, not linearly. Subsequent Sprints will require even more work to reach Done. This is due to a loss of context, code drift, and other factors. For example, 4 hours of undone work per Sprint for 6 Sprints won't be 24 hours of work but could be more like 80 hours. This "undone work" uncertainty has no place in a framework that promotes transparency. Every effort should be made to eliminate undone work and so-called "testing," "stabilization," and "hardening" Sprints. The team should slow down, adhere to the Definition of Done, deliver more Done functionality, and realize more outcome. Output should never be the goal.

The Scrum Values

Successful use of Scrum depends on people becoming more proficient in living these five Scrum Values: Commitment, Focus, Openness, Respect, and Courage. I will explain these values and provide some examples in this section.

The Scrum Team *commits* to achieving its goals and to supporting each other. Their primary *focus* is on the work of the Sprint to make the best possible progress toward these goals. The Scrum Team and its stakeholders are *open* about the work and the challenges. Scrum Team members *respect* one another to be capable, independent people and are respected as such by the people with whom they work. The Scrum Team members have the *courage* to work on tough problems and do the right thing—even when nobody is watching.

These values give direction to the Scrum Team with regard to their work, actions, and behavior. The decisions that are made, the steps taken, and the way Scrum is used should reinforce these values, not diminish or undermine them. The Scrum Team members learn and explore the values as they work with the Scrum events and artifacts. When these values are embodied by the Scrum Team and the people they work with, the empirical Scrum pillars of transparency, inspection, and adaptation come to life, building trust.

Table 1-13 lists some examples of how Developers can live the Scrum Values.

TABLE 1-13 Examples of how Developers can live the Scrum Values.

Scrum Value	Examples
Commitment	<ul><li>Committing to doing Scrum fully (according to the *Scrum Guide*)</li><li>Committing to collaborating, sharing, and learning</li><li>Committing to the achieving the Sprint Goal and delivering outcomes/value</li><li>Committing to continuous learning and improvement</li><li>Committing to adhering to the Definition of Done</li><li>Committing to the Scrum Values</li></ul>
Focus	<ul><li>Collaborating as a team to decide what to work on first</li><li>Creating a fit-for-purpose, Done increment without gold plating or technical debt</li><li>Collaborating as a team on a single PBI at a time (also known as swarming or mobbing)</li><li>Respecting the timebox to not overly plan, design, or develop</li><li>Respecting shared accountability to focus on overall outcome</li><li>Keeping the Product Goal or Sprint Goal in mind as work is planned/executed</li><li>Focusing on outcomes, not output</li></ul>
Openness	<ul><li>Being transparent about the team's progress (or lack thereof)</li><li>Asking for help as well as offering help</li><li>Collaborating with other Developers, the Product Owner, and stakeholders</li><li>Sharing and hearing others' perspectives</li><li>Admitting to being wrong and learning from mistakes</li><li>Being "teachable"</li></ul>
Respect	<ul><li>Acknowledging that people are naturally resourceful, creative, and capable</li><li>Accepting the diverse backgrounds, experiences, and range of skills</li><li>Knowing that people are motivated by autonomy, mastery, and purpose</li><li>Respecting the order of items in the Product Backlog</li><li>Assuming that people have good intentions and are doing their best</li><li>Listening, hearing, and acknowledging others' opinions and perspectives</li></ul>
Courage	<ul><li>Being transparent about the team's progress (or lack thereof)</li><li>Not releasing (or even showing) undone work to stakeholders</li><li>Able to say "I don't know" and ask for help</li><li>To hold each other accountable for not meeting commitments</li><li>Admitting to being wrong, making mistakes, or making wrong assumptions</li><li>Sharing a dissenting opinion or engaging in productive conflict</li><li>Developing something that hasn't been done before</li><li>Beginning development without having all of the "requirements" or details</li></ul>

This is just a partial list of ways that the Developers can live the Scrum Values on a daily basis. The rest of this book will list many complementary practices and associated ways to encourage the Scrum Values when applying those practices. Professional Scrum practitioners can quickly identify when the Scrum Values are being applied, encouraged, or hindered.

Professional Scrum

In this chapter I've described Scrum as it's defined in the *Scrum Guide*. When a team begins practicing Scrum, they will select their own complementary practices (user stories, Fibonacci estimation, story mapping, acceptance test-driven development, etc.). This will result in that team having created their own process. In other words, Scrum + complementary practices = a team's process. While custom-tailored to your team and organization, simply practicing *mechanical Scrum* like this is not enough. Ken Schwaber and Scrum.org offer something better: *Professional Scrum*. More than just a marketing term, Professional Scrum is a combination of several concepts and adherence to their principles.

Professional Scrum is defined as a combination of these elements:

- **Mechanical Scrum** Practicing Scrum according to the *Scrum Guide*

- **Continuously practicing empiricism** Relentlessly seeking improvement through regular inspection and adaptation and then being transparent about it

- **Continuously practicing the Scrum Values** Living the values of commitment, focus, openness, respect, and courage each day through interactions with individuals

- **Continuously practicing technical excellence** Uncovering better ways of working and developing the product and sharing those learnings with others on the Scrum Team

> **Note** I will refer to *"Professional Scrum"* in this book hundreds of times. I will mention *Professional Scrum* Teams, *Professional Scrum* Developers, *Professional Scrum* Masters, *Professional Scrum* Product Owners, and *Professional Scrum* Trainers. I do this repeatedly to differentiate over simple mechanical Scrum or even "certified" Scrum—both of which tell me nothing about the individual's or team's relentless pursuit of empiricism and improvement.

The Professional Scrum Developer

The *Scrum Guide* does not provide guidance on *how* to develop a product. In fact, during the time between Sprint Planning and the Sprint Review, the guide is intentionally vague. Other than requiring a Daily Scrum and suggesting regular Product Backlog refinement, not much guidance is provided at all. In fact, the rules only state that a Daily Scrum should occur, taking no longer than 15 minutes.

So, what about the other 7 hours and 45 minutes of the day? What should the Developers be doing during that time? That's the million-dollar question. I could simply say "it depends," at which point, you would throw this book in the recycle bin. Well, the short answer is, they should do the right thing, even when nobody is looking, while constantly improving in all they do. In other words, they should be practicing Professional Scrum. There are many longer answers, but it will take me another 10 chapters to go through it. I hope you have an interest as well as some patience.

Remember that developing a complex product like software is a risky endeavor for both the Scrum Team and the stakeholders (the customer or users). The development process is a complex undertaking consisting of analyzing, designing, coding, testing, and deploying. More things can go wrong than right. Any small mistake or fault can lead to wasted effort—if you are lucky. Some mistakes can lead to outright damage. Professional Scrum Teams understand this, and they make sure their stakeholders understand this too. Ideally the stakeholders will share in these risks. This means that the stakeholders as well as the Scrum Team understand that they are both equally responsible for identifying and mitigating these risks, as well as sharing responsibility if a risk evolves into waste or a disaster of some sort.

Let's drop the customer out of the discussion for a minute. Developers on a Scrum Team collectively own many things. They own their successes and failures, just as they collectively own the code, bugs, technical debt, and other issues. Professional Scrum Developers should learn to rely on their fellow team members and to trust them. They know that they must be resolute, forthright, transparent, and able to compromise in order to reach their goals.

When I'm meeting with a new team, I will often ask what they think the software developer's job is. "To write code," is the almost universal answer that I hear. Being a career developer myself, I used to give the same answer. As my understanding has evolved, this answer now irks me a bit. I believe that a better answer would be that a developer's job is *to provide value in the form of working product*. This answer encapsulates the attributes of a Professional Scrum Developer, which I've listed in Table 1-14.

TABLE 1-14 Attributes of a Professional Scrum Developer.

Attribute	Attribute
Collaborates whenever possible	Isn't hesitant to ask for help
Collectively owns all aspects of product development	Isn't afraid to work outside of their comfort zone
Doesn't know everything but is willing to learn	Knows that they are part of a team and that they have an equal voice
Doesn't release undone work	Looks for and minimizes waste in their practices
Has a stake in the success of the product	Makes commitments to their team members and keeps them
Has the right and responsibility to maximize self-managing capabilities	Obeys the Definition of Done without generating technical debt
Is honest in their estimates, goals, and forecasts	Only does work that provides value to the product
Is more than just a coder or programmer	Plans realistic goals and then commits to achieving them
Is not a hobbyist, but a professional	Reflects the Scrum Values as they work
Is responsible for the quality of the product	Respects the *Scrum Guide* and follows it rules
Is transparent in what they do and how they do it	Says "no" when appropriate

Chapter Retrospective

Here are the key concepts I covered in this chapter:

- **Scrum Guide**   The official definition of Scrum written by Ken Schwaber and Jeff Sutherland—the co-creators of Scrum. The *Scrum Guide* describes the Scrum framework and codifies the rules of Scrum. You should download it from *www.scrum.org/scrumguides* and read it now. Its updates will supersede any guidance I've provided in this chapter.

- **Empiricism**   Asserts that knowledge comes from experience and decisions should be made based on what is observed and known. Inspection, adaptation, and transparency enable empiricism.

- **Developers**   The individuals on the Scrum Team who do the work. The Developers are a cross-functional group, of typically three to nine professionals who develop the forecasted work during the Sprint. Developers are more than just coders/programmers. They also include testers, architects, database professionals, UI/UX designers, analysts, and IT professionals.

- **Product**   A vehicle to deliver value to stakeholders. It has a clear boundary and a well-defined consumer, and achieves some measurable value. A Product could be a service, a physical product or something more abstract—like software.

- **Product Goal**   Describes a future state of the product that can serve as a target for the Scrum Team to plan against. The Product Goal is evident in the Product Backlog.

- **Product Owner**   Represents the voice of the stakeholder (user or customer) and is responsible for maximizing the value of the product and the work of the Developers.

- **Scrum Master**   The Scrum Master is responsible for ensuring Scrum is understood and practiced correctly within the Scrum Team as well as the organization.

- **Stakeholders**   Anyone interested in the successful development of the product. Stakeholders can be managers, executives, analysts, domain experts, attorneys, sponsors, members from other teams, customers, and users of the software.

- **Sprint**   A fixed-length event of one month or less that contains the other Scrum events.

- **Sprint Planning**   The first event in the Sprint where the work to be performed is forecasted, along with a plan for developing it, and a Sprint Goal describing why the work is being performed.

- **Daily Scrum**   An opportunity for Developers to synchronize activities and create a plan for the next 24 hours.

- **Sprint Review**   An opportunity for stakeholders to inspect the Done PBIs in the releasable Increment in order to provide feedback.

- **Sprint Retrospective**   An opportunity for the Scrum Team to inspect its practices and adapt a plan for improvement in the next Sprint.

- **Product Backlog refinement**   An opportunity for the Scrum Team to discuss upcoming PBIs in order to add detail and estimates to make them ready for Sprint Planning. Product Backlog refinement is optional but should take up no more than 10 percent of the capacity of the Developers.

- **Product Backlog**   An artifact that contains an ordered list of everything that might be needed in the software product, including bugs or defects to fix. Items in the Product Backlog are called Product Backlog items (PBIs).

- **Sprint Backlog**   An artifact that contains the forecasted PBIs, the plan for developing them, and the Sprint Goal describing why the Scrum Team is doing this work.

- **The Increment**   An artifact representing the sum of all done PBIs during this and previous Sprints. An Increment serves as a concrete stepping-stone toward the Product Goal.

- **Definition of Done**   A shared understanding of what it means for the Developers to be done with the development of an individual PBI and, as a corollary, the Increment itself.

- **Undone work**   Any PBI that was started in a Sprint but not yet finished. The Developers should operate in a way to minimize undone work as it is considered waste.

- **Scrum Values**   Commitment, focus, openness, respect, and courage. Successful use of Scrum depends on people becoming proficient in living the Scrum Values.

- **Mechanical Scrum**   Simply following Scrum according to the *Scrum Guide*.

- **Technical excellence**   Experimenting with and uncovering better ways of developing and sharing those learnings with others on the team.

- **Professional Scrum**   Practicing mechanical Scrum but with a relentless and continuous application of empiricism, the Scrum Values, and technical excellence.

- **Professional Scrum Developer**   A Developer who operates under the principles of Professional Scrum.

CHAPTER 2

Azure DevOps

Since Microsoft first introduced what would become Visual Studio Team System (VSTS) back in 2004 (code name *Burton*), development teams have begun improving the way they plan, track, and manage their software development projects. No longer are they tracking code changes in meaningfully named zip files or Microsoft Visual SourceSafe, bugs, and requirements in Microsoft Excel, and performing automated builds using batch files. Team Foundation Server, a component of VSTS, integrated those pillars of software development and even tossed in reporting so that everyone could stay informed. The game of software development had changed forever; it had gone professional. In this chapter, I will provide a brief history of Microsoft's application lifecycle management (ALM) and DevOps products and introduce Azure DevOps Services, Azure DevOps Server, and Visual Studio.

A Brief History

In its first iteration, this stack of tools was marketed as only providing support for the software development lifecycle (SDLC). This lifecycle included everything to do with developing a software product, such as requirements, architecture, coding, testing, configuration management, and project management. In the subsequent release of VSTS 2008, Microsoft (and the rest of us) refactored its thinking to regard these tools as having much broader capabilities for supporting application lifecycle management (ALM). To further this point, Microsoft also added tools for database developers to the VSTS family. ALM includes everything that is part of the application lifecycle, not just development. ALM combines business management with software engineering. Scrum, or any other agile framework, process, or methodology, is encompassed by ALM.

With the 2010 version, Microsoft doubled down on their support for ALM. They introduced Microsoft Test Manager to allow teams to create and manage their testing effort. Hierarchical work items enabled a richer breakdown of the planned work. Product Backlog item (PBI) work items could now be linked to multiple child task work items. By using Lab Management, teams could configure environments of virtual machines and automate the build, deploy, test cycle against complex environments. It was during the 2010 product cycle that Microsoft gave us the first official Scrum process template, thus acknowledging and formalizing their support for Scrum. Also in 2010, the Microsoft Developer Division began development of a software-as-a-service (SaaS) offering based on Team Foundation Server.

The 2012 version continued the tradition of providing strong ALM tooling, especially in the area of agile (Scrum) planning and management capabilities. Feedback became a first-class concern with the introduction of Feedback Manager and storyboards. Microsoft also continued their migration of capabilities from desktop clients, like Team Explorer, to the browser. The web portal, Team Web Access, now included a team project landing page and rudimentary dashboard. The concept of a team was enhanced, allowing multiple teams per project—each with their own members, backlogs, and iterations. Scrum teams were given an interactive task board so that they could plan and track their tasks during a Sprint. Microsoft introduced their cloud-based Team Foundation Service preview, publicly revealing their intentions to start providing Team Foundation Server as a service.

In 2013 Microsoft introduced additional collaboration features such as portfolio management (hierarchical backlog levels), Kanban support, work item charts, and team rooms. Test management and execution continued to migrate from the Microsoft Test Manager desktop client to the browser, with the ability to create test plans, test suites, and test cases. Microsoft also saw the writing on the wall and started including Git support in both the Visual Studio client as well as Team Foundation Server. This reveals the world's (including Microsoft's) fascination with Git. Additionally, you could browse, comment on, diff, and view history of both Team Foundation Version Control (TFVC) as well as Git version control repositories. Release Management was finally added, due to Microsoft's acquisition of the InRelease product from InCycle Software. On the cloud front, Microsoft officially released Visual Studio Online—their set of Azure-hosted development services. Microsoft would rename this product a few more times in the coming years.

The 2015 version continued to add polish to the team experience by allowing multiple, customizable dashboards and enhanced Kanban board capabilities. Additionally, more and more functionality was moved, or at least duplicated, from the Visual Studio IDE (Team Explorer) and Microsoft Test Manager to the web portal. It was clear that Microsoft saw the browser as the future of their ALM tools. Developers also got a new build system with a completely rewritten architecture—one that was not based on XAML. Team Foundation Server extensions continued to emerge on the Visual Studio Marketplace. Microsoft, in an attempt to remove the confusion about whether Visual Studio Online is an IDE in the cloud or a set of development services, renamed Visual Studio Online to Visual Studio Team Services— and VSTS was back—at least in acronym form!

By 2017, Microsoft had essentially stopped referring to ALM in its marketing materials, in lieu of the new, trendy term DevOps. As such, they continued to invest in more and more engineering features, such as code search, package management, as well as Git, build, and release improvements. Microsoft also continued to provide love to agile teams, with a new work item user experience, the ability to follow a work item, Kanban board live updates, as well as dashboard and widget improvements. Additional extension points meant that the community could inundate us with awesome extensions. Microsoft continued their cadence of updating the cloud-hosted Visual Studio Team Services every three weeks, and their on-premises Team Foundation Server about every 3-4 months with new features and bug fixes.

The year 2018 saw many additional features added/moved to the web portal from the venerable desktop clients. More papercuts were fixed and small improvements made to the agile experience, including a new work item form with an optimized look and feel for mobile devices. The bulk of improvements applied to the engineering tools, however. Code repositories, building, testing, and releasing saw many enhancements. This version also saw the further deprecation and removal of features that had become antiquated or irrelevant, and/or didn't align with Microsoft's roadmap. Also, this was the first time that Team Foundation Server was released without a "companion" version of Visual Studio. This wasn't a big deal, but it just felt weird.

Microsoft renamed things again in 2019. The cloud-based Visual Studio Team Services was renamed *Azure DevOps Services* and the on-premises Team Foundation Server was renamed *Azure DevOps Server*. Although the industry wasn't ready for yet another name, when the dust settled, it was nice to be able to refer to both the on-premises and cloud-based products as simply Azure DevOps. Finally having "DevOps" in the product name made it easier to demonstrate to the industry that, yes, Microsoft did have a DevOps solution. At the same time, Microsoft also defined and named the Azure DevOps services themselves: Azure Boards, Azure Test Plans, Azure Repos, Azure Pipelines, and Azure Artifacts.

Over the past 15 years, the SDLC, then ALM, and now DevOps tools have proven themselves capable of helping organizations manage the entire lifespan of their application development, reduce their cycle times, and eliminate waste. Azure DevOps does this by integrating different teams, platforms, and activities with the goal of enabling a continuous flow of business value, which includes every aspect of the application's lifecycle, from when it first begins as an idea or need, all the way to its retirement. This lifecycle can include initiating the project, defining and refining requirements, design, coding, testing, packaging, releasing, deploying, and even operating, including monitoring.

In today's fast-paced, startup, open source, mobile, app store ecosystem, the lifecycle of a software idea can be quite short. Scrum can be used to start quickly and deliver the vision incrementally, and Azure DevOps can be used to safeguard that work and the quality of the product. Both can reduce risk and waste. On the other hand, some organizations and products have a very long lifecycle, such as line-of-business (LOB) systems. These may need more emphasis on governance and operations and less on time to market. Regardless, the tools you will learn about in Azure DevOps support both ends of this spectrum.

In this chapter, I will look more closely at the various Azure DevOps services, always with a focus on those tools that can empower a Scrum Team to deliver value continuously. I will also devote Chapter 3 to Azure Boards and Chapter 7 to Azure Test Plans.

Delivering Continuous Value

For the most part, our industry has emerged from building one-tier, two-tier, and n-tier applications that are internally managed by an organization. On the whole, we now build richer, more immersive applications powered by continuous services. These applications are delivered across a broad spectrum of connected systems—from mobile devices to traditional laptop and desktop computers. In parallel with this trend, software development practices are continuing to emerge that seek to enable a continuous *flow* of value.

> **Note** The term *flow*, which you will learn about in Chapter 9, is the movement and delivery of customer value through a process. Regardless of the underlying goals and motivations, teams want fast, smooth flow that creates value quickly while minimizing risk and avoiding cost of delay, and to do so in a predictable fashion.

Business conditions are demanding shorter and shorter cycle times, rapidly realizing the concept released to market. This is putting dramatic pressure on organizations and teams to deliver value rapidly and continuously without lowering quality or generating lethal technical debt. If your organization and team are not experiencing this yet, then you will be soon or you will be working for a competitor who is. In order to compete, companies must broaden their focus from merely improving their development practices to improving the entire value stream. This reality has been the impetus behind the main themes of the DevOps movement, as well as the tools in Azure DevOps.

Smell It's a smell when management expects that a tool (by itself) will be able to radically reduce cycle times or enable continuous delivery of value. It takes a combination of improving the existing process and practices as well as solid tooling. In fact, sometimes it takes the removal of a tool. A few years ago, one of my clients asked how establishing a specific branching strategy in Team Foundation Server would help them reduce their cycle time from 47 days to 7 days. I told them it wouldn't and that if they pursued this course of action, it would most likely increase their cycle time. The solution (as it almost always is) would be for them to radically change their processes, practices, and (most importantly) their culture to achieve such an extreme degree of improvement.

Here are the main themes for the tools in Azure DevOps:

- **Agile software development**   Agile techniques and methods, such as Scrum, have helped software development teams improve dramatically in the past two decades. They have fundamentally transformed the industry consensus as to the right way to develop software by implementing inspection and adaptation and shortening feedback loops. Azure DevOps continues to add and improve its tooling to support agile software development, regardless of the flavor of framework, methodology, or process you are following.

- **Quality enablement**   A shift away from traditional quality control (post-development testing) toward ensuring that quality is defined and delivered as a first-class requirement—even before coding begins.

- **DevOps**   An integration of development and operations to enable faster feedback cycles, reductions in time to resolve bugs and outages, and a focus on pushing smaller packages of functionality into production more frequently.

- **Continuous delivery (CD)**   The rapid flow of incremental business value through the entire technical value stream. CD is made possible through agile methods, quality enablement, proper tooling, and other practices.

The continuous delivery of value requires a tuned orchestration of people, practices, and tool usage. This goes beyond just managing changes to code using version control. The full value of the tools in

Azure DevOps cannot be realized unless work items, version control, builds, tests, and releases are used correctly. In other words, your team should consider using all of the Azure DevOps services together, not separately. This orchestration, resulting in a continuous delivery of value, can be seen in Figure 2-1.

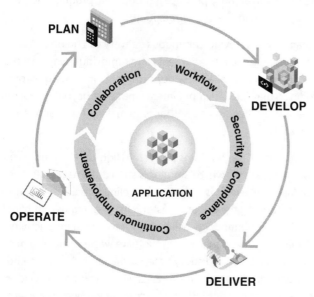

FIGURE 2-1 Achieving a continuous delivery of value.

When people view a diagram like Figure 2-1, they don't usually think about subsequent cycles. They tend to step through a diagram like this once, understand each phase, and file it away. In reality, teams rarely begin with a greenfield environment. Startups tend to be the exception and, even then, that field turns brown very quickly. After working software is deployed, it affects every part of the cycle. Maintenance work becomes integrated into the Product Backlog alongside new features and ideas. Needless to say, it can be challenging to build a responsive, modern application on top of existing technology, manage operations, and try to deliver value while keeping the existing application up and running.

Let's break down software delivery into three parts: plan, develop, and deliver. The Scrum framework helps with the planning part by instructing us on how to plan and manage work. Smart developers using modern tools and practices have no problem with developing the product. It's the *delivery* part of that value stream that has been the challenge—mostly because of impediments beyond the control of the team. For example, team members may deliver a potentially releasable Increment, but operations struggle to get it deployed and running in as timely a fashion. This is just one of many potential impediments.

Current agile and lean concepts have helped to improve the way that teams deliver software. As an industry we're getting better, but many gaps still exist. Fortunately, the tools in Azure DevOps, when employed by a Professional Scrum Team, can close these gaps.

Delivering the product is supported by a variety of functions that need to be integrated carefully in order to make delivery happen. For example, many organizations still consider quality an afterthought—something tested into a product after the developers are done. This bolting-on of last-minute fixes doesn't improve quality and can actually produce a suboptimal product and, worse, generate technical debt that must be repaid (with interest) at some point.

The list of impediments keeping a team from achieving continuous delivery can go on and on. The only solution is for such a team to start inspecting their impediments and removing them, one at a time on a regular basis. Teams must be relentless and take every opportunity to improve their practices to improve the entire value stream. Doing so may include impediments that exist beyond the Scrum Team. Organizations must give these teams the freedom and authority to make these changes and do what they can to ensure those changes persist.

The combination of Professional Scrum and Azure DevOps is powerful. High-performance Professional Scrum Teams know how to apply agile practices while using tools so that the net result is effective. Through constant inspection and adaptation, waste is identified and eliminated. This may mean starting to use a new feature in the tool or stopping the use of a wasteful feature. Teams that blindly use new, shiny tools without thinking and experimenting are just asking for a drop in productivity. At the same time, teams that think that all planning and management tools are wasteful are also missing the boat. For teams that only use manual whiteboards and sticky notes to manage their work, I hope they never have to "roll back" their task board or recover from a disaster, like a new janitor accidentally cleaning that board. In addition, these teams cannot trace PBIs to source code to releases, never mind the inability to obtain any metrics relating to their flow.

Azure DevOps Services

In September 2018, Microsoft renamed Visual Studio Team Services and Team Foundation Server to Azure DevOps Services and Azure DevOps Server, respectively. With this new Azure DevOps name, Microsoft now had an official "DevOps" tool, or suite of tools—even though they always did. Now it was official. Now it was in the name. At the same time, they productized the individual services, as you can see in Figure 2-2. For example, instead of referring to "the agile tools inside Visual Studio Team Services," people could now just refer to "Azure Boards." It made it easier to have a conversation with existing users as well as others in the marketplace who were looking to leave their existing DevOps tools.

Taken together, the various Azure DevOps services provide an integrated DevOps solution that enables software development teams of all sizes to quickly deliver continuous, high-quality value. It provides both individual developers, an entire Scrum Team, or a Nexus of Scrum Teams with the ability to build business and consumer applications—on any platform, using any language.

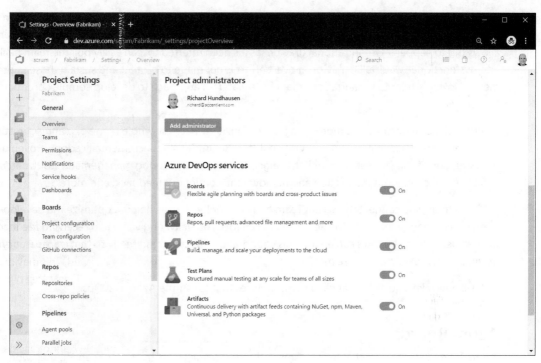

FIGURE 2-2 Enabling the various Azure DevOps services.

Note *Nexus* is a simple framework that implements Scrum at scale across multiple teams to deliver a single integrated product. It can be applied to 3–9 Scrum Teams that are working in a common development environment and are focused on producing a combined increment every Sprint with minimal dependencies. You can download and read the Nexus Guide at *www.scrum.org/resources/scaling-scrum*. I will cover Nexus in more detail in Chapter 11 , "Scaled Professional Scrum."

Azure DevOps assists everyone on the team in collaborating more effectively while building and sharing institutional knowledge. What's more, artifacts and data from each of the services are aggregated, stored, and made available for querying and reporting. Analytics provides a concise data model and reporting platform that you can use to answer quantitative questions about the past or present state of your product or process. These queries and reports can provide real-time transparency and traceability, as well as historical trending of the progress and quality of both the product and process.

Software products are developed and delivered by people, not by processes, practices, or tools. Processes and practices need to adapt and evolve to accommodate changes in scope and culture. Azure DevOps provides an environment that can be adapted to a Scrum Team's uniqueness and enhances it with proven agile practices that can be adopted at any pace. Over time, the team, as well as the organization, will become more productive by using these tools. This assumes that the culture allows this to happen. "Organizational gravity," as it's been referred to, can easily pull the team back into dysfunctional, waterfall behaviors if overt effort is not made to protect and sustain those improvements. I lovingly refer to organizations that take a waterfall approach to develop complex products as "waterfallian."

Professional Scrum involves more than just coding. It involves the whole range of planning, testing, and management activities. Azure DevOps enables a Scrum Team to incrementally adopt proven practices with out-of-the-box support for lightweight requirements, backlog management, task boards, code reviews, continuous integration and deployment, continuous feedback, and more.

These tools help connect the Scrum Team and stakeholders while helping optimize the development process and reduce risk. The Scrum Team can focus on delivering value and receiving feedback from stakeholders. The tools can enable this cooperation directly or indirectly so that transparency is maximized and expectations are met.

I will now take a quick look at each of the services that constitute Azure DevOps.

Azure Boards

Azure Boards is the service that helps teams and organizations plan, track, and manage their work. In other words, it's the service that teams use to visualize and manage their software development efforts. It provides a rich set of capabilities, including native support for Scrum and Kanban boards, customizable dashboards, and integrated reporting.

The Azure Boards service provides access to work items and related features like these:

- **Product Backlog** View and manage your Scrum Team's Product Backlog items in an ordered list, as in Figure 2-3.

- **Sprint Planning** Drag and drop items from the Product Backlog to the Sprint Backlog to create the forecast and plan the Sprint.

- **Sprint Backlog** View and manage the forecasted work and the plan for the current Sprint.

- **Task board** View and manage the task work items that constitute the plan for delivering the forecasted Product Backlog items.

- **Kanban board** Visualize and manage your team's workflow.

- **Queries and charts** Create work item queries and charts for reporting or bulk editing.

I will dive deeper into Azure Boards in Chapter 3, "Azure Boards."

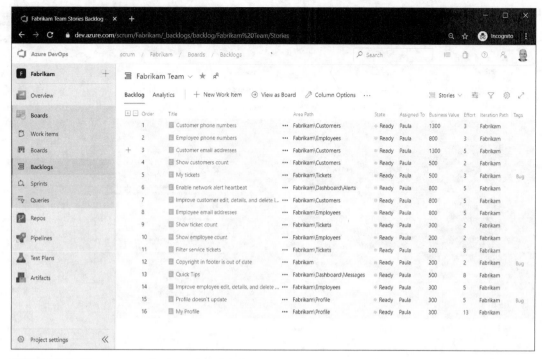

FIGURE 2-3 Backlogs page on the Azure Boards hub.

Azure Repos

Azure Repos is the service that provides version control tools that teams can use for collaborating and managing changes to their code and other files. Azure Repos supports Git in a standard way so that teams can use whatever clients and tools they choose, such as Git for Windows, Git for Mac, third-party Git services, and tools such as Visual Studio and Visual Studio Code.

The Azure Repos service provides features like these:

- **Browser-based** Easily visualize and manage repositories, branches, commits, pushes, and tags directly in the browser, as in Figure 2-4.

- **Multiple repositories** Create, import, and manage one or more repository per project.

- **Pull requests** Review code, add comments, and vote on changes prior to it being pulled into a branch.

- **Branch policies** Require pull requests, a minimum number of reviewers, linked work items, build validation, and other quality practices to further protect branches from low-quality changes.

- **Team Foundation Version Control (TFVC)** Support for the legacy, centralized version control until you are ready to migrate to Git.

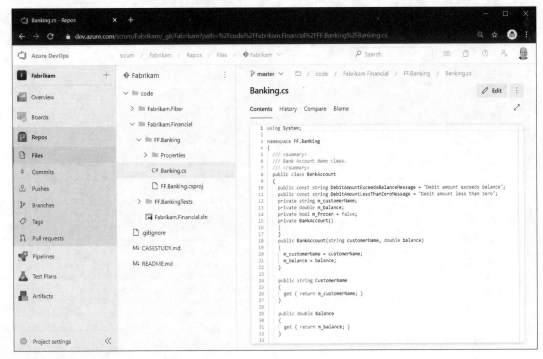

FIGURE 2-4 Files page on the Azure Repos hub.

I will not be covering Azure Repos in any depth in this book.

Azure Pipelines

Azure Pipelines is the service that a team can use to automatically build, test, and deploy their code. Azure Pipelines can pull code from many different repositories, as well as build almost any application type in almost any language and deploy it to almost any target platform. Through the use of visual designers or a straight coding experience, teams can create complex pipelines to automate the orchestration drudgery once and for all.

The Azure Pipelines service provides build and release features like these:

- **Robust pipelines**   Pick from hundreds of tasks available in Azure Pipelines and the Azure DevOps Marketplace to build, test, package, deploy, or perform other custom functions, as in Figure 2-5.

- **Hosted agents**   Use convenient, Microsoft-hosted agents on Windows, Unix, or macOS to build, test, and deploy applications as needed.

- **Continuous Integration**   Trigger a pipeline to run a build whenever a code change is detected.

- **Continuous Delivery**   Trigger a pipeline to create and deploy a release when a build successfully completes.

- **Configuration as Code**   Use YAML files to configure the pipelines, placing them under version control alongside code.

- **Queries and charts**   Use analytics to view pipeline reports and dashboard widgets such as duration, test failures, and pass rate reports.

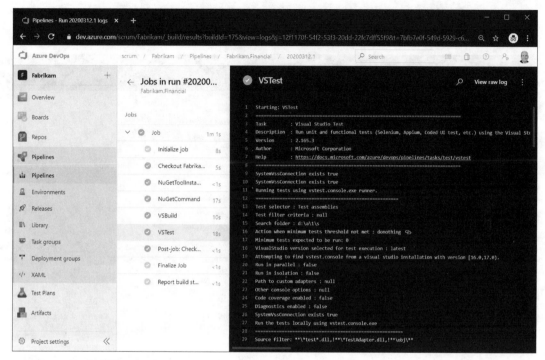

FIGURE 2-5 Pipelines page on the Azure Pipelines hub.

I will not be covering Azure Pipelines in any depth in this book.

Azure Test Plans

Azure Test Plans is the service that helps teams and organizations plan, manage, and execute their testing efforts. It supports teams of all sizes and types regardless of whether they are practicing manual, automated, or exploratory testing to drive collaboration and quality. The browser-based experience enables defining, executing, and charting the results of a team's tests.

The Azure Test Plans service provides access to testing-related work items and related features like these:

- **Test Case Management**   Organize your tests into test plans, test suites, and test cases, as in Figure 2-6.

- **Manual Testing**   Specify the specific steps and expected results in a test case to instruct the person executing the test and also to collect artifacts and telemetry of the test run.

- **Automated Testing**   Select a pipeline and test assembly to automatically run the automated test associated with the test case.

- **Exploratory Testing**   Use the browser-based extension to track your exploratory testing session and also to collect artifacts and telemetry.

- **Charts and reports**   Use the charts, runs, and progress reports to track quality metrics related to the testing effort.

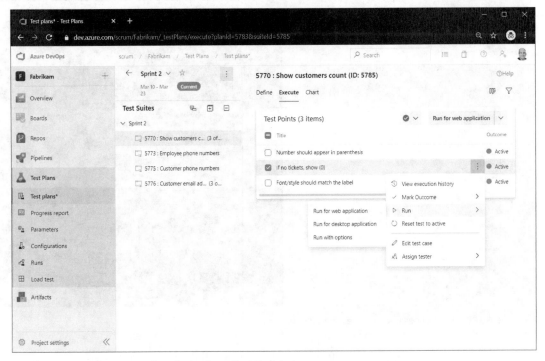

FIGURE 2-6 Executing a manual test in the Azure Test Plans hub.

I will dive deeper into Azure Test Plans in Chapter 7, "Planning with Tests."

Azure Artifacts

Azure Artifacts is the service that a team can use to create and share Maven, npm, NuGet, and Python package feeds from public and private sources. Formerly known as package management, Azure Artifacts can be fully integrated with Azure Pipelines to pull from or publish to, enabling the ultimate in reusability. By sharing binary packages, and not source code, teams limit complexity as well as the number of dependencies.

The Azure Artifacts service provides useful features like these:

- **Upstream sources**   Enable a single feed to store both the packages that your team produces and the packages your team consumes from other feeds in your organization as well as from "remote feeds" such as npmjs.com, nuget.org, Maven Central, and PyPI, as in Figure 2-7.

- **Release views** Filter the feed to a subset of packages that meet criteria defined by the view.

- **Universal packages** Store one or more files together in a single unit that has a name and a version.

- **Immutability** After a particular version of a package is published to a feed, that version number is permanently reserved.

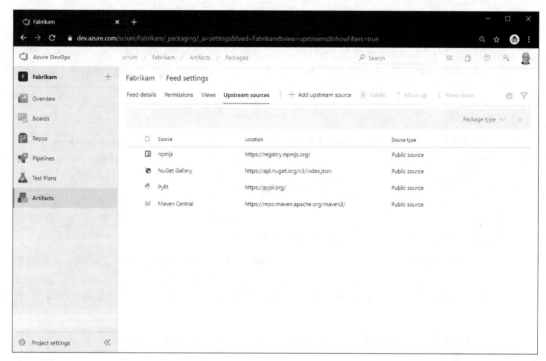

FIGURE 2-7 Specify upstream sources in Azure Artifacts.

I will not be covering Azure Artifacts in any depth in this book.

Azure DevOps Server

When Microsoft renamed Visual Studio Team Services to Azure DevOps Services in 2019, they also renamed Team Foundation Server to Azure DevOps Server. Azure DevOps Server is the on-premises version of the cloud-hosted Azure DevOps Services. It is updated regularly, including major updates every year or two. Both offerings provide the same great integrated, collaborative environment that supports the essential services, such as agile tools for work planning and tracking, Git, continuous integration, continuous delivery, test planning and execution, and package feeds.

Whereas Azure DevOps Services is an Azure-hosted SaaS offering, Azure DevOps Server is an on-premises offering that's built on SQL Server. Companies usually choose the on-premises Azure DevOps Server when they want their data to stay within their network.

> **Tip** Although both Azure DevOps offerings provide the same essential services, it is recommended that organizations and teams use the cloud-hosted Azure DevOps Services. Not only does the cloud-hosted offering provide a scalable, reliable, and globally available hosted service, but it's backed by a 99.9 percent SLA, monitored by Microsoft's 24/7 operations team and available in local data centers around the world. Your organization can transition from capital expenditures (servers and access licenses) to operational expenditures (subscriptions). It also means that you can give your administrator a break from having to continually upgrade, configure, and back up your on-premises instance. For more details on Azure DevOps Services data protection, visit *https://aka.ms/AzureDevOpsSecurity*. Only in situations where data cannot be stored offsite—for compliance or SLA reasons—should you consider Azure DevOps Server.

For the examples and guidance in this book, I will be talking about and using Azure DevOps *Services*. If your organization or team is using a current version of Azure DevOps Server, you should be fine following along with the guidance. If you are using an older Team Foundation Server instance, it's time to upgrade—and not just for the sake of the examples in this book!

Migrating to Azure DevOps Services

If your organization or team is using the on-premises Azure DevOps Server or the legacy Team Foundation Server, you will likely want to migrate to the cloud-based Azure DevOps Services. Personally, I think you should do so as soon as possible so that the flow of features and bug fixes comes much more quickly—and without effort on your part.

When your team decides to make the move to the cloud, you have a couple of choices:

- **Manually**   Start fresh by creating a new organization and new projects and by manually bringing over the relevant work items, code, and pipeline definitions from your existing environment.

- **Using the Data Migration Tool**   Use the migration tool created by Microsoft to perform a high-fidelity migration of an entire project collection.

A flexible way to move to Azure DevOps Services is to manually copy your most important assets and start relatively fresh. You can be picky about the projects you bring over, as well as the assets in each project. Unfortunately, this approach can be difficult if your team is in the middle of a large project. Advance planning can help. Open source tools or tools from the Azure DevOps Marketplace that leverage the public APIs can be helpful as well.

Another option is to use the high-fidelity data migration tool created and maintained by the Azure DevOps product team at Microsoft. This tool operates at the SQL Server database level so that it can provide a high-consistency migration to Azure DevOps Services. In other words, you will get everything migrated, as is. This is the best option if you want to move your existing on-premises instance to the cloud in its entirety. You may have to upgrade your existing Team Foundation Server or Azure DevOps Server instance to the latest version before using this migration tool. You can find requirements and more details in a downloadable Migration Guide available at *https://aka.ms/ AzureDevOpsImport*.

> **Tip** When migrating to Azure DevOps Services either manually or by leveraging tooling, your team should consider abandoning Team Foundation Version Control (TFVC) and moving to Git. If you haven't heard by now, Git has become the standard version control tool in our industry. It's what Microsoft uses internally and so should you and your team. Also, it's clear that all Azure Repos and Azure Pipelines innovation will focus on Git and that TFVC is on life support.

Visual Studio

A number of different client applications and tools can connect to and interact with Azure DevOps. Many are provided by Microsoft as well as the Azure DevOps partners and community. Even more can be created using Azure DevOps public APIs. The most popular clients from Microsoft are Visual Studio, Visual Studio Code, and Excel. I will cover Excel in Chapter 5, "The Product Backlog."

Many different audiences use Visual Studio. They differ depending on the makeup of the team, the software products they are building, and the speed at which they deliver and maintain those products. To support various audiences, Visual Studio is available in different editions. The editions range in price from free (Community edition) to thousands of dollars (Enterprise edition). For a thorough comparison of the Visual Studio editions, visit *https://visualstudio.microsoft.com/vs/compare*.

> **Smell** It's a smell when I see members of a professional software development team using the Community edition of Visual Studio. Although it is functionally equivalent to the Professional edition and the (free) price is tempting to management, it tends to reflect their inability to value what the team does or how they do it. This is not to mention the fact that the Community edition's EULA disallows enterprise organizational users from using it for developing or testing—except for a narrow list of situations. It also could be that the Community edition was temporarily installed for evaluation, training, or experimentation purposes, which are valid use cases.

All Visual Studio editions include the coding and testing tools that most teams need to develop, analyze, debug, test, collaborate, and deploy modern applications in a number of languages on a number of platforms. Each installation of Visual Studio includes Team Explorer, which is a plug-in that connects Visual Studio to projects in Azure DevOps Services (or Azure DevOps Server). It allows you to manage source code, work items, and other artifacts. The operations available to you depend on which version control system—Git or TFVC—was selected when the project was created. You can see Team Explorer in Figure 2-8.

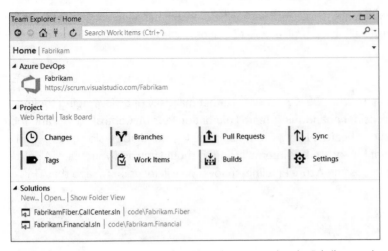

FIGURE 2-8 Here, Visual Studio Team Explorer is connected to the Fabrikam project.

Fabrikam Fiber Case Study

Most Developers on the Fabrikam Fiber Scrum Team have a Visual Studio Professional edition subscription, which allows them to develop, debug, and test as they progress through the Sprint. A couple of Developers, however, have a Visual Studio Enterprise edition subscription. They needed some more advanced features, such as IntelliTrace, Microsoft Fakes, and Live Unit Testing. Because Paula (the Product Owner) and Scott (the Scrum Master) aren't involved in the actual development effort, they won't need Visual Studio. Instead, they have chosen to interact with Azure DevOps using a browser, Excel, and Microsoft Power BI Desktop.

Visual Studio is not the only integrated development environment (IDE) that can connect to Azure DevOps. A number of other IDEs can interact with Azure DevOps using plug-ins, such as these:

- **Visual Studio Code** Use the Azure DevOps extension to connect to Azure DevOps Services to monitor builds and manage pull requests and work items.

- **Team Explorer Everywhere** Install a plug-in for Eclipse to connect to and interact with Azure DevOps Services or Azure DevOps Server.

- **Android Studio** Access and manage Azure DevOps Services or Azure DevOps Server Git repositories by installing the plug-in for Android Studio.

- **IntelliJ** Access and manage Azure DevOps Services or Azure DevOps Server Git repositories by installing the plug-in for IntelliJ IDEA.

- **Visual Studio for Mac** No plug-in or extension is required to connect to and interact with Azure DevOps Services or Azure DevOps Server Git repositories.

Visual Studio Subscriptions

The preferred way to obtain Visual Studio and access to Azure DevOps is through a Visual Studio subscription. Formerly known as MSDN subscriptions, they enhance an organization or team's investments by providing a comprehensive set of resources to create, deploy, and manage applications on most platforms and devices, including Android, iOS, Linux, macOS, Windows, web, and the cloud. Subscriptions also provide a cost-effective way for organizations to obtain software, services, training, and other resources for your development and testing needs.

Several Visual Studio subscription options are available:

- **Cloud subscription**   Allows you to essentially *rent* Visual Studio, Azure DevOps, and the subscriber benefits you need without a long-term contract

- **Standard subscription**   The familiar subscription model where you are buying Visual Studio and not simply renting like in the cloud subscription

- **Standalone license**   A one-time purchase of Visual Studio without access to major version upgrades or access to Azure DevOps

For most organizations, the standard subscription model of Professional or Enterprise edition makes the most sense. Both provide the latest versions of Visual Studio, with product updates for as long as the subscription is active. Both also provide access to Azure DevOps Services or Azure DevOps Server. Professional edition provides access to the basic features of all Azure DevOps services except for Azure Test Plans. Enterprise edition provides access to everything. For more information on the various subscription models, visit *https://visualstudio.microsoft.com/vs/pricing-details*.

The cloud subscriptions are particularly interesting because they allow the renting of Visual Studio and Azure DevOps on a month-to-month basis. Organizations and teams can now start and stop projects whenever they need, with flexibility and the convenience of having the development tools consolidated on the same bill with other Microsoft Azure cloud services such as virtual machines and storage. Although an annual cloud subscription may cost *less* than a perpetual license, the licensing rights end when you stop paying for the subscription. In other words, there is no perpetual use rights allowing you to continue using Visual Studio after the subscription has expired.

> **Note** Speaking of as-needed subscriptions, Microsoft also offers *Visual Studio Online*, a cloud-powered development environment for long- or short-term projects. You can work with these environments from Visual Studio, Visual Studio Code, or a browser-based editor. There is support for Git repos, extensions, and a built-in command-line interface so that you can edit, run, and debug your applications from any device. For more information on Visual Studio Online, visit *https://online.visualstudio.com*.

Azure DevOps Access Levels

Access levels grant or restrict access to select features in the web portal. The level of access you have is directly related to the type of license you have purchased. An administrator will manage who can access the Azure DevOps Services organization (or the Azure DevOps Server) by adding them as users, and also selecting their access level and permission.

Here are the supported access levels:

- **Stakeholder**   Provides partial access to features. Can be assigned to an unlimited number of users for free without the need for a license or subscription.

- **Basic**   Provides access to most features. Assigned to users with a Basic license or Visual Studio Professional subscription.

- **Basic + Test Plans**   Provides access to all of the Basic features plus those in Azure Test Plans. Assigned to users with a Basic + Test Plans license or Visual Studio Enterprise, Visual Studio Test Professional, or MSDN Platforms subscription.

There is also a Visual Studio subscription access level, which will automatically select the correct set of features based on the user's type of Visual Studio subscription. The system recognizes the user's subscription—Visual Studio Enterprise, Visual Studio Professional, Visual Studio Test Professional, or MSDN Platforms—and enables any other features that are included in their subscription level. These access levels will apply in each Azure DevOps organization in which that user is a member, whether they created the organization or were added by someone else.

> **Note**   Access levels are not the same as access or permissions. *Access* indicates that a particular user can sign into Azure DevOps and, at a minimum, view information. *Access levels* grant or restrict access to select features. They enable administrators to provide their users with access to the features they need and pay only for those features. *Permissions*, granted through security groups, provide or restrict users from performing specific tasks in Azure DevOps. Administrators need to understand each of these terms and how they relate to each other.

Each Azure DevOps Services organization gets five Basic licenses for free, which means that five users will be able to use Azure Boards, Azure Repos, Azure Pipelines, and Azure Artifacts without having to purchase a license. If, for example, an organization has five developers with Visual Studio Professional subscriptions and five developers using Visual Studio Code (free), then the five free Basic licenses can be used by the Visual Studio Code developers.

Fabrikam Fiber Case Study

All Developers on the Fabrikam Fiber Scrum Team who have a Visual Studio subscription will automatically be assigned their set of features when they sign in to Azure DevOps Services. Those with Visual Studio Professional subscriptions will be assigned the Basic access level. Those with Visual Studio Enterprise will also be able to use Azure Test Plans as well as some other features, such as self-hosted pipelines. Paula (the Product Owner) and Scott (the Scrum Master) will each use one of the five free Basic licenses to access Azure DevOps Services. They will consider upgrading to the Test Plans license if they start to become more involved in testing and using Azure Test Plans.

Stakeholder Access

Stakeholders are users with free but limited access to the features in Azure DevOps. The Stakeholder access level allows users to add and modify work items, manage build and release pipelines, and view dashboards. They can also check project status and provide direction, feedback, feature ideas, and business alignment to a team.

If a stakeholder needs access a feature that supports the daily work of the Scrum Team, they should have at least Basic access, if not Basic + Test Plans. This might include them needing to change the priority of an item within a backlog or creating queries or charts. If a stakeholder needs to perform these tasks, or even access code, tests, builds, or releases, then this would, according to Scrum's definition, make them a Scrum Team member and no longer a stakeholder. Regardless, they would need a proper license and access level at that point.

Fabrikam Fiber Case Study

Paula (the Product Owner) has opted to give certain users and managers Stakeholder access. Some of those individuals are a member of the Readers permission group and cannot change any information. In other cases, where there has been a history of collaboration and trust, those stakeholders have been added to the Contributors permission group, allowing them to directly add and edit work items. Paula knows that this is actually a way to take the strain off the Scrum Team by having stakeholders (users) create and update their own Product Backlog items.

GitHub and the Future

In 2018 Microsoft acquired GitHub, the world's leading software development platform, for $7.5 billion. GitHub has over 40 million users and is at the heart of the open source community. The synergies between GitHub and Azure DevOps are numerous. Together, GitHub and Azure DevOps provide an end-to-end experience for development teams to easily collaborate, build, and release code.

It's clear that Microsoft intends to make GitHub the home for every developer, not just open source developers. As such, they are intent on making integration with GitHub a first-class priority. To achieve this, Azure DevOps and GitHub share the same leadership, insights, and tooling inside Microsoft. It's also apparent that Microsoft is investing in and innovating GitHub at a higher rate than they are in Azure DevOps.

This acquisition raises several questions about the future of Microsoft owning two separate but similar products with a growing overlap of features. For example, are we to use Azure Repos or GitHub repositories? Today the two platforms are not at feature parity, so the messaging is about integration, not migration. I suspect that will flip in the future.

One thing is certain—all professional software developers need to learn Git, whether or not they plan on using GitHub.

Chapter Retrospective

Here are the key concepts I covered in this chapter:

- **DevOps** The union of people, process, and products to enable continuous delivery of value to our end users. Scrum is a process framework and fits within the boundaries of process-agnostic DevOps. High-performance Scrum Teams know that a balance of Scrum and DevOps practices are required to deliver complex software products efficiently.

- **Azure DevOps** Microsoft's suite of developer services that support teams to plan work, collaborate on code development, and build, test, and deploy applications.

- **Azure DevOps Services** Microsoft's cloud-based Azure DevOps offering. Formerly known as Team Foundation Service, Visual Studio Online, and Visual Studio Team Services.

- **Azure DevOps Server** Microsoft's on-premises Azure DevOps offering. Formerly known as Team Foundation Server.

- **Azure Boards** The Azure DevOps service that helps teams and organizations plan, track, and manage their work.

- **Azure Repos** The Azure DevOps service that provides version control tools that teams can use for collaborating and managing changes to their code and other files.

- **Azure Pipelines** The Azure DevOps service that a team can use to automatically build, test, and deploy their code.

- **Azure Test Plans** The Azure DevOps service that helps teams and organizations plan, manage, and execute their testing efforts.

- **Azure Artifacts** The Azure DevOps service that a team can use to create and share Maven, npm, NuGet, and Python package feeds from public and private sources.

- **Visual Studio Subscription**   The preferred way of licensing Visual Studio and gaining access to Azure DevOps.

- **Access levels**   Used to grant or restrict access to select features in the web portal. Access levels are different from access and permissions, which restrict access to Azure DevOps and what tasks can be performed.

- **Stakeholder access**   Users with free but limited access to the features in Azure DevOps. If appropriate, stakeholders (in Scrum) can make use of the free Stakeholder access level to increase transparency.

CHAPTER 3

Azure Boards

As I mentioned in the previous chapter, Azure Boards is the Azure DevOps service that helps teams plan and track their work. It's the service that provides the work items, backlogs, boards, queries, and charts—all the building blocks that a team needs to visualize and manage their work.

The look and feel of Azure Boards is partially driven by the process that a team selects when they create the project. This process defines the building blocks of the work item tracking system. It also serves as the basis for any process model customization that a team might want to perform.

In this chapter I will dive into Azure Boards and discuss the various processes that can be selected, focusing on the Scrum process. I will also show you how to create an inherited process to customize Azure Boards' behavior. In Part II of this book, "Practicing Professional Scrum," I will delve even deeper into how the backlogs and boards explicitly support Scrum.

Choosing a Process

Several processes are available out of the box. These system processes are designed to meet the needs of most teams. Some of them are more formal, like the Capability Maturity Model Integration (CMMI) process. Some of them are lightweight, like the Basic process. Some of them are intended to match the *Scrum Guide*, like the Scrum process.

Here are the system processes available when creating a new project:

- **Agile**   For teams that use agile planning methods, use user stories, and track development and test activities separately

- **Basic**   For teams that want the simplest model that uses issues, tasks, and epics to track work

- **CMMI**   For teams that follow more formal project methods that require a framework for process improvement and an auditable record of decisions

- **Scrum**   For teams that practice Scrum and track Product Backlog items (PBIs) on the backlog and boards

These system processes differ mainly in the work item types that they provide for planning and tracking work. Basic is the most lightweight and closely matches GitHub's work item types. Scrum is the next most lightweight. The Agile process is a bit "heavier" but supports many agile method terms. CMMI provides the most support for formal processes and change management.

When creating a project, a process must be selected, as you can see in Figure 3-1. After creation, the project will use the work item types, workflow states, and backlog configurations as defined by that process.

FIGURE 3-1 Selecting a process when creating a new project.

Note A *process* is different than a *process template*. A process defines the building blocks of the work item tracking system, supports the inheritance process model, and supports customization through a rich UI. It's available in Azure DevOps Services and Azure DevOps Server, but not for legacy Team Foundation Server versions. A process template is the legacy way of defining the building blocks of the work item tracking system. Process templates are expressed in XML and support customization through the modification and importing of XML definition files.

Work Item Types

Work items are the core elements of planning and tracking within Azure DevOps. They identify and describe requirements, tasks, bugs, test cases, and other concepts. Work items track what a team and team members have to do, as well as what they have done. Work items, and the metrics derived from them, can be visible within various queries, charts, dashboards, and analytics.

You can use work items to track anything that your team needs to track. Each work item represents an object stored in the work item data store. Each work item is based on a work item type and is assigned an identifier that is unique within an organization (or project collection in Azure DevOps Server). The work item types that are available to the project are based on the process used when the project was created, as you can see in Table 3-1.

TABLE 3-1 Work item categories available across the different processes.

	Scrum	Agile	CMMI	Basic
Work Item Category				
Requirement	Product Backlog Item	User Story	Requirement	Issue
Epic	Epic	Epic	Epic	Epic
Feature	Feature	Feature	Feature	-
Bug	Bug	Bug	Bug	Issue
Task	Task	Task	Task	Task
Test Case	Test Case	Test Case	Test Case	Test Case
Issue	Impediment	Issue	Issue	-
Change Request	-	-	Change Request	-
Review	-	-	Review	-
Risk	-	-	Risk	-

As you can see, the Agile process is very similar to the Scrum process. As far as work item types are concerned, the only differences are the type names of the *Requirement* and *Issue* work item categories. Agile refers to them as a *User Story* and *Issue*, respectively, whereas Scrum refers to them as a *Product Backlog Item* and *Impediment*, respectively. Figure 3-2 shows an example of this.

> **Note** Microsoft introduced work item *categories* in Team Foundation Server 2010. Categories are essentially a meta-type and enable the various processes to have their own names and behaviors of work item types, without breaking the functionality of Azure Boards. Examples of work item categories that have different names include Requirement, Bug, and Issue.

You can also see how heavy and formal the CMMI process is, with official Change Request, Review, and Risk work item types, as well as the antiquated Requirement work item type. I've helped hundreds of teams install, understand, and use Azure DevOps and Team Foundation Server and can count the number of CMMI projects I've run into on one hand. Conversely, the Basic process has only a few work item types—just barely sufficient to track work and also to more closely match how work is managed on GitHub. It is also the default process, so there are many Basic process projects in existence, if only by accident.

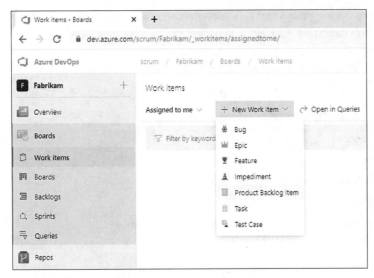

FIGURE 3-2 Work item types available to a Scrum project.

Another distinguishing feature of the different processes is the workflow states for the requirement category work item types. The workflow states define how a work item progresses upon its creation to its closure. You can see this natural progression by process in Table 3-2. Each state belongs to a state category (formerly known as a metastate). State categories enable the agile tools in Azure Boards to operate in a standard way regardless of the project's process.

TABLE 3-2 Requirement workflow states across the different processes.

Scrum	Agile	CMMI	Basic
New	New	Proposed	To Do
Approved	Active	Active	Doing
Committed	Resolved	Resolved	Done
Done	Closed	Closed	
Removed	Removed		

Hidden Work Item Types

Team Foundation Server 2012 introduced the concept of a *hidden* work item type. Work item types that are in this category are not able to be created from the standard user interfaces, such as the New Work Item drop-down list in Azure Boards. The reasoning behind this is that there are specialized tools for creating and managing these types of work items. Besides, creating these types of work items in an ad hoc way outside the context of the tooling doesn't make sense.

All processes, even the Basic one, support these hidden work item types:

- **Shared Parameter, Shared Steps, Test Plan** and **Test Suite**   Created and managed by the tools in Azure Test Plans. I will take a closer look at all the testing work item types in Chapter 7, "Planning with Tests."

- **Feedback Request** and **Feedback Response**   Used to request and respond to stakeholder feedback using the Test & Feedback extension.

- **Code Review Request** and **Code Review Response**   Used to exchange messages in legacy Team Foundation Version Control (TFVC) code review in the My Work page in Visual Studio Team Explorer. These code reviews are not to be confused with those related to Git pull requests.

Microsoft knew that teams typically wouldn't be creating these work item types outside the context of their dedicated tools. They actually did us a favor by hiding them from the various UIs where we create and manage work items. Referring back to Figure 3-2, notice that there weren't any of these hidden work item types listed.

The Scrum Process

Shortly after Microsoft released Team Foundation Server 2010, they made the Microsoft Visual Studio Scrum version 1.0 process template available for download. This new template was designed from the ground up to embrace the rules of Scrum as defined in the *Scrum Guide*. It was the result of collaboration between Microsoft, Scrum.org, and the Professional Scrum community. Everyone knew that Scrum had become the dominant agile framework in software development. Microsoft recognized this as well. They also knew that teams using Team Foundation Server and Scrum together wanted a lighter-weight experience, resulting in less friction. What resulted was a minimalistic process template that followed the rules of Scrum. There were over 100,000 downloads of this new process template in the first couple of years.

Over the years, through ongoing collaboration with the Professional Scrum community, Microsoft learned a thing or two about the Scrum process and the community using it. Primarily, they have learned that teams liked it! These teams appreciate its simplicity and straightforward support of Scrum. As you saw in Table 3-1, there are not a lot of extraneous work item types beyond what is needed to plan and track a project using Scrum. In fact, it's even more lightweight than the Agile process.

Many Scrum Teams evaluating Azure Boards currently use whiteboards and sticky notes to track their work. Since you can't get any lighter weight than that, any prospective software tool would need to be as lightweight as possible. We kept this guiding principle in mind as we created the Scrum process, and I still keep it in mind as I write this book.

Scrum Work Item Types

I want to spend some time talking specifically about the work item types in the Scrum process, and how a Scrum Team should (and shouldn't) use them. I will focus on just those items that directly relate to planning and executing work. The work items related to Azure Test Plans (test plans, test suites, test cases, etc.) will be covered in Chapter 7.

Product Backlog Item

In Scrum, the Product Backlog is an ordered (prioritized) list of the outstanding work necessary to realize the vision of the product. This list can contain new things that don't exist yet (features), as well as broken things that need to be fixed (bugs). In Azure Boards, the Product Backlog Item (PBI) work item type enables the Scrum Team to capture all of these various requirements with the least amount of documentation as is necessary. In fact, only the *title* field is required.

Later, as more detail emerges, the PBI can be updated to include business value, acceptance criteria, and an estimation of effort, as you can see in Figure 3-3.

FIGURE 3-3 Adding detail to a PBI work item.

As you create or edit PBI work items, consider the following Professional Scrum guidance while entering data in the pertinent fields:

- **Title (Required)**   Enter a short description that succinctly identifies the PBI.

- **Assigned To**   Select the Product Owner or leave blank, but don't assign to a Developer. This will reinforce the fact that the whole team owns the work on the PBI.

- **Tags**   Optionally add tag(s) to help find, filter, and identify the PBI. For example, some Scrum Teams opt not to use the Bug work item type in lieu of the PBI work item type and will simply tag those PBIs with "Bug."

- **State**   Select the appropriate state of the PBI. States are covered later in this section.

- **Area**   Select the best area path for the PBI. Areas must be set up ahead of time and can represent functional, logical, or physical areas or features of the product. If the PBI applies to all areas your team covers or you aren't sure of the specific area, then leave it set to its default value. For Nexus implementations, each team within a project can have its own corresponding areas as well as a default area. I will talk about Nexus in Chapter 11, "Scaled Professional Scrum."

- **Iteration**   Select the Sprint in which the Developers forecast that they will develop the PBI. If they have yet to forecast the PBI, then leave it set to the default (root) value.

- **Description**   Provide as much detail as necessary so that another team member or stakeholder can understand the purpose of the PBI. The user story format (As a <type of user>, I want <some goal>, so that <some reason>) works well here to ensure that the who, what, and why are captured. You should avoid using this field as a repository for detailed requirements, specifications, or designs.

- **Acceptance Criteria**   Describe the conditions that will be used to verify whether the team has developed the PBI according to expectations. Acceptance criteria should be clear, concise, and testable. You should avoid using this field as a repository for detailed requirements. Bulleted items work well. Gherkin (given-when-then) expressions work even better.

- **Discussion**   Add or curate rich text comments relating to the PBI. You can mention someone, a group, a work item, or a pull request as you add a comment. Professional Scrum Teams prefer higher-fidelity, in-person communication instead.

- **Effort**   Enter a number that indicates a relative rating (size) of the amount of work that will be required to develop the PBI. Larger numbers indicate more effort than smaller numbers. Fibonacci numbers (story points) work well here. T-shirt sizes (S, M, L, XL) don't, only because this is a numeric field. Effort can be considered the (I)nvestment in Return on Investment (ROI).

- **Business Value**   Enter a number that indicates a fixed or relative value of delivering the PBI. Larger numbers indicate more value than smaller numbers. Fibonacci numbers work well here. Business Value can be considered the (R)eturn in ROI.

- **Links**   Add a link to one or more work items or resources (build artifacts, code branches, commits, pull requests, tags, GitHub commits, GitHub issues, GitHub pull requests, test artifacts, wiki pages, hyperlinks, documents, and version-controlled items). You can see an example of linking a PBI to a wiki page in Figure 3-4. You should avoid explicitly linking PBIs to other PBIs, features, or epics using the Links tab. Instead, use drag and drop to establish hierarchical relationships within the backlogs. I will cover this in Chapter 5, "The Product Backlog."

- **Attachments**   Attach one or more files that provide more details about the PBI. Some teams like to attach notes, whiteboard photos, or even audio/video recordings of the Product Backlog refinement sessions and Sprint Planning meetings.

- **History**   Every time a team member updates the work item, Azure Boards tracks the team member who made the change and the fields that were changed. This tab displays a history of all those changes. The contents are read-only.

FIGURE 3-4 Adding a link to a wiki page.

While the PBI progresses on its journey to "ready" for Sprint Planning, the previous list of fields are really the only ones that need to be considered and completed. For the other fields on the PBI work item form, you should discuss as a team whether or not you should be using them because tracking data in those fields is most likely waste. When the PBI is forecast to be developed, additional fields and links will start to emerge, including links to task and test case work items, test results, commits, and builds.

Smell It's a smell when I see tasks created and associated with a PBI prior to Sprint Planning. Perhaps the Scrum Team knows what the plan will be, but what if it changes? The creation and management of those tasks will be wasted time and, what's worse, stubborn Developers may want to stick to their archaic plan, even though conditions might have changed. To avoid this pain and waste, don't create tasks until Sprint Planning where those PBIs are forecast or later in the Sprint.

A PBI work item can be in one of five states: New, Approved, Committed, Done, or Removed. The typical workflow progression would be New ⇒ Approved ⇒ Committed ⇒ Done. When a PBI is created, it is in the New state. When the Product Owner decides that the PBI is valid, its state should change from New to Approved. When the Developers forecast to develop the PBI in the current Sprint, its state should change to Committed. Finally, when the PBI is done, according to the Definition of Done, the state should change to Done. The Removed state is used for situations where the Product Owner determines that the PBI is invalid for whatever reason, such as it is already in the Product Backlog, has already been developed, has gone stale, or is an utterly ridiculous idea. Deleting the work item is another option for these situations.

Bug

A bug communicates that a problem or potential problem exists in the product. A bug can be found in a product that has already been delivered to production, in a done Increment from a previous Sprint, or in the Increment being developed in the current Sprint. A bug is not—repeat not—a failed test. Failed tests simply indicate that the team is not yet done. This will be covered more in Chapter 7.

By defining and managing Bug work items, the Scrum Team can track these problems, as well as prioritize and plan the efforts to fix them. A bug could be as small as a typo in a data entry form or as large as a vulnerability that allows credit card data to be exposed. Figure 3-5 shows a Bug work item.

FIGURE 3-5 An example of a Bug work item.

> **Note** In Scrum, a bug is just a *type of* Product Backlog Item, but Azure Boards defines a separate work item type to track bugs. The reason behind this is that the Bug work item type tracks additional, defect-specific information, such as severity, steps to reproduce, system information, and build numbers. Otherwise, the Bug and PBI work item types are fairly similar, with a few exceptions. Bug work items don't have a *Business Value* field, but they do have a *Remaining Work* field. The presence of the *Remaining Work* field allows the Bug work item to act like a task and be managed alongside tasks on the Taskboard. By default, the backlog includes both PBIs and bugs and the Taskboard—when used in accordance with my guidance—contains only tasks.

When you create a Bug work item, you want to accurately report the problem in a way that helps the reader understand its full impact. The steps to reproduce the bug should also be listed so that other

Developers can reproduce the behavior. There may be additional analysis (triage) required to confirm that it is an actual bug rather than a behavior that was by design. By defining and managing Bug work items, your team can track defects in the product in order to estimate and prioritize their resolution. As a general rule, bugs should be removed, not managed.

As you create or edit Bug work items, consider the following Professional Scrum guidance while entering data into the pertinent fields:

- **Title (Required)**   Enter a short description that succinctly identifies the bug.

- **Assigned To**   Select the Product Owner or leave blank, but don't assign to anyone else. This will reinforce the fact that the whole team owns the work on the bug.

- **Tags**   Optionally add tag(s) to help find, filter, and identify the bug.

- **State**   Select the appropriate state of the bug. States are covered later in this section.

- **Area**   Select the best area path for the bug. Areas must be set up ahead of time and can represent functional, logical, or physical areas or features of the product. If the bug applies to all areas your team covers or you aren't sure of the specific area, then leave it set to its default value. For Nexus implementations, each team within a project can have its own corresponding areas as well as a default area.

- **Iteration**   Select the Sprint in which the Developers forecast that they will fix the bug. If they have yet to forecast the bug, then leave it set to the default (root) value.

- **Repro Steps**   Provide as much detail as necessary so that another team member can reproduce the bug and better understand the problem that must be fixed. If you use the Test & Feedback extension to create a Bug work item, this information is provided automatically from your test session.

- **System Info**   Describe the environment in which the bug was found. If you use the Test & Feedback extension to create the Bug work item, this information is provided automatically from your test session.

- **Acceptance Criteria**   Describe the conditions that will be used to verify whether the team has fixed the bug according to expectations. Acceptance criteria should be clear, concise, and testable. Consider using this field to document the expected results, as opposed to the actual results.

- **Discussion**   Add or curate rich text comments relating to the bug. You can mention someone, a group, a work item, or a pull request as you add a comment. Professional Scrum Teams prefer higher-fidelity, in-person communication instead.

- **Severity**   Since the Bug work item type doesn't have a *Business Value* field, you will need to instead select the value that indicates the impact that the bug has on the product or stakeholders. The range is from 1 (critical) to 4 (low). Lower values indicate a higher severity. The default severity is 3 (medium).

- **Effort**   Enter a number that indicates a relative rating (size) of the amount of work that will be required to fix the bug. Larger numbers indicate more effort than smaller numbers. Fibonacci numbers (story points) work well here. T-shirt sizes don't, only because this is a numeric field. Effort can be considered the (I)nvestment in ROI.

- **Found In Build**   Optionally select the build in which the defect was found.

- **Integrated In Build**   Optionally, select a build that incorporates the bug fix.

- **Links**   Add a link to one or more work items or resources (build artifacts, code branches, commits, pull requests, tags, GitHub commits, GitHub issues, GitHub pull requests, test artifacts, wiki pages, hyperlinks, documents, and version-controlled items). You can link the bug to a related bug, to an article explaining the root cause, to the original PBI that failed, or even to a parent PBI that serves to gather several bugs into one collective "fix" user story.

- **Attachments**   Attach one or more files that provide more details about the bug. Some teams like to attach notes, whiteboard photos, or even audio/video recordings. This could also include screenshots, action recordings, and video, which the Test & Feedback extension can provide automatically.

- **History**   Every time a team member updates the work item, Azure Boards tracks the team member who made the change and the fields that were changed. This tab displays a history of all those changes. The contents are read-only.

Just like a PBI, a Bug work item progresses on its journey to "ready" for Sprint Planning, the previous list of fields are really the only ones that need to be considered and completed. If there are other fields on your Bug work item form, you should discuss as a team whether or not you should be using them because tracking data in those fields is most likely waste. When the bug is forecasted to be fixed, additional fields and links will start to emerge, including links to task and test case work items, test results, commits, and builds.

A Bug work item, like the PBI work item, can be in one of five states: New, Approved, Committed, Done, or Removed. The typical workflow progression would be New ⇒ Approved ⇒ Committed ⇒ Done. When a bug is reported and determined to be genuine (that is, it's not a feature, a duplicate, or a training issue), a new Bug work item is created in the New state. When the Product Owner decides that the bug is valid, its state should change from New to Approved. When the Developers forecast to fix the bug in the current Sprint, its state should change to Committed. Finally, when the bug is done, according to the Definition of Done, the state should change to Done. The Removed state is used for situations where the Product Owner determines that the bug is invalid for whatever reason, such as it's already in the Product Backlog, it's actually a feature, it's a training issue, it's not worth the effort, or it has already been fixed. Deleting the work item is another option for these situations.

Fabrikam Fiber Case Study

Because the Bug work item type does not have a *Business Value* field and also contains several extraneous fields, Paula has decided not to use that work item type. This is not to say that the Product Backlog won't contain bugs, but rather that the Scrum Team will use the PBI work item type to track them. They will tag the PBIs accordingly, and put the repro steps and system information into the *Description* field. By doing this, the Product Backlog will contain only Product Backlog Item work items and each will have a *Business Value* and a *Size* field to compute ROI.

Epic

In Scrum, there is only one Product Backlog for a product and it contains only Product Backlog items. Some PBIs are quite small, deliverable in a single Sprint or less. Other PBIs are larger and may take more than one Sprint to complete. Huger PBIs may take many Sprints, even up to a year or more to complete. In Scrum, regardless of size, each item is simply called a Product Backlog item.

Organizations and teams prefer to have more specific language. They also prefer to have separate backlogs for these different-sized items, and that's why Azure Boards provides hierarchical backlogs. With hierarchical backlogs, an organization or team can start with "big picture" ideas called *epics* and break them down into more releasable-sized items called *features*, and finally into smaller, more executable-sized items.

An epic represents a business initiative to be accomplished, like these examples:

- Increase customer engagement

- Improve and simplify the user experience

- Implement microservices architecture to improve agility

- Integrate with SAP

- Native iPhone app

Note Epics and features are managed on their own backlogs. In Azure Boards, each team can determine the backlog levels that they want to use. For example, Scrum Teams may want to focus only on their Product Backlog and the higher-level Features backlog. Leadership may want to only see epics and maybe how they map to features. By default, the Epics backlog is not visible in Azure Boards. A team administrator must enable it before you can view and manage epics on that backlog, as you see in Figure 3-6.

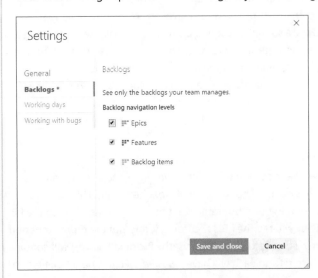

FIGURE 3-6 Enabling the Epics backlog.

Epic work items are similar to PBI work items. As you create or edit Epic work items, consider the following Professional Scrum guidance while entering data into the pertinent fields:

- **Title (Required)**   Enter a short description that succinctly identifies the epic.

- **Assigned To**   Select the Product Owner or leave blank. Alternatively, you can assign it to the stakeholder advocating for the epic.

- **Tags**   Optionally add tag(s) to help find, filter, and identify the epic.

- **State**   Select the appropriate state of the epic. States are covered later in this section.

- **Area**   Select the best area path for the epic. Areas must be set up ahead of time and can represent functional, logical, or physical areas or features of the product. If the area applies to all areas your team covers or you aren't sure of the specific area, then leave it set to its default value. For Nexus implementations, each team within a project can have its own corresponding areas as well as a default area.

- **Iteration**   Optional, but you can select the Sprint in which the Developers forecast that they will either begin or complete the development of the epic. If they have yet to begin work, then leave it set to the default (root) value.

- **Description**   Provide as much detail as necessary so that another team member or stakeholder can understand the purpose and goal of the epic.

- **Acceptance Criteria**   Describe the conditions that will be used to verify whether the team has developed the epic according to expectations.

- **Discussion**   Add or curate rich text comments relating to the epic. You can mention someone, a group, a work item, or a pull request as you add a comment. Professional Scrum Teams prefer higher-fidelity, in-person communication instead.

- **Start Date**   Optional, but you can set the date that work will commence on the epic. This could be the start date of the Sprint when the first PBI related to the epic is forecast for development. This field is key to using Delivery Plans.

- **Target Date**   Optional, but you can set the date that the epic should be implemented. This field is key to using Delivery Plans.

- **Effort**   Enter a number that indicates a relative rating (size) of the amount of work that will be required to develop the epic. Larger numbers indicate more effort than smaller numbers. Fibonacci numbers (story points) work well here. T-shirt sizes don't, only because this is a numeric field. Effort can be considered the (I)nvestment in ROI.

- **Business Value**   Enter a number that indicates a fixed or relative value of delivering the epic. Larger numbers indicate more value than smaller numbers. Fibonacci numbers work well here. Business Value can be considered the (R)eturn in ROI.

- **Links**   Add a link to one or more work items or resources (build artifacts, code branches, commits, pull requests, tags, GitHub commits, GitHub issues, GitHub pull requests, test artifacts, wiki pages, hyperlinks, documents, and version-controlled items). You should avoid explicitly linking epics to other epics, features, or PBIs using the Links tab. Instead, use drag and drop to establish hierarchical relationships on the backlog using the Mapping pane.

- **Attachments**   Attach one or more files that provide more details about the epic. Some teams like to attach notes, whiteboard photos, or even audio/video recordings.

- **History**   Every time a team member updates the work item, Azure Boards tracks the team member who made the change and the fields that were changed. This tab displays a history of all those changes. The contents are read-only.

Epic work items can be in one of four states: New, In-Progress, Done, or Removed. The typical workflow progression would be New ⇒ In-Progress ⇒ Done. When an epic is created it is in the New state. Once work begins, you should move it to the In Progress state, as I'm doing in Figure 3-7. Finally, when the epic is finished, because the last related feature is complete, then the state should change to Done. The Removed state is used for situations where the Product Owner determines that the epic is no longer needed, for whatever reason. Deleting the work item is another option in this situation.

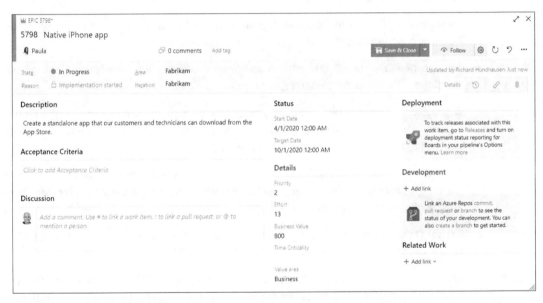

FIGURE 3-7 Setting an epic to In Progress.

Refining an epic means to break it down, or *decompose*, into one or more features. Feature work items are then created and linked back to the parent epic. This can be done manually by using the links in the work item form; inline on the Epics backlog; or by using the Mapping feature, as I'm doing in Figure 3-8. Refining is an ongoing process, with the features changing, merging, and splitting again as the Scrum Team learns more about the domain, the product, and the stakeholders.

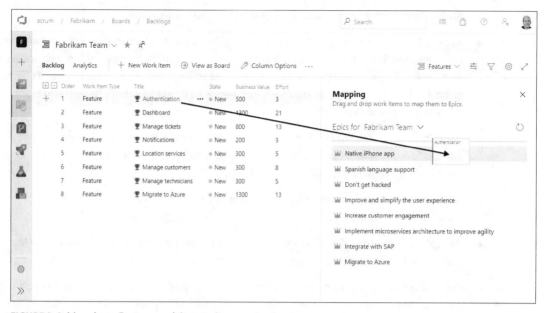

FIGURE 3-8 Mapping a Feature work item to its parent epic.

Feature

Whether or not you plan on using Epic work items, your team may still want to track features. Features are typically what stakeholders request and also what they expect to be delivered. If a feature is larger than can be delivered in a Sprint, then it must be broken down further—into other features or into more executable-sized items that are tracked and managed at the backlog level. Most teams I work with refer to these lowest, leaf-level items as user stories, or simply stories.

A feature typically represents a releasable component of software, like these examples:

- View technician details on the dashboard

- Ability to reassign tickets

- Support text alerts

- Find and filter tickets

Feature work items are similar to Epic work items, as you can see in Figure 3-9. As you create or edit Feature work items, consider the following Professional Scrum guidance while entering data into the pertinent fields:

- **Title (Required)** Enter a short description that succinctly identifies the feature.

- **Assigned To** Select the Product Owner or leave blank, but don't assign to anyone else. This will reinforce the fact that the whole team owns the work on the feature.

- **Tags** Optionally add tag(s) to help find, filter, and identify the feature.

- **State** Select the appropriate state of the feature. States are covered later in this section.

- **Area** Select the best area path for the feature. Areas must be set up ahead of time and can represent functional, logical, or physical areas or features of the product. If the area applies to all areas your team covers or you aren't sure of the specific area, then leave it set to its default value. For Nexus implementations, each team within a project can have its own corresponding areas as well as a default area.

- **Iteration** Optional, but you can select the Sprint in which the Developers forecast that it will either begin or complete the development of the feature. If they have yet to begin work, then leave it set to the default (root) value.

- **Description** Provide as much detail as necessary so that another team member or stakeholder can understand the purpose and goal of the feature.

- **Acceptance Criteria** Describe the conditions that will be used to verify whether the team has developed the feature according to expectations.

- **Discussion** Add or curate rich text comments relating to the feature. You can mention someone, a group, a work item, or a pull request as you add a comment. Professional Scrum Teams prefer higher-fidelity, in-person communication instead.

- **Start Date** Optional, but you can set the date that work will commence on the feature. This could be the start date of the Sprint when the first PBI related to the feature is forecasted for development. If using epics, then the epic's start date might coincide with the first related feature's start date. This field is key to using Delivery Plans.

- **Target Date** Optional, but you can set the date that the feature should be implemented. If using epics, then the epic's target date might coincide with the last related feature's start date. This field is key to using Delivery Plans.

- **Effort** Enter a number that indicates a relative rating (size) of the amount of work that will be required to develop the feature. Larger numbers indicate more effort than smaller numbers. Fibonacci numbers (story points) work well here. T-shirt sizes don't, only because this is a numeric field. Effort can be considered the (I)nvestment in ROI.

- **Business Value** Enter a number that indicates a fixed or relative value of delivering the feature. Larger numbers indicate more value than smaller numbers. Fibonacci numbers work well here. Business Value can be considered the (R)eturn in ROI.

- **Links** Add a link to one or more work items or resources (build artifacts, code branches, commits, pull requests, tags, GitHub commits, GitHub issues, GitHub pull requests, test artifacts, wiki pages, hyperlinks, documents, and version-controlled items). You should avoid explicitly linking features to other features, epics, or PBIs using the Links tab. Instead, use drag and drop to establish hierarchical relationships on the backlog using the Mapping pane.

- **Attachments**   Attach one or more files that provide more details about the feature. Some teams like to attach notes, whiteboard photos, or even audio/video recordings.

- **History**   Every time a team member updates the work item, Azure Boards tracks the team member who made the change and the fields that were changed. This tab displays a history of all those changes. The contents are read-only.

Feature work items can be in one of four states: New, In-Progress, Done, or Removed. The typical workflow progression would be New ⇒ In-Progress ⇒ Done. When a feature is created it is in the New state. Once work begins, you should move it to the In-Progress state. Finally, when the feature is finished, because the last related PBI is complete, then the state should change to Done. The Removed state is used for situations where the Product Owner determines that the feature is no longer needed, for whatever reason. Deleting the work item is another option in this situation.

FIGURE 3-9 Creating a Feature work item (notice the related epic in the lower right).

Refining a feature means to break it down, or *decompose*, into one or more PBI work items. PBI work items are then created and linked back to the parent feature. This can be done manually by using the links in the work item form; inline on the Features backlog; or by using the Mapping feature, as I'm doing in Figure 3-10. Refining is an ongoing process, with the PBIs changing, merging, and possibly splitting again as the Scrum Team learns more about the domain, the product, and the stakeholders.

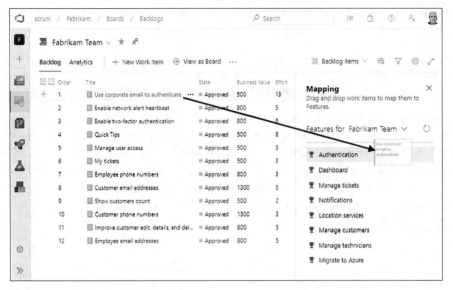

FIGURE 3-10 Mapping a PBI work item to its parent Feature work item.

Tip Azure Boards provides a few ways of visualizing and filtering by parent work items in the hierarchical backlogs. One option is to show parents in the backlog as nested read-only rows in the backlog. This option also includes an *unparented* section for those work items that don't have parents. These extra rows, while informational, can make your backlog view quite messy, especially if you have many parent rows. Another option is to select Column Options and add a *Parent* column, which will simply show the parent work item's title as a virtual field, as you see in Figure 3-11.

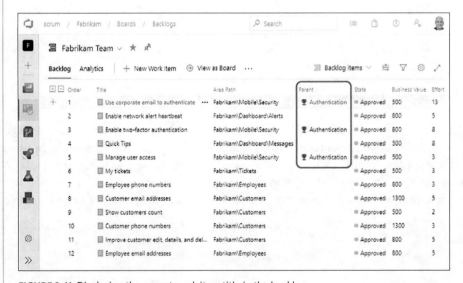

FIGURE 3-11 Displaying the parent work item title in the backlog.

Task

A Task work item represents a piece of detailed work that Developers must accomplish when developing a PBI. All tasks form the Sprint *plan* for achieving the Sprint Goal. These tasks, along with their associated PBIs, constitute the Sprint Backlog.

A task can be analysis, design, development, testing, documentation, deployment, or operations in nature. For example, the team can identify and create Task work items that are development focused, such as implementing an interface or creating a database table. They can also create testing-focused tasks, such as creating a test plan and running tests. A deployment-focused task might be to provision a set of virtual machines for hosting the deployed application. Figure 3-12 shows an example Task work item.

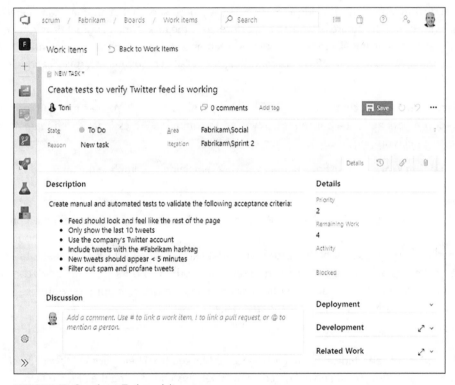

FIGURE 3-12 Creating a Task work item.

As you create or edit Task work items, consider the following Professional Scrum guidance while entering data in the pertinent fields:

- **Title (Required)** Enter a short description that provides a concise overview of the task. The title should be short but descriptive enough to allow the team to quickly understand what work is to be performed. Some teams have adopted a simple verb-noun naming convention (e.g., Create tests, Write code, Deploy app, etc.).

- **Assigned To** Select the team member who is responsible for ensuring that the task is completed. A task can be assigned to only one person at a time, so if two people pair up on a task, or the team mobs on a task, just pick one of them to be the owner. Leave it blank until someone starts working on it.

- **Tags**   Optionally add tag(s) to help find and identify the task.

- **State**   Select the appropriate state of the task. States are covered later in this section.

- **Area** (optional)   Typically matches the associated PBI that you are working on. When tasks are created from the Taskboard, the Area is automatically populated with the parent PBI's area.

- **Iteration**   Select the Sprint in which your team will be working on the task. The Sprint should be the same as the associated PBI. When tasks are created from the Taskboard, the Iteration is automatically populated.

- **Description** (optional)   Provide as much detail as is necessary so that another team member can understand the nature of work to be performed in the task. A meaningful title might be sufficient. Some teams like to track task-level acceptance criteria for particularly complex tasks in this field. Avoid using this field as a repository for detailed requirements, specifications, or designs.

- **Discussion**   Add or curate rich text comments relating to the task. You can mention someone, a group, a work item, or a pull request as you add a comment. Professional Scrum Teams prefer higher-fidelity, in-person communication instead.

- **Remaining Work**   The estimated hours of work remaining to complete the task.

> **Tip**   Initially, during Sprint Planning, the Remaining Work value should be an estimated pro-vided by the entire team. Later, after a team member begins working on the task, it should be updated by that person, who has more up-to-date knowledge of the work. Ideally, tasks should be 8 hours or less. If a task is going to take longer than 8 hours, it should be decom-posed into smaller tasks, in order to reduce risk and enable more collaboration options. Remaining work estimates should be updated daily.

- **Blocked** (optional)   Indicates whether the task is blocked from being accomplished. Blocked work should be identified and mitigated immediately. Instead of using the Blocked field, some teams have opted to use a "Blocked" tag.

- **Links**   Add a link to one or more work items or resources (build artifacts, code branches, commits, pull requests, tags, GitHub commits, GitHub issues, GitHub pull requests, test arti-facts, wiki pages, hyperlinks, documents, and version-controlled items). In general, you should avoid manually linking tasks to other PBIs, preferring to use the Sprint Backlog or Taskboard instead. Linking tasks to other tasks can help visualize dependencies, but it also has the smell of a command-and-control work breakdown structure.

- **Attachments**   Attach one or more files that provide more details about the task. Some teams like to attach notes, whiteboard photos, or even audio/video recordings.

- **History** Every time a team member updates the work item, Azure Boards tracks the team member who made the change and the fields that were changed. This tab displays a history of all those changes. The contents are read-only.

As your team uses tasks to plan, visualize, and manage its Sprint work, the previous list contains the only fields that you need to consider and complete. If there are other fields on your Task work item form, such as Priority or Activity, you should discuss as a team whether or not you should be using them because tracking data in those fields is most likely waste. That said, at the end of the day, *how* the team works, which includes how they will use Azure Boards, is up to them—which is an example of self-management.

> **Smell** It's a smell when I see that a team is using the *Activity* field on tasks. Professional Scrum Teams know that everything they do is considered a *development* activity, so using this field seems like waste. There is also a risk that Developers will become conditioned to look for their favorite type of task. For example, someone with a background in testing may only look for unassigned testing tasks, which is not necessarily what is best for the team's productivity, let alone achieving the Sprint Goal. An even greater fear is that others outside the Scrum Team will begin using the activity type for resource planning or assignment of work!

A Task work item can be in one of four states: To Do, In Progress, Done, or Removed. The typical workflow progression would be To Do ⇒ In Progress ⇒ Done. When a task is created, it is in the To Do state. When a team member begins working on a task, the state should be set to In Progress. When the task is finished, the state should be set to Done. The Removed state is used for situations where the Developers determine that the task is invalid for whatever reason, such as it doesn't apply anymore or it was a duplicate. Deleting the work item is another option for these situations.

Impediment

An Impediment work item is a report of any situation that blocks the team or a team member from completing work efficiently. By defining and managing Impediment work items, a Scrum Team can identify and track problems that are blocking it. More importantly, they'll have a backlog from which to work on improvements.

Impediments can be identified and, optionally recorded, at any time. They should be made transparent at least once a day, perhaps during the Daily Scrum. Professional Scrum Teams, however, don't wait until the Daily Scrum to raise and/or fix impediments. If the impediment is something that can be removed immediately, that's what should be done. If not, then the impediment could be recorded as an Impediment work item. The Scrum Team may also record impediments on a physical board or on a wiki page. Regardless, it's better to remove impediments than to track and manage them. The Scrum Master is responsible for facilitating the resolution of impediments—that the team cannot resolve themselves—as well as improving team productivity.

> **Tip** Having a transparent, prioritized backlog of impediments and improvement ideas at a team's fingertips can be very beneficial. If and when management comes offering to help, an item can be pulled from the top and discussed. Even if the budget is tight and management can't afford new hardware, software, or services, you may still have impediments that don't require an expenditure. For example, management may not have any money for faster laptops, but they could ask the Project Management Office (PMO) to ease off on the weekly status report requirements.

Figure 3-13 shows you an example of an Impediment work item.

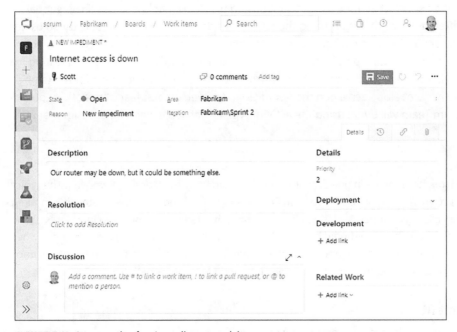

FIGURE 3-13 An example of an Impediment work item.

As you create or edit Impediment work items, consider the following Professional Scrum guidance while entering data in the pertinent fields:

- **Title (Required)** Enter a short description that accurately and succinctly describes the impediment.

- **Assigned To** Select the team member or stakeholder who will be responsible for resolving the impediment. Don't assume that the Scrum Master will always own/remove impediments.

- **Tags** Optionally add tag(s) to help find and identify the impediment.

- **State** Select the appropriate state of the impediment. States are covered later in this section.

- **Area** (optional) Select the best area for the impediment. If the impediment applies to all areas your team covers or you aren't sure of the specific area, then leave it set to its default value.

- **Iteration** (optional)   Typically a team selects the Sprint in which the impediment occurred, but Iteration could also represent the Sprint in which the impediment was removed. Leaving it set to its default value is fine as well.

- **Description**   Provide as much detail as necessary so that another person can understand the impediment and its impact.

- **Resolution**   Provide as much detail as necessary to describe how the impediment was resolved. Over time, these resolutions could establish a "lessons learned" reference.

- **Discussion**   Add or curate rich text comments relating to the impediment. You can mention someone, a group, a work item, or a pull request as you add a comment. Professional Scrum Teams prefer higher-fidelity, in-person communication instead.

- **Priority**   Select the level of importance for the impediment on a scale of 1 (most important) to 4 (least important). The default value is 2.

- **Links**   Add a link to one or more work items or resources (build artifacts, code branches, commits, pull requests, tags, GitHub commits, GitHub issues, GitHub pull requests, test artifacts, wiki pages, hyperlinks, documents, and version-controlled items). For example, you may want to link the impediment to one or more blocked tasks or PBIs or other impediments.

- **Attachments**   Attach one or more files that provide more details about the impediment. Some teams like to attach notes, whiteboard photos, or even audio/video recordings.

- **History**   Every time a team member updates the work item, Azure Boards tracks the team member who made the change and the fields that were changed. This tab displays a history of all those changes. The contents are read-only.

> **Note** Impediments may seem similar to tasks, and vice versa. To add further confusion, impediments are referred to as *issues* in other processes, and issues in the Basic process represent work to be done. To keep it straight in your head, consider this simple definition of an impediment—which is anything that hinders or prevents you or your team from achieving the Sprint Goal. In other words, Impediment work items are used to track unplanned situations that *block* work from getting done, whereas Task work items represent the plan for developing the forecasted PBIs in the Sprint Backlog and achieving the Sprint Goal.

An Impediment work item can be either Open or Closed. When an impediment is created, it is in the Open state. When the impediment is resolved/removed, the state should be set to Closed. Deleting the impediment after it has been removed is another option.

Impediments can be configured to show on the Boards. You can also track and manage them using a work item query. A team administrator could create a shared query looking for Work Item Type of Impediment and State of Open sorted by Priority. This query could then be surfaced on a dashboard or a wiki page, as I've done in Figure 3-14.

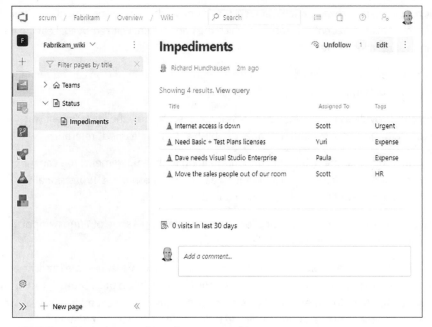

FIGURE 3-14 Displaying open impediments on a wiki page.

Scrum Work Item Queries

Work item queries allow you to view, understand, and manage your workload. By running the appropriate query, you can return lists of work items showing you the PBIs, bugs, tasks, impediments, test cases, and other work items that pertain to you or your team. You can filter and sort those items in many ways. You can then decide on which of these work items to take action. Queries can also be used to perform bulk work item updates. For example, the Product Owner can query those PBIs in a specific area in order to make bulk changes to the *Business Value* field.

With queries, you can perform these functions:

- Review work that's planned, in progress, or recently done

- Perform bulk updates, such as assigning new PBIs to the Product Owner

- Create a chart to get a count of items or the sum of a field

- Create a chart and add it to a dashboard

- View a tree of parent-child-related work items

Note Queries can be executed from Microsoft Excel and other clients. When you have many work items to add or modify, Excel can really save you time. Simply create a flat list query of epics, features, PBIs, bugs, or tasks and open it in Excel. You must first install the (free) Azure DevOps Office Integration plug-in, which supports Microsoft Excel 2010 or later versions, including Microsoft Office Excel 365.

When saving a query, you can save it to *My Queries* or, if you have permissions, save it to *Shared Queries*. As you might guess, only you can view and run queries saved under *My Queries*. Queries that you and others save under *Shared Queries* can be viewed by everyone with access to the project. Queries can be organized within folders and even marked as favorites.

Here are some queries that your Professional Scrum Team may want to create:

- **Open impediments**   The Scrum Team and especially the Scrum Master should be mindful of these.

- **PBIs assigned to someone other than the Product Owner**   In Scrum, the Product Owner, not anyone else, "owns" the Product Backlog items. Empty assigned-to values are okay.

- **New or approved PBIs with tasks**   It is wasteful to create tasks ahead of Sprint Planning.

- **Approved PBIs without acceptance criteria**   How will the team members know what the expectations are or when development is done?

- **New or approved PBIs with a root-level area**   Are these items really cross-cutting, or did someone just forget to assign an area?

- **New or approved PBIs assigned to a Sprint**   Either someone entered the wrong iteration or someone forgot to set the state.

- **Committed or done PBIs without a Sprint**   Either someone forgot to set the iteration or someone goofed up the state.

- **PBIs without links to features**   Assuming you are using features, it may be useful to see the unparented items.

- **Features without links to epics**   Assuming you are using epics and features, it may be useful to see the unparented items.

- **Features without links to PBIs**   Assuming you are using features, it may be useful to see the items without children.

- **Epics without links to features**   Assuming you are using epics and features, it may be useful to see the items without children.

Here are some additional query ideas that pertain to the current Sprint:

- **Committed PBIs without tasks**   Perhaps the plan for delivering these items really is that simple. It's more likely that the team forgot to create a plan or a PBI was snuck into the Sprint after Sprint Planning.

- **Committed PBIs with associated tasks from other Sprints**   These PBIs were either rolled over from a previous Sprint with a part of the plan remaining in that Sprint, or there are some serious problems with your iteration values or planning practices.

- **Committed or done PBIs with no business value**   How will the Product Owner explain the investment in something with no value? More likely, someone forgot to enter the business value.

- **Committed or done PBIs with no effort**   Well, that was easy. Somebody probably forgot to enter the effort.

- **Committed or done PBIs without acceptance criteria**   How will the team members know what the expectations are or when development is done?

- **To-do or in-progress tasks outside of current Sprint**   Looks like a previous Sprint plan was not cleaned up correctly.

- **To-do tasks are assigned to a team member**   It's better to leave to-do tasks unassigned so that any team member with capacity can help out in order to increase the chances of meeting the Sprint Goal.

- **To-do or in-progress tasks without remaining work**   Assuming your team has a working agreement to estimate hours for tasks, this query can show those tasks that were overlooked.

- **To-do or in-progress tasks with remaining work > 8**   Assuming your team has a working agreement that no task should take longer than 8 hours, this query can show those tasks that need to be decomposed.

- **Tasks not linked to a PBI**   Not all work in the Sprint Backlog needs to pertain to developing the forecasted PBIs, but it can be a smell if there are "free-floating" tasks in there.

- **Tasks with activity set**   Assuming you follow my advice in this chapter and don't see any value in using this field, then it can be helpful to see which tasks may have this accidentally set. You can then turn this into a learning opportunity.

- **Blocked tasks**   Whether the Scrum Team uses the *Blocked* field or sets a tag or both, it can be useful to know which tasks are currently blocked.

- **In-progress tasks not assigned to a team member**   Who's working on these tasks?

- **One team member has multiple in-progress tasks**   Don't you know that multitasking is a myth and attempting it will damage your brain? Perhaps one of the tasks is actually done or blocked.

- **Team members without tasks**   With the exception of the Product Owner and Scrum Master (unless they are also a Developer), everyone should be working out of the Sprint Backlog. Be careful with this query—it could become a weapon in the wrong hands.

- **Done tasks have remaining work > 0**   How can you be done with a task if there is still remaining work? More than likely, this was just an oversight.

- **Done PBI has new or in-progress tasks**   How can you be done with a PBI if one or more associated tasks are not done?

If the Scrum Team is using Azure Test Plans, there are a few more query ideas to consider, specifically related to testing:

- **New or approved PBIs with test cases**   It is wasteful to create test cases ahead of Sprint Planning.

- **Committed PBIs without test cases**   Perhaps you are proving acceptance in some other way, such as exploratory testing.

- **No test plan for current Sprint** Maybe this Sprint is exceptional and doesn't require any acceptance testing, or more likely, someone hasn't created a test plan for it yet.

- **Test cases not linked to a PBI** It's a smell to see test cases in a test plan that are not explicitly linked to one or more PBIs. Perhaps it is a cross-cutting acceptance test, but it could also be an oversight.

Scrum Guide Drift

When the Scrum process (formerly known as the Visual Studio Scrum process template) was introduced at Microsoft's TechEd North America conference in New Orleans in 2010, it *exactly* matched the *Scrum Guide*. Over the years, however, the two have drifted apart. The *Scrum Guide* evolved while the Scrum process template did not. For example, in late 2014 Microsoft went crazy and added support for the Scaled Agile Framework (SAFe) to *all* of their process templates—even our beloved Scrum one. Although this was good in that users now had additional hierarchy support in the backlog, it also added extraneous fields.

Also, in 2014 the *Scrum Guide* was moved off Scrum.org and posted to the neutral ScrumGuides.org. At the same time, all the major Scrum organizations in the world acknowledged this as the official definition of Scrum. Unfortunately, Microsoft didn't get the memo. Sure, they still have a Scrum process, but it no longer matched the *Scrum Guide* and it was no longer "barely sufficient."

> **Note** For more than a decade, survey after survey has demonstrated that agile is and remains the most popular and successful way to develop software. Those same surveys also show that Scrum is the most popular framework to become agile—always in the 80–90 percent range of agile organizations. With this in mind, Microsoft should make the Scrum process the *default* process when creating a project. It used to be.

Over the years the Professional Scrum community has maintained in close relationship with Microsoft, and we've done what we could to keep the Scrum process from drifting too far from the *Scrum Guide*. In this section, I will explore the current differences between the two.

Work Item Types

Azure DevOps offers more than a dozen work item types—most of which don't particularly relate to planning and managing work. Therefore, I will focus only on those work item types that I have previously listed in the Scrum process section.

- **Bug** The *Scrum Guide* does not mention bugs at all. That's because a bug is a *type of* PBI. The confusing part is that the Scrum process also includes a PBI work item type. In my opinion, the only reason the Bug work item type exists is so that tooling such as the Test & Feedback extension can create a specific work item with repro steps and system information—both of which could be tracked in a PBI work item's description field.

- **Epic** and **Feature**   Again, the *Scrum Guide* only mentions Product Backlog items. It doesn't mention epics and it doesn't mention features. Microsoft did this back in 2014 to support SAFe. Professional Scrum Teams using Azure Boards have since become comfortable with hierarchical backlogs, even though they could have engineered these backlogs to use the existing PBI work item type on all backlog levels.

Backlog Levels

As I have mentioned, Microsoft introduced hierarchical backlogs to support scaled agile practices. If they had kept the PBI work item type at each backlog level, that would have kept in alignment with Scrum. But since they didn't, we now have epics, features, and PBI work item types, and the result is a goofy mix of terminology.

If organizations and teams want to use the hierarchical backlogs, and most of the ones I consult with do, then perhaps Microsoft could rename the lowest leaf-level "Backlog items" to something like "Stories"—which is the most popular term I see used. In this way, it's made clearer that the names of the backlog levels are all *not* Scrum but more industry-standard names for types of PBIs. You and your organization can refer to the items in this lowest level however you'd like.

PBI Work Item Fields

Over the years, Microsoft has added many fields to the "barely sufficient" PBI work item type. In this section, I will take a look at those fields in the PBI work item type and give my Professional Scrum opinions, including why the use of some may be considered waste.

- **Assigned To**   Sounds very command-and-control. The label and underlying field should be changed to something that sounds more like a tool for self-managing teams, such as *Owned By*. Also, Azure DevOps should let you tag which user in the project is the "Product Owner" and make that the default value of this field.

- **Reason**   For the Scrum process, the reasons are all read-only and weak. This field should just be removed or hidden from the form.

- **Iteration**   Should always default to the root level when adding a PBI. Rarely would a Scrum Team add a PBI directly to an existing Sprint. Microsoft got carried away with the use of a team's default iteration in this regard.

- **Priority**   Product Owners don't necessarily need a field for priority, as the position of the PBI in the ordered Product Backlog suggests its "priority." If a Product Owner wants to track an individual PBI's priority, it would be better to express business priority using the *Business Value* field so that all PBIs can be compared relative to each other on a common field and using a common scale like Fibonacci.

- **Effort**   Invokes thoughts of specifying hours and classic project management, instead of something more abstract and better suited for complex work like Fibonacci numbers or story points. *Size* would be a better label and underlying field name.

- **Value area**   Product Backlog items can have value for a number of reasons, well beyond the two options in this drop-down. What's more, teams may think that *architecture* work has value, which is rarely the case. Architectural work is required to deliver the kind of value that a stakeholder is looking for, but it's rarely of direct value itself. It's better to not use this field so that all items can have a measure of value relative across the same stratification.

- **Business Value**   As a corollary to what I mentioned earlier, I think *Value* would be a much simpler and better label and underlying field name.

PBI Work Item Workflow States

One of the most controversial updates to the 2011 *Scrum Guide* was the removal of the term "commit" in favor of "forecast" in regard to the work selected for a Sprint. Prior to this change, practitioners used to say that the Development Team *commits* to the Product Backlog Items that it will deliver by the end of the Sprint. Scrum now calls that selection and practice a *forecast*—because it better reflects the reality of doing complex work in a complex domain.

Well, as you can guess, Microsoft never updated the Scrum process. The Scrum community has had to put up with the Committed workflow state, as you can see in Figure 3-15. I would welcome the change to *Forecasted*, or even *Planned*.

FIGURE 3-15 The Committed workflow state, another example of misalignment with the *Scrum Guide*.

Another small nit I have with the workflow states is the state of Approved. It's not bad, but *Ready* would be preferred. Although "Ready" is not an official thing in Scrum, it is mentioned in older *Scrum Guides:* "Product Backlog items that can be 'Done' by the Development Team within one Sprint are deemed 'Ready' for selection in a Sprint Planning."

Process Customization

As you've learned, each Azure DevOps project is based on a process that defines the building blocks for tracking work. Of the out-of-the-box *system* processes, the Scrum process most closely matches the *Scrum Guide*, but not entirely. There has been some drift over the past 10 years.

Fortunately, you can customize the Scrum process to make it more closely match the *Scrum Guide*, and even your own organization or team's specific needs. Achieving this requires you to create an *inherited* process first and then make customizations to that. Any changes you make to the inherited process automatically appear in the projects that use that process. You cannot make changes to the system processes.

You primarily customize a process by adding or modifying its work item types. This is done through an administrative user interface in the web portal.

The general sequence for process customization looks like this:

- **Create an inherited process**   Select a system process (for example, Scrum) and create an inherited process (such as Professional Scrum) based on it.

- **Customize the inherited process**   Add or modify work item types, work item fields, work item workflow states, and work item form UIs. You can also update backlog behavior.

- **Apply inherited process to project(s)**   Create new projects using the inherited process or change existing projects to use the new inherited process.

- **Refresh and verify**   Refresh the web portal and explore the changes to the work items and backlogs.

Note This section covers the inheritance process model, which is available in Azure DevOps Services and Azure DevOps Server. Legacy Team Foundation Server instances used an XML process model, which provided support for customizing work tracking objects and agile tools for a project. With this older model, you had to update the XML definitions of work item types, process configuration, categories, and more. On-premises XML process configuration is beyond the scope of this book.

Professional Scrum Process

If your organization or team cares about the *Scrum Guide* and wants to address the drift between it and the system Scrum process, you should consider following the instructions in this section to create a custom, inherited Professional Scrum process. Doing so is completely optional but may result in a better experience for Scrum teams. It will also help organizations and teams that are just adopting Scrum where precise language and terms are important for establishing a new mental model.

Process customization takes place at the organization level (or at the collection level for the on-premises Azure DevOps Server). You can select any of the system processes and create an inherited process, as I've done with the Scrum process in Figure 3-16.

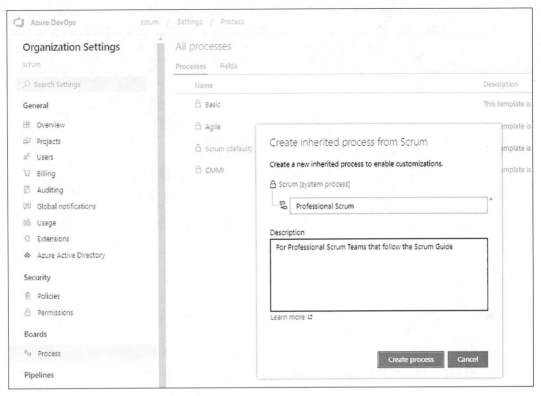

FIGURE 3-16 Creating a Professional Scrum process from the Scrum system process.

After I create the inherited Professional Scrum process and set it to be the default process, I then disable the Bug work item type. This allows the Scrum Team to use the PBI work item type for all work in the Product Backlog. Teams can add a "Bug" tag to those PBIs if they wish.

Next, I update the Product Backlog Item work item type, making the following changes:

- **Hide the *Priority* and *Value Type* fields from the layout**   I would like to remove these fields altogether, but that customization isn't allowed in Azure Boards.

- **Change the Effort label to Size**   I could also create a new *Size* field behind the scenes, but I will leave the Effort field in use.

- **Change the Business Value label to "Value"**   I could also create a new value field behind the scenes, but I will leave the *Business Value* field in use.

- **Rename the Details group to "ROI"**   The only two fields in this group now are related to ROI. It would be awesome to include a computed ROI field, but that functionality is not available outside of using an extension.

Next, I make changes to the workflow states by adding two new states: Ready (which maps to the Proposed category) and Forecasted (which maps to the In Progress category). I keep the default colors for these new states. Next, I hide the Approved and Committed states, replacing them with the Ready and Forecasted states that I just created, as you can see in Figure 3-17.

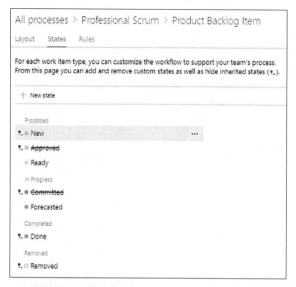

FIGURE 3-17 Customizing PBI workflow states in the Professional Scrum process.

For organizations and teams that use the Epic and Feature work item types, you can make similar customizations by hiding those extraneous fields that you don't use (e.g., *Priority, Time Criticality, Value Area*, and even *Start Date* and *Target Date*). You could also rename the labels and normalize the work-flow states, as I did for the PBI work item type.

I also hide the *Priority* and *Activity* fields from the Task work item type. The last customization I do is to rename the lowest leaf-level backlog from "Backlog items" to "Stories" (or whatever the organiza-tion/team would like it to be called). Leaving it named *Backlog items* is confusing, because in actuality, all backlog levels contain "backlog items."

After these changes are made, I can start creating projects based on the Professional Scrum process. If I have any existing projects, I can also change them to use the new Professional Scrum process, which you can see in Figure 3-18. Later, if I make any changes to the Professional Scrum process, all projects based on it will instantly reflect that change.

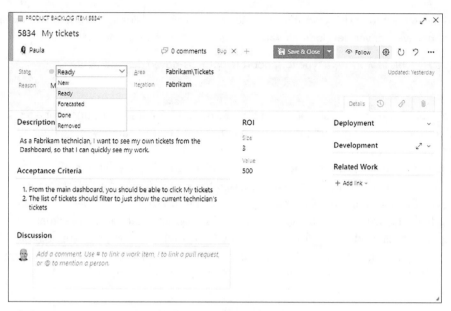

FIGURE 3-18 PBI work item form after applying the Professional Scrum process.

Other Customizations

Beyond matching the Scrum process to the *Scrum Guide*, organizations and teams may want to make additional customizations to work items and backlogs. Here are some examples I've collected over the years from various teams and other consultants:

- Add a *Team* field to the PBI work item type to indicate which team owns it, rather than using the *Area* field gimmick.

- Add a *Value Stream* work item type and corresponding top-level backlog that sits above Epic.

- Add a *Planned Sprint* field to the Impediment work item type for planning when the improvement will be done. The system *Iteration* field can be used for the Sprint where the impediment was discovered.

- Add a new workflow state to the Impediment work item type to indicate which improvement(s) are currently *in-progress*.

- Add an *Improvement* work item type to plan and track any improvement experiments being performed.

- Add a *Remaining Work* field to the PBI work item type to store the rolled-up sum of any child task *Remaining Work* values. External automation, such as an extension, would be required to do the rolling up.

- Add default user story description text ("As a *<type of user>*, I want *<some goal>* so that *<some reason>*") to the *Description* field of the PBI work item type.

- Add default text to the PBI work item type *Acceptance Criteria* fields to suggest a "given-when-then" behavior-driven development (BDD) or "given-when-then-fail" format.

- Add a *Hypothesis* work item type to support hypothesis-driven development.

Tip Use your Scrum or Professional Scrum process the way that it was designed for a few Sprints before customizing anything. I've seen teams want to immediately make their new project look and behave like their old project or culture. For example, if after reading this book you decide to abandon your Agile process project, creating a new Professional Scrum project, don't immediately add the fields that used to be in your old project (for example, the *Original Estimate* and *Completed (hours)* fields in the Task work item type—which we removed from the Scrum process for a reason; tracking original estimates and actual hours are generally considered waste). Just know what you are doing and why you are doing it before making any "improvements" and weigh the benefits against potential waste and misuse. Don't inadvertently change the rules of Scrum by customizing the tool!

Fabrikam Fiber Case Study

The Scrum Team has decided to follow the guidance in this section and create and use the Professional Scrum process, inherited from the Scrum process. They will apply this new process to their existing Fabrikam project. This will be the process that is referenced in the coming chapters. For that reason, you may want to take a moment and create a Professional Scrum process yourself so that you can better follow along.

Chapter Retrospective

Here are the key concepts I covered in this chapter:

- **Process** When creating a project, you will need to select a process. Microsoft provides several out-of-the-box processes, referred to as system processes.

- **Scrum process** A Scrum-centric process created through a collaboration of Microsoft and the Professional Scrum community.

- **Work item types** Although there are more than a dozen Azure DevOps work item types, including a number of hidden ones, the ones that apply to planning and managing work are Product Backlog Item, Bug, Epic, Feature, Task, and Impediment. Task and Test Case work items should be created and linked only during the Sprint in which you are working on their parent PBIs.

- **Queries**   There are a number of queries that a Scrum Team could create and share to track and manage the work in the Scrum development effort.

- **Scrum Guide drift**   Over the years, the *Scrum Guide* has evolved while the Scrum process has not. This issue can be overcome by creating and customizing an inherited process.

- **Inherited process**   A child of one of the system processes that can be customized in a structured way, inherited processes can be used to create new projects as well as applied to existing projects. Future changes to the inherited process are instantly visible in all the projects that use that process.

Practicing Professional Scrum

In this part of the book, I will begin demonstrating how to practice Professional Scrum and Azure DevOps together effectively. The previous part established a baseline understanding of the three areas of knowledge required before proceeding: Scrum, Azure DevOps, and Azure Boards. Over the next several chapters, you will see how these three fit together and how a team can optimize their use to deliver business value in the form of working software.

I will begin with the discussion and activities surrounding product planning. This will take us up to the beginning of the first Sprint. I refer to this collection of activities as the *pre-game*—everything from envisioning the product, provisioning the Azure DevOps environment, setting up the project, organizing the team, building and refining the Product Backlog, and preparing for the first Sprint falls into the pre-game. As you can imagine, a lot is involved in the pre-game, with numerous opportunities to get sidetracked. We will stay focused on the intersection of Scrum and Azure DevOps; other books are available that explain the intricacies of product planning.

The remaining chapters will follow the rules of Scrum very closely as I establish how a Professional Scrum Team works within a Sprint using the relevant tools found in Azure DevOps. At times, I will focus on using Azure Boards to plan and track a Sprint and manage the daily work. I will also delve into the other Azure DevOps services to demonstrate how team members can collaborate effectively to maximize their flow as well as the quality of their product. I will continue to use the Fabrikam Fiber case study to give examples of how a team might use the many options that are available.

> **Tip** High-performance Scrum Teams take the "let the team decide" mantra seriously, and they don't abuse it. These teams have learned to effectively live within the balance of increasing value in the product with decreasing waste in the process. In short, they are not distracted by the shiny features but take an experimental approach to any new tool or practice.

CHAPTER 4

The Pre-game

In the game of rugby, or any professional sport for that matter, many activities must be performed prior to kickoff: prior games are analyzed, sponsors are secured, stakeholder input is provided, rules get clarified, playing fields get selected, calendar dates get negotiated, teams get selected, and player positions get designated. Scrum development efforts also have a pre-game, where many of these same types of activities are performed. The Scrum pre-game is the time period when the vision is established all the way up to the start of the first Sprint. The pre-game is not timeboxed, and not all development efforts make it out of the pre-game.

Many important activities can be performed during the pre-game (in no particular order):

- Establish the vision, scope, and goals of the product.

- Identify sponsors and stakeholders.

- Establish the Scrum Team (Product Owner, Scrum Master, and Developers).

- Establish the software development environment (e.g., provision Azure DevOps, build and release pipeline agents, deployment environments).

- Educate individuals on the rules of Scrum.

- Educate individuals on Azure DevOps.

- Define the high-level product requirements.

- Create the initial Product Backlog.

I recognize that some of the activities I outline in this chapter are considered to be *execution* in nature as opposed to *preparation*. An example of an execution activity would be provisioning Azure DevOps. Some Scrum Teams prefer to do these kinds of activities during an actual Sprint, within a timebox and while collaborating with an engaged Product Owner. Since many of the activities I outline in this chapter are executed one time only and must be performed *before* development using Azure DevOps can occur, I have lumped them together into the pre-game.

Smell It's a smell when I see a team spending too much time setting up their environment. Developers do not need the most perfectly awesome configuration of tools prior to their first Sprint. In fact, not until they begin work will they actually know what they need. Just as their software product will evolve, so will their tools and practices. If a team has historically procrastinated getting started on a new development effort, consider executing these activities in Sprint 1. The timebox will force the team to produce an increment of working functionality in the same Sprint that they set up their environment. In other words, their tooling will be barely sufficient.

Note The concept of the pre-game (or "Sprint 0" as some call it) does not exist in the *Scrum Guide*. Whatever the team wants to calls it, they are just not yet practicing Scrum. Because of this, most of the pre-game activities I've listed here will be out of the scope of this chapter. I will focus only on those activities directly related to provisioning the Azure DevOps environment.

Setting Up the Development Environment

It should go without saying that before a Scrum Team can begin using Azure Boards to implement Scrum, someone must provision Azure DevOps. For the cloud-hosted Azure DevOps Services, this is as simple as signing up, creating a new organization, and entering payment details. For the on-premises Azure DevOps Server, this means installing and configuring.

This section assumes that you have a properly provisioned Azure DevOps Services organization available for the team to use. I'll also assume that you have access to a *helpful* organization owner or project collection administrator to serve your Scrum Team as needed. Maybe this is you, but if not, hopefully it's someone who is a friend of the team. In my experience, this is a recipe for success. If the administrator understands software development, that's great. If the administrator understands Scrum, that's awesome. In my experience, if the administrator only has an IT/operations background, be prepared for some friction.

Tip Having a Scrum Team member also be the Azure DevOps administrator is not ideal. A Scrum Team should be able to focus on building great product, not administering their DevOps tools. High-performance Scrum Teams are ones whose team members can avoid being distracted by activities that don't directly contribute to delivery of business value.

Creating an Azure DevOps Organization

The Azure DevOps organization is a mechanism for organizing and connecting groups of related projects. An organization might be a business division, a regional division, or other enterprise structure. You can choose one Azure DevOps organization for your entire company, or separate organizations for specific business units, or even teams. In other words, your business structure should act as a guide to the number of organizations that you create in Azure DevOps.

Before you begin using Azure DevOps Services, someone in your organization or team will need to sign up and create a new organization. You can use an Azure Active Directory (AAD), a Microsoft Account (MSA), or a GitHub account to sign up. It's easy enough to create these, if you don't happen to have one. Even if your company doesn't yet have an AAD instance, one can be created for free by using the Azure portal. Double-check first, because having an AAD is required if your organization is using Azure or Microsoft 365 (formerly known as Office 365), so you may already have one.

AAD is the recommended choice because it's so convenient to have your Azure DevOps Services organization backed by AAD and to be able to have everyone's names available to select from without having to remember if someone is a live.com, hotmail.com, outlook.com, Gmail, etc. Connecting to AAD will map existing Azure DevOps users in the organization to their corresponding identities in AAD.

The Azure DevOps Services organization name will become part of the URL that individuals use to access the services. The organization URL uses the format *https://dev.azure.com/{organization}*. For example, since my organization name is *Scrum*, the URL to access the services is *https://dev.azure.com/scrum*. This is something to consider if your company's name is *Fabrikam Fiber and Cable Management Limited*. Perhaps just shrinking it to *Fabrikam* would be preferred. The trick, like with web domains, is finding a short organization name that is not taken. At any time, an organization administrator can rename the organization to a better name and, thus, a better (hopefully) simpler URL.

> **Note** A default *project collection* is created when you sign up with Azure DevOps Services. However, unlike the on-premises Azure DevOps Server (or Team Foundation Server), the existence of the collection has been abstracted out of the URL and UI. In other words, you don't even have to worry about it. The cloud-based Azure DevOps Services URLs are shorter and you probably won't miss what you never knew existed.

For a larger company, you can create multiple organizations. For example, the Fabrikam company might create three Azure DevOps organizations: *Fabrikam-Marketing*, *Fabrikam-Engineering*, and *Fabrikam-Sales*. The organizations are all for the same company but are mostly isolated from each other. Each organization has a separate URL, a separate list of users, a separate subscription. Setting up boundaries like this is not required, but it is helpful if and when doing so makes sense.

When you create an organization, you can also choose the Azure region where your organization is hosted. You may choose your organization's region based on locality and network latency, or because you have sovereignty requirements for specific data centers. You can see an example of this in Figure 4-1. A default location is selected, based on the closest Microsoft Azure region where Azure DevOps Services is available.

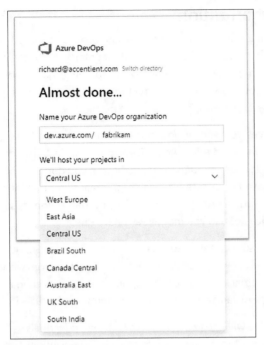

FIGURE 4-1 Select the region where a new Azure DevOps organization will be created.

Note Azure DevOps Services maintains all of your data within the selected Azure region, including work items, source code, and test results. Geo-redundant mirrors and offsite backups are stored within the region as well. Azure DevOps Services stores information that is more global in nature, such as user identities and profile information, in either the U.S. data center (for U.S.-based users) or the EU data center (for EU-based users). Profile data for users from all other countries are stored in the U.S. data center.

Providing Access to the Organization

After the organization has been created, with the best name, in the best region, it's time to make additional settings so that your teams, team members, and stakeholders can access it. This step will involve adding them as users, specifying their AAD or MSA credentials, selecting the type of license, and providing access to one or more projects. This step also involves setting up billing to pay for the user licenses, self-hosted and additional pipeline jobs, and other services.

To review, the following types of users can join your organization for free:

- Five users who get Basic features

- An unlimited number of Visual Studio subscribers who get Basic features, and possibly Basic + Test Plans and self-hosted parallel jobs

- An unlimited number of users who get Stakeholder features

Users not listed here must have a Basic or Basic + Test Plans license purchased for them. You do this through user assignment–based billing, where you will pay for the users who are assigned a specific access level. When a user is removed, those charges stop. You can set the organization default to Stakeholder access so that once you run out of user assignment–based licenses, each new user will be assigned a free Stakeholder license. This way, there aren't any users who are not assigned a license, and every new user gets at least some access to Azure DevOps. You can see user license assignment starting to emerge in Figure 4-2.

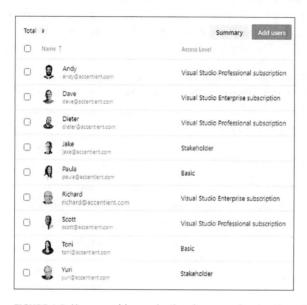

FIGURE 4-2 You can add organizational users and assign them licenses.

> **Tip** Although it is possible to add users to the organization directly, it makes more sense to do so once a project is created. This way, you can add them and assign a project role all in one step. In fact, most of the interesting configurations I will look at, as they relate to Professional Scrum Teams, are done at the project level, as you will see later in this chapter.

Before you can purchase a Basic or Basic + Test Plans license or any other paid service, you must set up billing. All billing is done through Azure; therefore, you will have to set up an Azure subscription. Even though you may not be using Azure or any other Azure service, this is how Azure DevOps is billed. Payment options include credit card as well as invoiced billing through the Enterprise Agreement, Cloud Solution Providers, and other programs. All charges appear on a monthly Azure bill. In addition to being an Azure DevOps organization owner or collection administrator, you must be added as an owner or contributor to the Azure subscription that will be used for billing.

Tip Check out the Azure calculator to predict monthly Azure DevOps costs. It can help management understand the costs and help with their expenditure decisions. You can find it at *https://azure.microsoft.com/pricing/calculator*.

Microsoft offers a *multi-organization* billing option, which allows you to pay once per user, regardless of the number of organizations they belong to. Multi-organization billing simply groups the per user charges at the subscription level, so only organizations that share a common Azure subscription can be billed together.

Multi-organization billing does not make sense for all customers. For example, each organization gets five free Basic users, which apply to the billing subscription, not the organization. If most of the users access only one organization, then five free users may be more cost-effective. If many users access multiple organizations, then multi-organization billing is likely the better option.

Smell It's a smell when I see members of a professional software development team being assigned Stakeholder access. Although it may be adequate for some stakeholders, every member of a Scrum Team requires at least Basic, if not Basic + Test Plans access. The (free) price of Stakeholder access is tempting to management, but it tends to reflect their inability to value what the team does or how they do it.

Other Organization Configurations

Creating the organization, setting up billing, and providing user access are the primary steps in provisioning an Azure DevOps Services environment. Some of those activities are ongoing, such as when team members join or leave the development effort or new stakeholders desire access.

Here are some of the other configuration and administrative activities that may be pertinent to setting up your Azure DevOps environment:

- **Auditing**   Access, export, and filter audit logs that list all changes that occur throughout the organization. Changes occur when a user or service identity edits the state of an artifact, like changing permissions, deleting resources, changing branch policies, and so on.

- **Global notifications**   Enable organization-wide notifications by configuring default subscriptions, subscribers, and other settings.

- **Usage**   Investigate and view high usage levels by you and other users in the organization.

- **Extensions**   View and manage extensions that have been installed, requested, or shared. Extensions will be discussed later in this chapter.

- **Security - Policies**   View and manage various application connection, security, and user policies.

- **Security - Permissions**   View and manage organization-level permission groups and permissions, such as specifying which users can help with organization- and collection-level administrative activities.

> **Tip** The user who created the Azure DevOps organization becomes the organization *owner*. The organization owner is also a member of the Project Collection Administrators group and cannot be removed. The owner can be changed, however. Also, to reduce the risk of losing admin access, you should add additional users to the *Project Collection Administrators* permission group. As roles and responsibilities change, you can change owners.

- **Boards - Process**   View and manage system and inherited processes. I covered this in Chapter 3, "Azure Boards."

- **Pipelines - Agent pools**   View and manage the agent pools, which are containers that logically group one or more build/release pipeline agents, for your whole organization.

- **Pipelines - Settings**   View, enable, and disable various pipeline settings, such as limiting variables that can be set at queue time.

- **Pipelines - Deployment pools**   View and manage the deployment pools, which are containers that logically group one or more deployment targets, such as the servers in a test environment.

- **Pipelines - Parallel jobs**   View and manage the number of in-progress jobs as well as the maximum number of parallel jobs available.

> **Note** For each parallel job listed, you can run a single job at a time in your organization. Microsoft-hosted job limits are separate from self-hosted (on-premises) job limits. The free tier limits include one Microsoft-hosted pipeline and one self-hosted pipeline (with no hour limits). Additional parallel jobs can be purchased for both Microsoft-hosted and self-hosted environments.

- **Pipelines - OAuth configurations**   View and manage OAuth-based service connections to third-party services, such as GitHub, GitHub Enterprise Server, and Bitbucket Cloud.

- **Artifacts - Storage**   View and manage your usage of pipeline artifacts and packages stored in Azure Artifacts feeds.

Azure DevOps Marketplace Extensions

Microsoft knew early on that they would not be able to keep up with the development community's demand for Azure DevOps features and customizations. Shadow groups within Microsoft created and blogged freely downloadable tools. Even Microsoft Research gave us some great toys over the years, a couple of which ended up inside Visual Studio Enterprise edition. We also had CodePlex and GitHub, which were good ways of sharing open source plug-ins and extensions. The support was great, but we lacked a centralized marketplace, as Visual Studio had with its Visual Studio Gallery.

Seeing the success of the Eclipse marketplace and numerous other app stores, Microsoft decided to offer something similar. At the Connect() event in November 2015, Microsoft launched the Visual Studio Marketplace, which replaced the Visual Studio Gallery. At the same time, they also launched the Visual Studio Team Services Marketplace. They added support for the on-premises Team Foundation Server the following spring. Since then, we've enjoyed a unified marketplace full of wonderful extensions published by Microsoft, Microsoft partners, and the community at large.

Today, the Azure DevOps Marketplace boasts more than 1,200 extensions across all the services: Azure Artifacts, Azure Boards, Azure Pipelines, Azure Repos, and Azure Test Plans. You can search for extensions that work in the cloud (Azure DevOps Services) or on-premises (Azure DevOps Server) as well as by other factors, such as price and certification. For example, Figure 4-3 shows the first page of Azure Boards extensions for Azure DevOps Services sorted by number of installations.

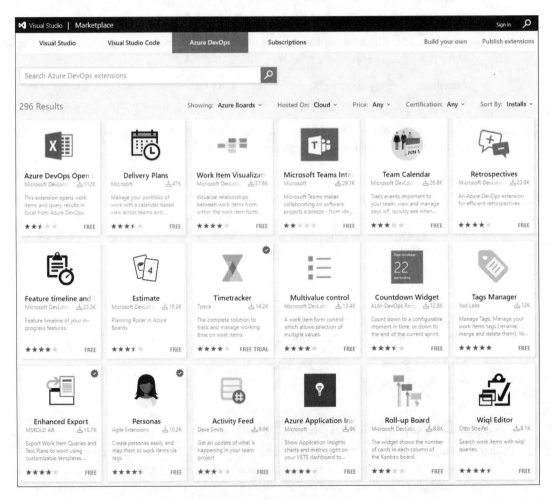

FIGURE 4-3 There are many great Azure Boards extensions in the Azure DevOps Marketplace.

Extensions for Professional Scrum Teams

Ultimately, which extensions a Scrum Team wants to install and use should be up to the team. In the spirit of experimentation, inspection, and adaptation, they should install and evaluate an extension, using it minimally for one to three Sprints in order to form an opinion on its value. At the next Sprint Retrospective, the Scrum Team can share their findings and decide if they want to keep using it, change the way they are using it, or drop it. Although this applies particularly well to extensions, this should be the approach that a Scrum Team takes with all tools and practices they experiment with.

Over the many years training, consulting, and coaching Scrum Teams, I've come up with a list of some extensions that have proven themselves valuable. I'm not saying that my list should be *your* list. In fact, you may think some of these are less useful than I do. You may also find other extensions that are awesome and can really enable your team. Just be sure to inspect their value and adapt their usage accordingly.

Here's my partial list of extensions that have proven themselves valuable in the visualization and management of work for Scrum Teams (in no particular order):

- **Azure DevOps Open in Excel**   Opens work items and query results in Excel from Azure DevOps. Requires Excel and Azure DevOps Office Integration to be installed.

- **CatLight**   Notifies Developers in the system tray when pull requests, builds, bugs, and tasks need their attention.

- **Decompose**   Allows you to quickly break down work items from epics to features to PBIs to tasks. Use keyboard shortcuts to easily promote and demote work items between different hierarchy levels, as shown in Figure 4-4.

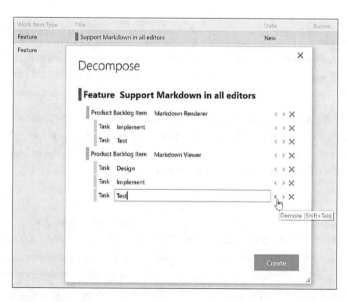

FIGURE 4-4 Use the Decompose extension to quickly create work item hierarchies.

- **Definition of Done**   Lets you view and modify your team's Definition of Done. Visualize this definition on each PBI work item, and even use it as a confirmation step before setting the work item to the Done state.

- **Feature timeline and Epic Roadmap**   Allows you to plan or track work items in progress by visualizing them on a Sprint calendar. This tool helps you visualize the portfolio-level work items (epics and features) as they are worked over multiple Sprints. Microsoft won't be investing further in this extension, especially once their Delivery Plans 2.0 extension is published.

- **Product Vision**   Allows the Product Owner to easily set the product vision and make it visible to team members and stakeholders.

- **Retrospectives**   Enables the collecting, grouping, voting, and visualizing of feedback obtained during a Sprint Retrospective.

- **SpecMap**   Lets you create engaging story maps directly from the work items in Azure Boards. Use these story maps to visualize user journeys, refine PBIs, and plan releases.

- **Sprint Goal**   Enables the Scrum Team to display a Sprint Goal on the Sprint page.

- **Tags Manager**   Lets you manage work items tags by viewing, renaming, merging, and even deleting them in a central place.

- **Team Calendar**   Allows you to track the events important to the team, view and manage days off, quickly see when Sprints start and end, and more.

- **ActionableAgile Analytics**   Lets you analyze and display many aspects of a team's development process. This extension uses the team's own data to predict when specific PBIs might be completed, when a release might be ready, or how many PBIs might be in a release.

- **Test & Feedback**   Supports running manual or exploratory tests. This browser-based extension can capture notes, screenshots with annotations, screen recordings, user actions (as an image action log), page load data, and other system information. This rich data can be associated with work items that are created, such as issues, tasks, and feedback response work items.

As you can tell, this list of extensions is primarily Azure Boards and Azure Test Plans related. This is not to say that Scrum Teams can't make use of some great extensions in other categories. There are plenty available too. I just wanted to focus on those extensions that dealt with viewing and managing work, and not any particular development or engineering practices.

Note Only organization owners and Project Collection Administrators can install an extension. If you don't have those permissions, you can *request* an extension instead. You must be a contributor for your organization in order to request an extension. You can view and monitor your requested extensions on the Requested tab in the Extensions page. An administrator can then approve (and thus install) the extension. Hopefully the delay in requesting and approving is minimal. Delays like this are proportional to how agile your organization is (or isn't).

Setting Up Product Development

This section explores those activities related to setting up software product development within Azure DevOps. Some of these activities are one-time events, whereas others are ongoing, such as configuring areas and iterations. Before proceeding, it's assumed that the following activities have already been completed:

- You have a product with goals and stakeholders.

- The Scrum Team has been formed and roles identified.

- All team members are in the company's Azure Active Directory (AAD) or have Microsoft Accounts.

- An Azure DevOps Services organization has been created and backed by AAD, and billing details have been entered.

- Appropriate Azure DevOps licenses have been purchased for the team members.

- Azure DevOps extensions have been installed.

- Appropriate client software has been installed (for example, Visual Studio, Visual Studio Code, Microsoft Office).

Creating a Project

The Azure DevOps project is the container for the software product's development lifecycle. All work items, code and other version-controlled artifacts, test cases, test results, pipelines, builds, and releases are stored in a project. Technically, they are stored as a combination of Azure SQL table data and blob storage. To look at it from a Scrum perspective, the Azure DevOps project represents the product being developed and is a container for the Product Backlog, the Sprint Backlog, the source code, and the tests that form the Increment and thus the product. The project also contains queries, charts, and other visualizations that enable a team to assess their progress and the quality of their work.

Note You may have noticed that I don't refer to "team project," just "project." For whatever reason, the "team" prefix has been disappearing from usage over the past few years. You can still find "team project" and "team project collection" (Azure DevOps Server) referenced in the product and docs, but for the most part, the world seems to have dropped the "team" prefix. If you find out why, let me know.

You can create a new project on the Projects page of Organization settings. You will need to provide the name and description of the project and select the visibility, version control system, and process, as I'm doing in Figure 4-5. By default, only organization owners and Project Collection Administrators can create a new project.

The name of the project should be short and simple, and relate to the product being developed. The name does not need to include the version, release, Sprint, team, feature set, area, or component. All of these items can be tracked within the project using areas, iterations, teams, and repositories. For example, if I were creating a project to plan and track development of the Fabrikam application, I would consider naming it something simple like *Fabrikam* rather than *FabrikamV1, FabrikamBeta, FabrikamSprint1, FabrikamDev,* or *FabrikamWeb*.

FIGURE 4-5 We create the Fabrikam project using a custom Professional Scrum process.

Creating a project is the easy part, but planning it can be more difficult. It's imperative that you know what product—or component of a larger product—that your team will be developing. If the product has a name, you should consider using that for the name of the project. If the product doesn't yet have a name, or the name is something like "The web app that consumes the web service from our financial partner," you'll want to give it an actual name first. This is the first step in it becoming a product. It sounds trivial, but having a clear, meaningful name will begin the process of focusing less on *how* the software works and more on *what* it should be doing. If you are at a complete loss, then consider Wikipedia's page listing fictional computer names at *https://en.wikipedia.org/wiki/List_of_fictional_computers* for inspiration.

> **Smell** It's a smell if the Developers don't know the name of the product they are developing. Maybe it doesn't have a name, or maybe they just don't care to know it. Either way, this demonstrates a lack of product-minded thinking. For a successful adoption of Scrum to occur, this must change. The Scrum Master and Product Owner can help with this.

Determining How Many Projects You Will Need

The answer is almost always one, as I'll explain.

The scope of an Azure DevOps project is a function of the product being developed, its components, the number of Developers, and whether they are dedicated to that one product. Developers include anyone who performs analysis, design, coding, testing, deployment, operations, or other activities that help turn a PBI into done, working software. The ideal formation is a Scrum Team of no more than 10 members dedicated to working on a single product. This would simply require one project containing both the Product Backlog and the engineering aspects (repositories, pipelines, and feeds) of the product being developed.

Unfortunately, I don't see this very often. More often I come across tiny teams, huge teams, or teams having to context-switch across multiple products and domains. Good'ish news here: Azure DevOps can support all of these environments as well.

On a micro-team of only one or two Developers, they won't necessarily be practicing Scrum, but they could still make use of an Azure DevOps project. I would hope they would take advantage of the Product Backlog, but as far as planning and tracking work within a Sprint, they may not need tooling for that. For medium-large teams with 10 or more developers, they would want to split into smaller, correctly sized Scrum Teams to work more efficiently within the rules of Scrum. They can still all work within the same Azure DevOps project. Table 4-1 shows a summary of this discussion.

TABLE 4-1 Azure DevOps Projects Based on Teams and Products.

Developers	Products	Azure DevOps projects	Notes
1–2	1	1	■ Won't be using Scrum ■ Will want a backlog
	> 1	1	■ Won't be using Scrum ■ Will want a single backlog ■ Can use separate code repositories and pipelines

Developers	Products	Azure DevOps projects	Notes
3–9	1	1	■ Ideal for a Scrum Team
	> 1	1	■ Will be using Scrum
			■ Will use a single backlog
			■ Can use separate code repositories and pipelines
10–18	1	1	■ Two Scrum Teams
			■ Will use a single backlog, partitioned by team
			■ Can use separate code repositories and pipelines
10+	> 1	Varies	■ Each Scrum Team can cover a portion of the products and then refer to the above
			■ May need to redefine what the "product" is
19+	1	1	■ Ideal for Nexus Scaled Scrum framework

I've seen large products with 80+ developers working within the same Azure DevOps project and using a single backlog. This increased complexity demands using work item areas to designate the responsible team, not to mention being deliberate about using—and naming—the corresponding repositories and pipelines. Even developers in the unfortunate situation of having to develop or support multiple products at the same time can still do so within one project—by having a single backlog from which to order and plan the work.

Fabrikam Fiber Case Study

Scott created an Azure DevOps project named *Fabrikam* based on the custom, inherited Professional Scrum process. The Fabrikam project is a private project using Git version control.

One organization I worked with claimed to be using Scrum. It had multiple product managers, each vying for the team's time. Needless to say, its queue of work was quite full, always changing, and always being reprioritized. Focus was low. Chaos was high. One product manager wanted a set of improvements on their system and asked the IT director when she could have them. The IT director performed some high-level analysis and provided a rough estimate, telling the manager that work could be started in about 9 months and, once started, shouldn't take longer than about a month. The manager was displeased. She asked if the team could start sooner and work on her features in between their other tasks. The IT director replied, "We can do that, but then it will take us 12 months total, put other projects behind, and you won't like the quality of our work."

The IT director was describing the fundamental result of *Little's law*, which is that for a given process, in general, the more things that you work on at any given time (on average), the longer it is going to take for each of those things to finish (on average). The IT director realized that the manager, as well as the organization, had a belief in magic. They thought that the team could work efficiently on multiple things at once; could work at an unsustainable pace when it suited the organization; and could give accurate estimates for unknown, complex problems. Eventually, the IT director became the Product Owner and was able to order the work in a way where the team could focus, spending an entire Sprint working on the most important work—as decided by the Product Owner, not the cadre of managers.

Tip If your development organization is broken and can't be fixed, rest assured that Azure DevOps can implement whatever dysfunctional processes you can throw at it. For a situation where a small number of Developers must service a large number of products in a prioritized, just-in-time way, then I suggest they take a look at Kanban. Kanban is a method for developing software with an emphasis on just-in-time delivery. Kanban instructs developers to pull work from a queue using a visual board that shows the work in progress by state. It can better support unplanned work.

Adding Project Members

When a project is first created, only the creator is a member. They will need to add other team members. This can be done in one step as a new Azure DevOps user is added from the organization's Users page. For an existing user, a Project Administrator (e.g., the person who created the project) will need to add them directly to the project.

Project members can be added via the Permissions page, assigning them to one of the built-in groups. You can see these groups in Figure 4-6. Members can also be added to a team (which in turn is a member of a permission group). I'll talk about teams in a bit. If you're adding a user to Azure DevOps for the first time, you'll need to add them as an Organization user first.

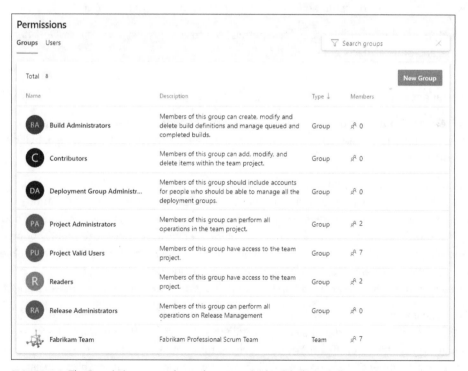

FIGURE 4-6 The Permissions page shows the groups in the Fabrikam project.

Projects contain several project-level permission groups—each with a default set of permissions. These are created at the time the project is created, and they are all empty. The exceptions are the Project Administrators and Project Valid Users groups, which contain the user who created the project. These groups are populated by the user who created the project, who is also typically a Project Collection Administrator. The Project Administrator must decide who else needs access to the project and what level of permission they need.

Here are details about the built-in permission groups:

- **Build Administrators** Members of this group have permissions to administer build resources and build permissions for the project. Members can manage test configurations/environments, create test runs, and manage builds.

- **Contributors** Members of this group have permissions to contribute fully to work items, code, and pipelines. Contributors don't have permissions to manage or administer resources.

- **Deployment Group Administrators** Members of this group have permissions to administer deployment groups and agents.

- **Project Administrators** Members of this group have permissions to administer all aspects of the project and teams.

- **Project Valid Users** This group contains all users and groups that have been added anywhere within the project. You cannot directly modify the membership of this group.

- **Readers** Members of this group have permissions to view project information, work items, code, and other artifacts but not modify them.

- **Release Administrators** Members of this group have permissions to manage all release operations.

- **<Project> Team** The project's default team is also considered a permission group. It can have membership and be assigned permissions like the other groups. I'll get into more details on teams in a bit.

According to Microsoft's design and documentation, the obvious choice would be to add all the members of the Scrum Team (including the Product Owner and Scrum Master) to the Contributors group, stakeholders to the Readers group, and a select few to the Project Administrators group. Although I acknowledge that this is a valid choice, I believe that it can lead to impediments during a Sprint. For example, if the Project Administrator is unavailable, the team members might be blocked from adding a new work item area, creating a shared query, or adding a new team member. These sound like minor things, but they can add up to a fair amount of waste during the course of a Sprint.

I believe that if a team has a certain level of proficiency in the tool, and the members trust one another, they should all be added to the Project Administrators group. This epitomizes the self-managing qualities of the Scrum Team. You can easily do this by making the team (for example, Fabrikam Team) a member of the Project Administrators permission group, after which you can remove the team from the Contributors group. You can see the final result in Figure 4-7.

FIGURE 4-7 We can make the Fabrikam Team a member of the Project Administrators group.

Fabrikam Fiber Case Study

After Scott created the Fabrikam project, he made the default Fabrikam Team a member of the Project Administrators group. He also removed the Fabrikam Team as a member of the Contributors group, since that was no longer required. Scott then added all the Scrum Team members to the Fabrikam Team and then added two key stakeholders, Jake and Yuri, to the project's Readers permission group. During upcoming Sprint Retrospectives, the team will discuss and decide if any permission levels need to be adjusted for any team members or stakeholders.

Teams

Teams are similar to permission groups but with added capabilities when it comes to using the agile planning tools in Azure Boards. Like groups, teams allow the organizing of collaborating individuals who will be working on similar areas of the product. Teams differ from groups in that they enable their members to access and use the agile planning tools in order to define and manage their Product Backlog, Kanban board, Sprint Backlog, and task board. By default, a new team has the same permissions as the Contributors permissions group.

When a project is created (such as Fabrikam), a default team is created with the same name as the project. Team members can be added to this team and start using the agile planning features right away, without the need for additional configuration. In scaled product development situations, like with the Nexus Scaled Scrum framework, a Project Administrator can create additional teams, giving them each a custom view of the overall Product Backlog. I cover this in Chapter 11, "Scaled Professional Scrum."

Azure DevOps allows a user to belong to more than one team. Before considering this as an option, make sure that it would be prudent. Rarely do team members belong to more than one Scrum Team, especially within the same product. If the team member is shared between a couple of teams, then this might make sense, but let's think about it. Forget about Azure DevOps supporting it or not and consider it from a commitment and focus perspective. Is it really the best use of this person's time to be physically and mentally switching teams and context like this? You'll learn more about this in Chapter 8, "Effective Collaboration." Perhaps other alternatives could be considered during the next Sprint Retrospective. If you are thinking that you should add the Product Owner and Scrum Master to multiple teams, don't. This is exactly the purpose of the default team. Members of the default team can see all the items in the Product Backlog. I cover this in Chapter 11.

Note When you create a new team, Azure DevOps can also create a similarly named area path at the same time. This ensures a strong connection between teams and areas, as you can see in Figure 4-8. You can skip this when creating a new team by clearing the Create An Area Path With The Name Of The Team check box. Also, keep in mind that if you rename your team, it will not rename the associated area path. You must do that manually. The default team's area is the root enabling team members to view and manage the entire Product Backlog.

FIGURE 4-8 We can create a corresponding area path when creating a new team.

After creating a team, its iterations (Sprints) and areas can be specified. Iterations that the team selects will appear in the backlogs as Sprints for the sake of forecasting and planning. The areas that a team specifies determine what work items show up in the team's backlog. Teams can also designate one of those areas as its default. This is then suggested when creating new work items.

The user who created the team becomes the *Team Administrator*. Team Administrators can manage the team and team membership, as well as configure the agile tools. Essentially, Team Administrators can configure, customize, and manage all team-related activities for their team. Each team must have at least one Team Administrator. To remove a Team Administrator, you must first add a second Team Administrator before you can remove the first Team Administrator.

Here are some of the activities that a Team Administrator can perform:

- Create and manage team alerts
- Select team area paths
- Select team Sprints
- Configure team backlogs
- Customize the Kanban board
- Manage team dashboards
- Set working days off
- Show bugs on backlogs and boards

Note It's easy to confuse Project Administrator with Team Administrator. Whereas the Project Administrator is a permission group with a set of permissions, the Team Administrator is a role that is tasked with managing team assets. A team member can be both a Team Administrator and a member of the Project Administrator role, which (as you can probably guess) is what I recommend for every member of the Scrum Team.

Fabrikam Fiber Case Study

To further limit impediments and reinforce self-management, Scott makes sure that every member of the Scrum Team is not only a Project Administrator, but also a Team Administrator of the Fabrikam Team. You can see this in Figure 4-9. During upcoming Sprint Retrospectives, the team will discuss and decide if any permission levels need to be adjusted for any team members or stakeholders.

(Case Study continued on the next page.)

Administrators

All teams need to have an Administrator.

richard@accentient.com ✕   Dave ✕   Toni ✕   Andy ✕   Dieter ✕   Paula ✕   Scott ✕

Add

FIGURE 4-9 Each member of the Fabrikam Scrum Team has been made a Team Administrator.

Other Project Configurations

Creating the Azure DevOps project, configuring permissions, and adding team members are the most important provisioning steps. Some of those activities are ongoing, such as when team members join or leave the development effort or new stakeholders need access. You should rarely have to futz with the groups and permissions, however.

Here are some of the other configuration and administrative activities that may be pertinent to setting up a project:

- **Azure DevOps services**   By default, all of the Azure DevOps services (Boards, Repos, Pipelines, Test Plans, and Artifacts) are enabled. A Project Administrator may want to disable, or later reenable, some of these services if they won't be used during the development effort.

- **Notifications**   Enable project-wide notifications by enabling/disabling subscriptions, adding new subscriptions, or changing delivery settings.

- **Service hooks**   View and manage integrations with other Microsoft or third-party services, such as App Center, Azure, Teams, Microsoft 365, or UserVoice.

- **Boards - Project configuration**   View and manage the iterations (Sprints) and areas for the project. Figure 4-10 shows the Sprints set up for this project. Figure 4-11 shows the areas set for this project. Since there is only one team—the Fabrikam Team—it will use all iterations and areas.

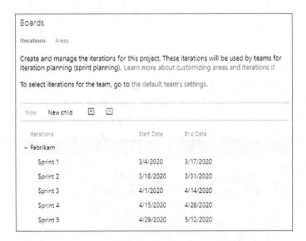

FIGURE 4-10 The first five Sprints have been created, each with a two-week duration.

FIGURE 4-11 Area paths reflect the logical or functional areas of the product.

- **Boards - Team configuration**   View and manage the backlog behavior, working days, bugs behavior, iterations, areas, and work item templates for each team.

- **Boards - GitHub connections**   Manage GitHub integration with Azure Boards.

- **Repos - Repositories**   View and manage permissions, options, and policies for the project's repositories.

- **Repos - Cross-repo policies**   View and manage pull request completion requirements for branches that match a pattern across all of your repositories.

- **Pipelines - Agent pools**   View and manage agent pools, which are containers that logically group one or more build/release pipeline agents, for your project.

- **Pipelines - Parallel jobs**   View and manage the number of in-progress jobs as well as the maximum number of parallel jobs available.

- **Pipelines - Settings**   View, enable, and disable various pipeline settings, such as limiting variables that can be set at queue time and retention policies.

- **Pipelines - Test management**   View and manage flaky test detection. A flaky test is a test that provides different outcomes, such as pass or fail, even when there are no changes in the source code or execution environment.

- **Pipelines - Release retention**   View and manage release retention policies.

- **Pipelines - Service connections**   View and manage connections and authentication details to a number of services, such as Azure, Chef, Docker, GitHub, Jenkins, Kubernetes, Nuget, Npm, and more.

- **Test - Retention**   View and manage automated and manual test retention policies.

Establishing Information Radiators

An *information radiator* is a generic term for any kind of display—physical or electronic—placed in a highly visible location that shows process or product information. Information radiators can be for the Scrum Team and/or stakeholders. Azure DevOps offers several information radiator solutions. Every Azure DevOps project has an Overview hub, which contains a summary page, one or more dashboards, and a wiki, which can contain multiple pages. Even if a project has all the Azure DevOps services (Azure Boards, Azure Repos, etc.) disabled, the Overview hub and those pages remain. Initially, the Overview hub is pretty boring. It's up to the team to decide what interesting and valuable content should be added.

Here's a quick look at each of these pages:

- **Summary**   This simple welcome page for the project shows high-level information such as project stats and team members.

- **Dashboards**   These customizable, interactive signboards provide real-time information. Dashboards can be associated with a team and display configurable charts and widgets.

- **Wiki**   The team can create wiki pages and share information with your team in order to better understand and contribute to your project.

Summary Page

The Summary page isn't all that interesting. It cannot be customized to the extent of being an effective information radiator, but it does have its purpose. It is the landing page for the project, so it should effectively communicate the purpose of the project and development effort.

To customize the Summary page, you basically have two options. First, if you have Azure Boards enabled, you can simply add a project description (yawn). Second, if you have Azure Repos enabled, you can point the Summary page to the home page of your wiki or to a readme file full of Markdown goodness, as I've done in Figure 4-12.

Dashboards

Dashboards are a great way for a team or stakeholders to visualize real-time information about the product or process. They are visually appealing, interactive, and customizable. A project can host several dashboards, each serving a specific purpose.

Each dashboard hosts an array of *widgets*. Widgets display specific information, such as query results, charts, team members, statistics, shortcuts, Markdown, or even an embedded webpage. You can see an example of this in Figure 4-13. Widgets can be resized and configured in a number of ways.

Over two dozen widgets are available out-of-the-box and more than a hundred available through the Azure DevOps Marketplace. Dashboards can be configured to periodically refresh their contents.

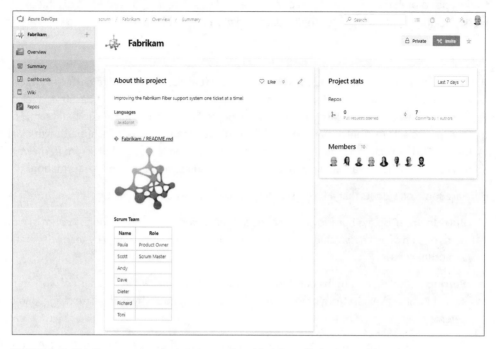

FIGURE 4-12 This Summary page points to a readme Markdown file in the repository.

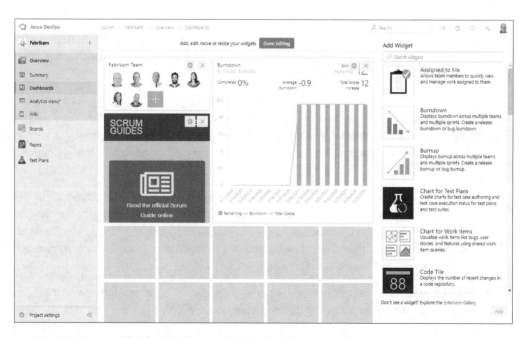

FIGURE 4-13 You can add widgets to the Overview dashboard.

Note Widgets that pertain to a specific Azure DevOps service are disabled if the service they depend on has been disabled. For example, if the Azure Boards service is disabled, work tracking–related widgets are disabled and won't be selectable.

When you create a dashboard, you can choose to make it a project dashboard or one specific to a team. Project dashboards display information or status about the project. Project dashboards also have configurable permissions. Team dashboards provide focused information specific to that team. For Azure DevOps projects that only have one team—the default team—project dashboards work just fine. For Azure DevOps projects in a scaled environment (e.g., using the Nexus Scaled Scrum framework), it might make sense for each team to have their own dashboard radiating their own information.

Here are some of widgets that a Professional Scrum Team may be interested in:

- **Burndown** Displays burndown by backlog level or work item type, either by count of work items or sum of Business Value, Effort, or Remaining Work. Configurable by team, date range, and other criteria.

- **Burnup** Displays burnup by backlog level or work item type, either by count of work items or sum of Business Value, Effort, or Remaining Work. Configurable by team, date range, and other criteria.

- **Chart for Test Plans** Tracks the progress of test case authoring or status of test execution for tests in a test plan. Several chart types and settings are available; this widget is intended for use by teams using Azure Test Plans.

- **Chart for Work Items** Displays a progress or trend chart that builds off a shared work item query, such as showing investment by product area. Several chart types and settings are available.

- **Cumulative Flow Diagram** Displays the flow of work through the system for a specified team, backlog, and timeframe. Primarily for Kanban teams, but since Scrum Teams can practice Kanban as well, this widget can be very helpful for visualizing and improving a team's flow. Configurable by team, backlog, swimlane, column, and date range. I cover this in Chapter 9, "Improving Flow."

- **Cycle Time** Used to monitor the flow of work through a system by computing the amount of time a team spends actually working on a PBI from start to finish. Primarily for Kanban teams, but since Scrum Teams can practice Kanban as well, this widget can be very helpful for visualizing and improving a team's flow. Configurable by team, backlog, swimlane, column, date range, and other criteria. I cover this in Chapter 9.

- **Markdown** Enables custom text, links, images, and more by using Markdown syntax. The widget can also be configured to point to a file stored in a repository.

- **Sprint Burndown** Displays work burndown for a Sprint by backlog level or work item type, either by count of work items or sum of Remaining Work. Configurable by team, date range, and other criteria.

- **Velocity** Helps to improve forecasting by displaying a team's velocity either by count of work items or sum of Effort, Business Value, or Remaining Work. Configurable by team, backlog level, work item type, date range, and other criteria, this widget is intended for use by teams using velocity.

Fabrikam Fiber Case Study

Paula has decided to simplify the Overview dashboard to show only the product vision. She installed the Product Vision extension by Agile Extensions. This allows her to simply enter the vision and then configure a size and color scheme. You can see the results in Figure 4-14. The Product Vision can also be viewed on a Product Vision page in the Boards hub.

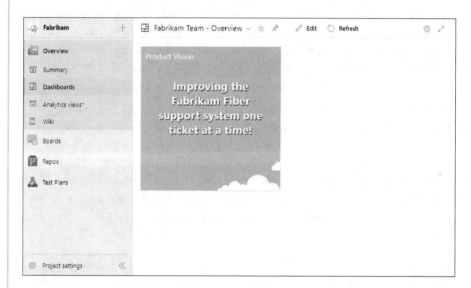

FIGURE 4-14 This Overview dashboard shows the Product Vision widget.

Wiki

Every Azure DevOps project has its own wiki. The wiki behaves like most other wikis you have probably used. You can use it to share information and collaborate with your team. Even stakeholders (with Stakeholder access) can view information in the wiki. The project wiki must be created before you can use it.

Behind the scenes, each wiki is supported by a Git repository. When you create the project wiki it will, in turn, provision a hidden Git repository to store the corresponding Markdown files and related artifacts. Even if your project uses TFVC for source control, it will still create a hidden Git repository to support the wiki. This is why users with Stakeholder access can't create or edit wikis; they have no capability to work in Azure Repos.

Here are some items that a Professional Scrum Team might consider publishing in the wiki:

- Sprint Goals

- Sprint Retrospective notes

- Definitions (Done, ready, bug, etc.)

- Development standards and practices

- Stakeholder contact list

- Technical documentation and notes

For some of these items, sticky notes or writing on a whiteboard would work just fine, and possibly be even more visible. Physical boards fall apart when it comes to broad visibility, searching, and fault tolerance (e.g., the nightly cleaning person). Also, if you want to track changes or control access of who can change these items, an electronic solution is preferred.

> **Tip** A team should not hide its Sprint Goal or its definitions. Because transparency is important, make these broadly available so that all stakeholders know what you are working on and what your definitions are—especially your Definition of Done. In addition, teams may want to increase transparency by publishing their metrics (burndowns, velocity, cycle time, cumulative flow diagram, etc.) charts and other visualizations. For the ultimate in transparency—and courage—you could even publish these on a large flat screen in the break room or cafeteria. Now that's confidence!

The wiki makes a great place to store your team's definitions, such as the Definition of Done. You can see an example of this in Figure 4-15. By making this and other definitions big and visible, a Scrum Team's transparency increases, as well as a shared understanding of the work that the Scrum Team is performing. Stakeholders, knowing what the team's Definition of Done is, will be informed the next time a Developer says that they are "done" or "not done" with something. There's an actual checklist, on the wiki, available to all, that describes what "done" means.

> ### Fabrikam Fiber Case Study
>
> The Scrum Team has decided to use the wiki for sharing information with one another as well as stakeholders. As part of that, they have created a wiki entry to track their Definition of Done. They will also create separate wiki entries for Sprint Goals and Sprint Retrospectives.

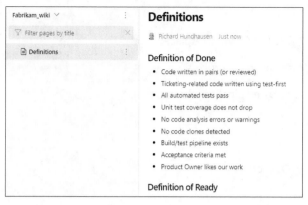

FIGURE 4-15 The wiki lets you share your team's Definition of Done.

Pre-game Checklist

Each team and organization's preparation will be different. Those just getting started with Scrum and/or Azure DevOps may have more work to do in their pre-game. Other teams can basically just start with their first Sprint. Regardless of where you and your team are, this checklist can be helpful in preparing your environment for an awesome first Sprint.

Scrum

- ❏ Identify the product.
- ❏ Establish the vision, scope, and business goals of the product.
- ❏ Identify product sponsors and stakeholders.
- ❏ Define the high-level product requirements.
- ❏ Establish the Scrum Team (Product Owner, Scrum Master, and Developers).
- ❏ Educate individuals on the rules of Scrum (if necessary).
- ❏ Locate the Product Backlog (if one exists).

Azure DevOps Organization

- ❏ Create the organization.
- ❏ Connect to Azure Active Directory (if necessary).
- ❏ Set up billing in Azure.
- ❏ Acquire user licenses.
- ❏ Add the users to the organization.
- ❏ Add a backup Collection Administrator.

- ☐ Configure global notifications.
- ☐ Install any extensions from the Azure DevOps Marketplace.
- ☐ Create a custom, inherited Professional Scrum process.
- ☐ Purchase additional parallel or self-hosted pipeline jobs.
- ☐ Create the Azure DevOps project.

Azure DevOps Project

- ☐ Enable/disable specific Azure DevOps services.
- ☐ Configure notifications.
- ☐ Create additional teams (if in a scaled environment).
- ☐ Add stakeholders to the Readers role (or, optionally the Contributors role).
- ☐ Configure at least the first three Sprints (iterations with consistent lengths and no gaps).
- ☐ Configure areas.

Azure DevOps Team

- ☐ Make each team a member of the Project Administrators group.
- ☐ Remove each team from the default Contributors group.
- ☐ Add users to their appropriate team(s).
- ☐ Make each user a Team Administrator.
- ☐ Choose an awesome image/avatar for the team.
- ☐ Configure what backlogs the team will use (Epics, Features, etc.).
- ☐ Configure team working days.
- ☐ Configure team iterations (at least the first three Sprints).
- ☐ Set team's Default iteration and Backlog iteration to root.
- ☐ Set the team's default area (root if only one team).
- ☐ Select the team's areas (root, including subareas if only one team).

Azure DevOps User

- ☐ Set full name and contact email in profile.
- ☐ Choose an awesome image/avatar.
- ☐ Set time and locale information.
- ☐ Configure notifications.
- ☐ Select theme (light or dark).

Chapter Retrospective

Here are the key concepts covered in this chapter:

- **Provisioning Azure DevOps Services**   Before being able to use the cloud-hosted Azure DevOps, you will need to create an organization, set up billing, and acquire user licenses.

- **Azure Active Directory**   Azure DevOps Services can be connected to Azure Active Directory to simplify the onboarding and managing of users from your organization.

- **Extensions**   Adds additional functionality to Azure DevOps. Available through the Azure DevOps Marketplace, an active community of extension providers.

- **Azure DevOps project**   This serves as the container for your product's lifecycle. One project can support multiple product areas, teams, Sprints, and releases.

- **Azure DevOps team**   This is similar to a permission group in an Azure DevOps project, but with additional context and support for using the agile planning tools.

- **Areas**   This hierarchy is useful for classifying work into different logical or functional areas of the product. Areas are used to help organize work items. For scaled development, each Azure DevOps team in a project can be associated with one or more areas. I cover this in Chapter 11.

- **Iterations**   Synonymous with Sprints, Iterations have a name, a start date, and an end date and are used to help organize and plan work.

- **Summary page**   This is the Azure DevOps project's landing page. The Summary page contains basic information by default. It can be customized by pointing at a readme file or a wiki.

- **Dashboards**   These interactive signboards can display a variety of static and dynamic content through the arrangement of widgets. A number of widgets are available out-of-the-box, with many more available in the Azure DevOps Marketplace.

- **Wiki**   This contains one or more pages of content authored by the team in order to collaborate, share information, and better understand the development effort. Wikis use the Markdown language, which is a lightweight markup language with plain-text-formatting syntax. Wikis are backed by Git repositories.

CHAPTER 5

The Product Backlog

The Product Backlog is an ordered list of everything that might be required of the product. It is the single source listing all requirements for any changes to be made to the product. It includes features to be added, changes to be made, and bugs to be fixed. Each item in the Product Backlog is called a Product Backlog item (PBI). PBIs can range from extremely important and urgent to quite trivial.

The Product Owner is responsible for the Product Backlog, but they may have others create and update its items. However, it is the Product Owner's responsibility to ensure that the items in the Product Backlog are clearly defined, understood by everyone, assigned a business value, and ordered (prioritized) correctly. The Developers collaborate with the Product Owner and other stakeholders (such as domain experts) as needed during Product Backlog refinement, Sprint Planning, and the Sprint Review to understand and size the items in the Product Backlog.

This chapter will focus on how to use Azure Boards to create and refine a Product Backlog. You will also see how to foster a healthy Product Backlog, how it can enable forecasting and release planning, and how to do so using these tools. If you are more interested in the concept of the Product Backlog and less on *how* to use Azure Boards to interact with it, you may want to reread Chapter 1, "Professional Scrum."

> **Note** In this chapter, I am using the customized Professional Scrum process, not the out-of-the-box Scrum process. Please refer to Chapter 3, "Azure Boards," for information on this custom process and how to create it for yourself.

Creating the Product Backlog

You create the Product Backlog by adding work items—one at a time or in bulk. In the default Scrum process, both PBI and Bug work item types appear in the backlog. In the custom Professional Scrum process, I have disabled the Bug work item type.

By default, only a title is needed in order to save a new work item to the Product Backlog. The Product Owner will probably require more information to assign the item a business value and order it in relation to other items. The Developers will definitely need more information than just a title to be able to size it and ensure that it is ready for Sprint Planning and development. Having a title is a good start, though.

According to the *Scrum Guide*, the Product Owner is responsible for the Product Backlog. This doesn't necessarily mean that they are the one *doing* the actual data entry. It just means that the Product Owner is responsible for ensuring that each PBI is clearly defined and understandable. In Chapter 3, I demonstrated that, in Azure Boards, anybody with the appropriate permission can create work items. A Project Administrator can assign these permissions in Project Settings.

As stated previously, I feel that everyone on the Scrum Team should be able to contribute to the Product Backlog—at a minimum. The Product Owner may also want other stakeholders, such as business analysts, customers, or the users themselves, to be able to create work items. Why not? The fewer people in the conversation, the less diluted the story will be. That said, if someone other than the Product Owner creates a work item, a conversation should take place so that person can explain its context, purpose, and business value. If possible, this conversation should take place face to face or over the phone, and not using the discussion tools.

When someone creates a PBI, they should focus on its value (the "what") and avoid descriptions of "how" it should be developed. When it comes time, the Product Owner can order the Product Backlog based on each item's value, risk, priority, dependency, learning, necessity, votes, or any other criteria the Product Owner desires.

The Product Backlog can emerge and evolve quickly if the business requirements of the product and other conditions change. To minimize waste, detailed requirements should be avoided except for the highest-ordered items (the ones at the top of the backlog). Even then, only enough details are required in order to size the item. The teams that I have worked with preferred having two to three Sprints' worth of refined PBIs at the top of their Product Backlog.

Smell It's a smell when I see tasks, tests, impediments, statements, gripes, or guidance in a Product Backlog. What I'm talking about is a PBI titled, "We should back up the database each night" or "Find a better place to meet for our Daily Scrum." The Product Backlog should contain only items that represent a potential change in the product being developed. This is not to say that a valid PBI couldn't have some goals, guidance, or gripes attached to it.

Creating a Product Backlog in Azure Boards

As previously mentioned, Azure Boards is just one of the services in Azure DevOps. Azure Boards is the service that provides the planning, tracking, visualizing, and managing of work. Using Azure Boards, a Scrum Team can manage backlogs at several levels (such as Epics, Features, and Stories). Developers can also plan their Sprints and create their Sprint plan by using the Sprint Backlog and related Taskboard.

Azure Boards enables a Scrum Team to perform the following Product Backlog–related activities:

- Add items to the Product Backlog.

- Import items into the Product Backlog.

- Establish hierarchical backlogs of Feature work items.

- Map PBI work items to Feature work items.

- Establish hierarchical backlogs of Epic work items.

- Map Feature work items to Epic work items.

- Report bugs.

- Manage and refine items in the Product Backlog.

- "Bulk-modify" items in the Product Backlog.

- Reorder the Product Backlog.

- Calculate velocity to be used for forecasting.

- Forecast how many PBIs might be deliverable in upcoming Sprints (also known as release planning).

- Track progress using the burndown, burnup, and other analytics.

- Move undone PBIs from a Sprint back to the Product Backlog.

- Calculate flow metrics such as CFD (Cumulative Flow Diagram), cycle time, and lead time. I will cover flow metrics in Chapter 9, "Improving Flow."

A Scrum Team typically uses the Backlogs page to add and manage PBIs. The Boards page will work as well, but it is better suited for teams practicing Kanban. You can use queries, Microsoft Excel, or other client applications to manage the backlogs, but the Backlog page should be the Scrum Team's primary UI for managing the backlog. Regardless of how it's created and managed, the ordered (prioritized) list of work items in the Product Backlog represents the vision, Product Goal(s), and release plan (roadmap) for the software product. It is the single source of truth.

 Tip Although you can add, view, and edit work items with the (free) Stakeholder access, I recommend that anyone who contributes to the Product Backlog have *Basic* access. Stakeholders can't change the backlog priority order, can't assign items to an iteration, and can't use the Mapping pane or the Forecasting tool. On the Boards page, Stakeholders can't add work items or update fields displayed on cards, although they can drag and drop work items to update status. Basic access provides full access to all backlog and Sprint Planning tools.

When you view the Product Backlog, only New, Ready, and Forecasted PBI work items are displayed. Done and Removed PBIs are not displayed. You would need to create and run a work item query or use the search feature to find Done and Removed PBIs. PBI work items in the Forecasted state and assigned to a Sprint are also not displayed in the Product Backlog by default. You would need to enable the *In*

Progress Items display option, or go to the respective Sprint Backlog, to see those PBIs. This setting is meant to keep the Product Backlog lean by removing excess noise.

If your organization has multiple Scrum Teams working on a single product, then you are in a *scaling* situation and should consider using the Nexus Scaled Scrum framework. When Azure Boards is configured for multiple teams, you will see only those PBI work items assigned to the area path(s) specified as belonging to your team. For example, let's assume there is a team *Red* that owns areas R1, R2, and R3 and a team *Blue* that owns areas B1 and B2. If a user on team Red changes the area of one of their PBIs to B1, it will disappear from their Product Backlog and appear on team Blue's Product Backlog. The team Red user would need to use the search feature or create a custom query to locate that work item in the future. Members of the Default team will, by default, see all items in the Product Backlog. I will dive deeper into scaling and how Azure Boards supports multiple teams in Chapter 11, "Scaled Professional Scrum."

Keyboard Shortcuts

Microsoft provides several time-saving keyboard shortcuts to quickly navigate within the Boards hub and manage work items. Shortcuts work with either lowercase or uppercase, as long as the focus is not on an input control. Shortcuts can be disabled as well if they become bothersome. At any time, from any screen, you can press the question mark on your keyboard to display the list of global and page-specific shortcuts. Table 5-1 lists the keyboard shortcuts that a Professional Scrum Team might be interested in.

TABLE 5-1 Azure DevOps keyboard shortcuts.

Scope	Shortcut	Description
General and project navigation	[g][h]	Goes to the organization home page (list of projects)
	[g][w]	Goes to the Boards hub
	[g][c]	Goes to the Repos hub
	[g][b]	Goes to the Pipelines hub
	[g][t]	Goes to the Test Plans hub
	[g][s]	Goes to the Settings page
	[s]	Goes to search
Boards hub	[w]	Opens the Work Items page
	[l]	Opens the Backlogs page
	[b]	Opens the Boards page
	[i]	Opens the Sprints page
	[q]	Opens the Queries page
	[z]	Toggles full screen
Backlogs page	Ctrl+Shift+F	Filters results
	Ctrl+Home	Moves the PBI to top

Scope	Shortcut	Description
	[m][b]	Removes the PBI from a Sprint (set its Iteration to the root)
	[m][i]	Moves the PBI to the current Sprint
	[m][n]	Moves the PBI to the next Sprint
	[ins]	Adds a child task
	[del]	Deletes the PBI

Azure DevOps also offers page-specific shortcuts for Boards, Repos, Pipelines, Test Plans, Artifacts, and the wiki. Page-specific shortcuts only work when on that specific page.

Adding Product Backlog Items

In Azure Boards, each backlog allows you to add a new work item. This is a quick way of adding a PBI to the Product Backlog (or to the Stories backlog if you're using the custom Professional Scrum process). The same capability exists in the Features and Epics backlogs as well. You can add the new item just *above* the current selection, to the *top* of the backlog, or to the *bottom* of the backlog. You can see these options in Figure 5-1. This method is fast, because you only have to provide a title, which you should keep short and meaningful. This approach allows you to quickly add several items in rapid succession, which can be helpful during Sprint Review, when feedback and ideas are flying about the room.

FIGURE 5-1 You can quickly add a PBI to the Product Backlog in Azure Boards.

Tip Clicking the *Add* button without specifying a title will open the work item form. This time-saving trick allows you to specify more than just the title when adding a PBI to the Product Backlog.

Besides the title, the rest of the fields will be assigned their default values, some of which might need to be changed right away, such as the *Assigned To* field. Most Scrum Teams like to assign PBI work items to the Product Owner, since they are the *owner* of that item. It is blank (unassigned) by default, which is also a valid choice. Over time, more details about the PBI will emerge, such as its area, description, value, acceptance criteria, and size. Some of these details may be known at the time of creation. Others may never materialize, especially if the Product Owner decides the item has no value. There are many ways to open a work item, but the easiest—if you are on the Backlogs page—is to just click its title.

Smell It's a smell when I see a PBI work item assigned to a team member and not to the Product Owner. This suggests that either (a) the team doesn't understand Scrum and its beautiful separation of the "what" from the "how" across the roles, or (b) the team isn't working as a team, and that individual plans on "owning" that item and doing all of the work to develop it.

Even if the PBI work item is assigned to the Product Owner, the smell can resurface in the Sprint Backlog if a forecasted PBI work item's *Assigned To* field changes to an individual team member to indicate who is working on it. In Professional Scrum, the *team* should be working on the PBI, not any individual. Individuals may work on tasks associated with the PBI but leave the PBI assigned to the Product Owner throughout its lifecycle, or simply leave it blank.

In a future version of Azure DevOps, I'd like to be able to specify more information about the Scrum Team specifically. I'd like to be able to indicate which team member is the Product Owner and which team member is the Scrum Master. By doing this, Azure Boards could auto-assign PBI work items to the Product Owner, and maybe even auto-assign Impediment work items to the Scrum Master. I hope this is on Microsoft's backlog, but if not, then maybe you could add it for me. Just visit *https://developercommunity. visualstudio.com*, select Azure DevOps Services, select Suggest A Feature, and add a feature idea.

If you find that a PBI work item's Iteration defaults to the current Sprint (such as Fabrikam\Sprint 1), it means that your team's Default iteration is set to @CurrentIteration or a specific iteration. You should change this by navigating to *Project > Settings > Team Configuration > Iterations*, and change the Default iteration to the root iteration (such as Fabrikam). This way, all PBI work items will be added to the Product Backlog *without* being assigned to any particular Sprint.

Fabrikam Fiber Case Study

Since the Product Backlog contains only PBI work items, Paula has decided to remove the *Work Item Type* column. She also removed the *Value Area* column knowing that, in Scrum, there is only business value, which is to say value as perceived by the stakeholders. She then added the *Area Path* and *Business Value* fields. This minimal set of columns allows her to focus on the basics, including ROI (BV/Effort).

The initial placement of a PBI on the Product Backlog is determined according to where you have added the item or moved the item on the page. A PBI's position in the Product Backlog is tracked behind the scenes in the *Backlog Priority* field (or in the *Stack Rank* field in the Agile and Capability Maturity Model Integration [CMMI] processes). As you add items or drag and drop items within the Product Backlog, a background process updates that field, which is used by Azure Boards to track the order of items on the backlog. By default, this field doesn't appear on the work item form.

These hidden numbers are not simply 1..n, but large integers with large gaps between them. Also, these gaps are not computed by halving logic, which can lead to problems if several items are added in one location (like at the top) at a time. Instead, the gaps are computed through optimized sparsification logic, which starts with the current item and "walks" up or down until it finds a gap. All the items in that range are then redistributed to create gaps for new backlog items without affecting others. If you are curious, you can view the *Backlog Priority* column and inspect those integers and gaps.

If you add a new PBI at the top of the backlog, its Backlog Priority will be lower than the item in row #2. If you add a new PBI at the bottom of the backlog, its Backlog Priority will be higher than the item in the second-to-last row. If you add a PBI through the Work Items page or an external client, such as Excel or Visual Studio, or via the REST API, the *Backlog Priority* column will be empty. PBI work items without a Backlog Priority appear at the bottom of the Product Backlog, as you can see in Figure 5-2.

Simply adding a new item to the *bottom* of the Product Backlog will cause the sparsifier—the logic that performs sparsification—to run and assign Backlog Priority values to those other PBIs. Also, a Backlog Priority will be assigned if you drag one of the new items or drag another item to the bottom.

Order	Title	Area Path	State	Business Value	Effort	Backlog Priority
1	Use corporate email to authenticate	Fabrikam\Mobile\Security	Ready	500	13	111100571
2	Enable network alert heartbeat	Fabrikam\Dashboard\Alerts	Ready	800	5	124994730
3	Enable two-factor authentication	Fabrikam\Mobile\Security	Ready	800	8	138888888
4	Quick Tips	Fabrikam\Dashboard\Messages	Ready	500	8	159722222
5	Manage user access	Fabrikam\Mobile\Security	Ready	500	3	170138888
6	My tickets	Fabrikam\Tickets	Ready	500	3	180555555
7	Employee phone numbers	Fabrikam\Employees	Ready	800	3	222222222
8	Customer email addresses	Fabrikam\Customers	Ready	1300	5	333333333
9	Show customers count	Fabrikam\Customers	Ready	500	2	444444444
10	Customer phone numbers	Fabrikam\Customers	Ready	1300	3	722222222
11	Improve customer edit, details, and delet...	Fabrikam\Customers	Ready	800	5	777777777
13	Added from Work Items page	Fabrikam	New			
14	Added from Microsoft Excel	Fabrikam	New			
15	Added from Visual Studio Team Explorer	Fabrikam	New			

Backlog · Analytics · + New Work Item · View as Board · Column Options · ...

FIGURE 5-2 PBIs added outside the Backlogs page won't have a Backlog Priority.

Tip Don't try to manually manipulate the *Backlog Priority* field. If you set it through Microsoft Excel, once you add or reorder the items in the backlog, the sparsification logic kicks in and sets the values to big numbers to introduce big gaps. This is by design—to improve performance. The *Backlog Priority* field was always intended to be a system field, and Microsoft's recommendation is to not update it directly. So, if you want to use Excel to set *Backlog Priority*, avoid the losing fight with the sparsifier and use the Backlogs page instead.

If you are curious, you can see the *Backlog Priority* field value changes made by the sparsifier. Simply select the History tab when you edit the work item and review the changes. You can compare the current value with the previous value, as shown in Figure 5-3.

FIGURE 5-3 Work item history shows the current and previous *Backlog Priority* values.

Since the Product Backlog order is persisted in a static field, other lists and queries can use it. For example, when the Developers forecast PBIs for a Sprint and then view the forecast in the Sprint Backlog, the PBIs will display in the same order. This is good, because if the Product Owner has signaled that the *Two factor authentication* is the most important item in the Product Backlog, then the team should see it at the top of the Sprint Backlog and consider executing its work first.

> **Note** Each time you drag a PBI in the Product Backlog, history is generated—and not just for that one item. Be mindful of this as you order the backlog.

Handling Epic PBIs

An epic PBI is any PBI that is too large to be completed in a single Sprint. For example, if the Developers have a velocity of 18 points, they should not forecast a PBI that is 21 points. Even if the Developers are not using velocity as a forecasting tool, they should never take on more work than they *feel* they can accomplish in a Sprint. In either case, the PBI is an epic and must be split—preferably vertically (by acceptance criteria) and not horizontally (by technology layer). Suffice it to say that splitting a PBI is a combination of science, art, magic, and a bit of luck at times.

I want to take a step back and differentiate epic PBIs from Epic work items found in the Epics backlog. When I say "epic PBI," I'm using the term "epic" as the agile community uses it, not the way Microsoft does. An epic PBI is a generic term for a large PBI, as opposed to Epic work items, which are at the top of the hierarchical backlog of work items, sitting above Features. An Epic work item can certainly hold an epic PBI, but it might be that a Feature work item would be a better container. It just depends on how epic the epic PBI is!

Speaking of epics and features, not all organizations and teams create their top-level Epic work items in the Epics backlog and then create associated, medium-level Feature work items in the Features backlog, and then the associated, lowest-level PBI work items in the Stories backlog. Even though it seems logical to plan work this way, no organization or team typically does it like that. In practice, PBI work items often get added to the (lowest-level) Stories backlog. During refinement, if the team realizes that a PBI is, in fact, a very large piece of work (also known as an epic PBI), they can either split it into two or more, smaller PBIs or "promote" the PBI to a Feature work item and then split it into two or more smaller PBIs.

> **Smell** It's a smell when I see an epic PBI near the top of an ordered Product Backlog. It's not ready for Sprint Planning. It should be split into smaller PBIs well in advance of the Sprint in which it is forecasted to be developed.

In addition to splitting the epic into two or more PBIs, the Product Owner should decide if a parent Feature work item should be created in the Features backlog and linked to the new PBIs in order to establish a hierarchy. Let's assume that you have an epic PBI to "Improve mobile UX," as shown in Figure 5-4. UX stands for *user experience*. Since this PBI's size is larger than our velocity of 18, it must be split.

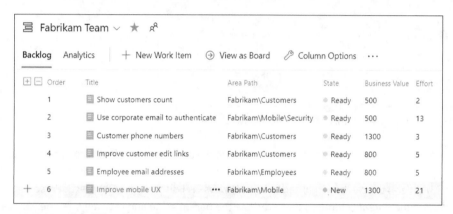

FIGURE 5-4 Epic PBIs are ones that are too large to forecast and develop in a Sprint.

First, you must split the epic PBI into two or more PBIs. Each of these would be of a smaller size (effort)—achievable in a single Sprint. Ideally, you'd want the titles to maintain some logical reference to each other, as well as the original epic PBI. For example, the epic PBI titled "Improve mobile UX" might yield three smaller PBIs titled "Improve mobile *dashboard* UX," "Improve mobile *ticketing* UX," and "Improve mobile *reporting* UX." You could also use the *Area* or *Tag* field to help keep these items associated.

What should be done with the original epic PBI? You have three options:

- Remove or delete it.

- Rename it to become one of the new, smaller PBIs.

- Promote it to a *Feature* work item, making it a parent of the new, smaller PBIs.

I want to focus on the last option. When I say "promote," I mean to change the work item type from PBI to Feature, as I'm doing in Figure 5-5, essentially moving it up one level so that I can make it a parent of the new, split PBIs. The Developers would never directly forecast or develop the Feature work item; it would just be used for organization and planning. When all the child PBIs were done, the Feature work item's state could be manually set to Done. The advantage to this approach is that it establishes a visual hierarchy and additional context. You can learn more about changing a work item type here: *https://aka.ms/change-work-item-type*.

Change work item type
Type
Feature
Add a reason for changing the type
This PBI was too large, so we are promoting it to a feature.
Learn more about changing the type
OK
Cancel

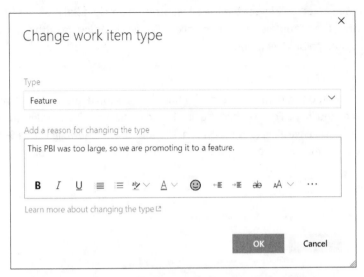

FIGURE 5-5 You can change a PBI work item to a Feature work item.

After changing the PBI work item to a Feature work item, you'll need to refresh the backlog, show the Mapping pane, and drag the child PBIs to the new feature. These actions will create a parent-child relationship. This hierarchy can be visualized in several ways, as I discussed in Chapter 3. Personally, I like to just include the *Parent* column so that I can see the parent's title and click it if need be. You can see this at the bottom of Figure 5-6.

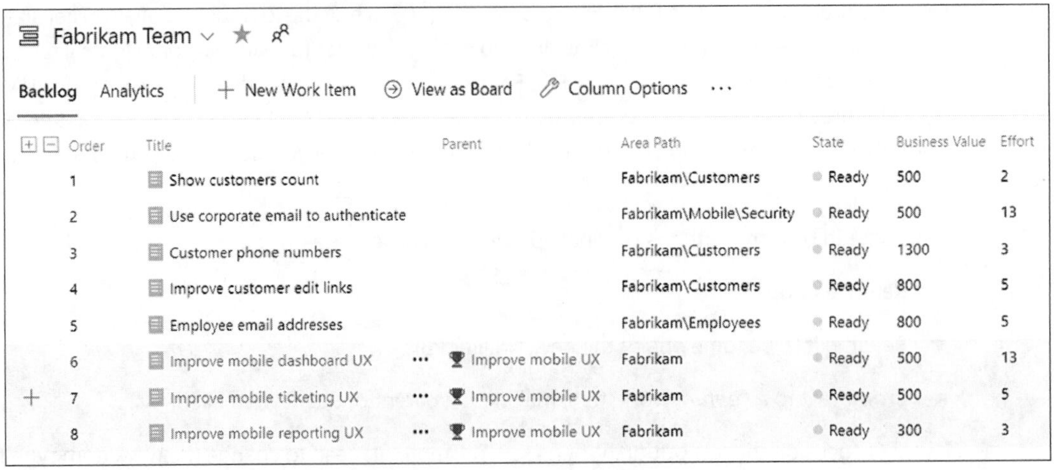

Fabrikam Team
Backlog
Analytics
New Work Item
View as Board
Column Options
Order
Title
Parent
Area Path
State
Business Value
Effort
1 Show customers count Fabrikam\Customers Ready 500 2
2 Use corporate email to authenticate Fabrikam\Mobile\Security Ready 500 13
3 Customer phone numbers Fabrikam\Customers Ready 1300 3
4 Improve customer edit links Fabrikam\Customers Ready 800 5
5 Employee email addresses Fabrikam\Employees Ready 800 5
6 Improve mobile dashboard UX Improve mobile UX Fabrikam Ready 500 13
7 Improve mobile ticketing UX Improve mobile UX Fabrikam Ready 500 5
8 Improve mobile reporting UX Improve mobile UX Fabrikam Ready 300 3

FIGURE 5-6 You can display the parent feature of newly split and linked PBIs.

> **Tip** You should avoid creating same-category links. These are parent-child links among work items of the same type, such as PBI to PBI. In Azure Boards, when a backlog contains same-category nested work items like this, the system cannot properly reorder work items. The system disables the drag-and-drop reorder feature and not all items will display under these circumstances. You may also run into error messages.
>
> Instead of nesting PBIs within the Stories backlog, Microsoft recommends that you maintain a flat list. In other words, only create parent-child links one level deep between items that belong to a different category. Use the Feature work item type when you want to group PBIs, and use the Epic work item type when you want to group features.

You can also use the bulk-modify feature to link several PBIs to a feature at once. Just select the related PBIs, right-click the selection, and select *Change Parent* (Figure 5-7). Here you can select the parent feature (or parent epic if you're working in the Features backlog).

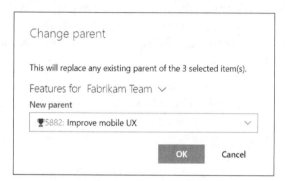

FIGURE 5-7 Azure Boards allows you to set the parent of many PBIs at once.

Establishing a feature-to-PBI hierarchy keeps the Product Backlog flat. There are no placeholder rows in there to distract you. This allows items to be reordered completely independent of one another. In other words, you are not forced to keep all the related work together. For example, maybe the Product Owner wants to improve the mobile dashboard UX and mobile ticketing UX now, but improving the mobile reporting UX can wait a few Sprints.

So, which option for handling epic PBIs is best? You can probably guess my answer. Let the Product Owner decide. Some Product Owners don't like to create hierarchies and don't like to use portfolio backlogs. Others do. Different Product Owners adopt naming conventions, or use areas or tags to group items together.

If you have different stakeholders who want to see different views and levels of the Product Backlog, then establishing a hierarchy can be beneficial for their sake. Doing so helps everyone visualize and understand the breakdown of work and be able to visualize the plan and assess progress at various levels. It also helps the Scrum Team explain to stakeholders the physics of complex product development and its trade-offs (such as "We had to do it this way because we couldn't physically build the whole feature in one Sprint").

Fabrikam Fiber Case Study

Paula has decided that when epic PBIs arrive in the Product Backlog, they will be split into smaller PBIs and have a Feature work item established as their parent. She recommends that each team member add the *Parent* column to their backlog to visualize this hierarchy.

Importing Product Backlog Items

Prior to your team using Azure Boards to maintain its Product Backlog, it's likely that your organization will maintain one or more lists of work. One list might track high-level requirements, another might contain feature requests from the users, and yet another will track bugs. These lists can range from an exotic arrangement of sticky notes, to an Excel spreadsheet, to a Microsoft SharePoint list, even to a dedicated ticketing tool.

Merging all this data into a common Product Backlog can be difficult, and I'm not just talking about navigating the people, politics, and permissions. Meaningful data must be extracted, transformed, and loaded. Information gaps must be filled. Duplicates must be resolved. Fortunately, Azure Boards offers a few options for importing and bulk-modifying work items.

Although you can continue to use Excel for bulk import and updates, doing so is no longer required. The native way of importing work items is to import directly from a CSV file. This CSV file must contain the *Work Item Type* and the *Title* fields, at a minimum. You can include other columns as needed. You can see an example of a CSV import file in Figure 5-8. All work items you import are created in a *New* state. This rule means that you can't specify field values that don't meet the field rules for the New state.

```
Import.csv - Notepad                                                                                    —    □    ×
File  Edit  Format  View  Help
Work Item Type,Title,Area Path,Assigned To,Business Value,Effort,Tags
Product Backlog Item,Enable network alert heartbeat,Fabrikam\Dashboard\Alerts,Paula <paula@accentient.com>,800,5,
Product Backlog Item,Quick Tips,Fabrikam\Dashboard\Messages,Paula <paula@accentient.com>,500,8,
Product Backlog Item,Show customers count,Fabrikam\Customers,Paula <paula@accentient.com>,500,2,
Product Backlog Item,"Improve customer edit, details, and delete links",Fabrikam\Customers,Paula <paula@accentient.com>,800,5,
Product Backlog Item,Employee phone numbers,Fabrikam\Employees,Paula <paula@accentient.com>,800,3,
Product Backlog Item,Employee email addresses,Fabrikam\Employees,Paula <paula@accentient.com>,800,5,
Product Backlog Item,Customer phone numbers,Fabrikam\Customers,Paula <paula@accentient.com>,1300,3,
Product Backlog Item,Customer email addresses,Fabrikam\Customers,Paula <paula@accentient.com>,1300,5,
Product Backlog Item,My tickets,Fabrikam\Tickets,Paula <paula@accentient.com>,500,3,Bug
Product Backlog Item,Use corporate email to authenticate,Fabrikam\Mobile\Security,Paula <paula@accentient.com>,500,13,
Product Backlog Item,Enable two-factor authentication,Fabrikam\Mobile\Security,Paula <paula@accentient.com>,800,8,
Product Backlog Item,Manage user access,Fabrikam\Mobile\Security,Paula <paula@accentient.com>,500,3,
                                                          Ln 1, Col 1        100%   Windows (CRLF)   UTF-8 with BOM
```

FIGURE 5-8 Here is a sample CSV import file of Product Backlog items.

The Import Work Items feature is found on the Work Items page as well as the Queries page. Simply choose the CSV file and import it. The process loads the imported work items into a query-style view. Imported items are in an unsaved state, allowing you to verify and tweak the results before saving them. Any import issues are highlighted so that you can resolve them. You can fix an issue by opening the work item directly or by using bulk-edit to fix several work items with the same issue.

Azure Boards also allows you to *export* work items to CSV. You'll need an existing query—one that contains all the columns you are interested in. Simply run the query and export to CSV. The exported file format looks similar to the import file format. If the exported file contains (at least) the *ID*, *Work Item Type*, *Title*, and *State* fields, you can reimport it later—after making changes—to update the work items in Azure Boards.

> **Smell** It's a smell to continue importing items into the Product Backlog after it's been created. Once Sprints begin and regular feedback yields a flow of new PBIs, they should be added to the Product Backlog, not some other list. The Product Owner and stakeholders should avoid storing those items and then dumping a batch of them into the backlog. It could be that the product definition has been broadened or a new scope of work has arrived. It could also be that the Scrum Team has dependencies on other teams or individuals in the organization that is batching work as well. As with all dependencies, these should be identified and mitigated.

Using Microsoft Excel

As everyone knows, Excel is extremely easy to use, and everyone in the organization seems to have a copy. I've used it to create dozens of Product Backlogs over the years. What many people don't know is that it can be used as an extract, transform, and load (ETL) tool as well. Okay, maybe it's not advertised as such, but it's true.

Using Excel, you can extract the data from an existing list by using copy/paste, one of the *Get Data* functions, or some form of automation. The data can then be transformed (normalized) and finally loaded (published) to Azure DevOps. Excel really is the Swiss Army knife of data manipulation.

There are a number of ways to import existing backlog items using Excel. Here is a high-level, step-by-step approach that I recommend:

1. Download and install the (free) Azure DevOps Office Integration.

2. Launch Excel.

3. Open a blank worksheet.

4. Rename the Sheet1 worksheet to **Source**.

5. Add a second worksheet named **Target**.

6. Load your data using the clipboard or one of the *Get Data* functions into the Source worksheet.

7. Select the Target sheet.

8. On the Team ribbon, click *New List*, connect to the Azure DevOps project (if necessary), and click *Input list*.

9. On the Team ribbon, click *Choose Columns*.

10. Select the Product Backlog Item work item type, add any additional columns you might need, and remove any extraneous columns. As you can see in Figure 5-9, *Area Path*, *Description*, *Business Value*, *Effort*, and *Tags* are good columns to consider adding.

11. From the Source sheet, copy the relevant columns and paste them into the respective columns in the Target sheet. You may need to set some fields manually (such as *Work Item Type* and *Product Owner*). You can also use the bulk-edit capabilities in Azure Boards to set these fields after import.

12. Clean up the data, especially if you are importing Area Path, State, or any numeric data.

13. On the Team ribbon, click *Publish*. If an error occurs, correct it and then publish again.

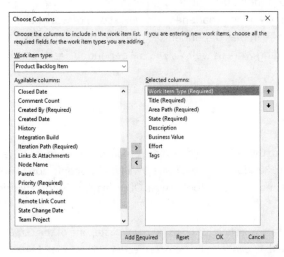

FIGURE 5-9 Select work item columns when importing PBIs from Excel.

The normal state transition workflow of a PBI work item is New ⇒ Approved ⇒ Committed ⇒ Done. For our custom, *Professional Scrum* process, I changed it to New ⇒ Ready ⇒ Forecasted ⇒ Done. When importing historical items, you must initially import the work item in the New state, but then you can switch to any of the other states after that. For example, even though the normal workflow is New ⇒ Ready ⇒ Forecasted ⇒ Done, you can import items as New and then change them directly to Done. This approach avoids the tedium of having to "walk the states," or publishing back to Azure Boards each time.

After you've published the work items, you can continue to use the Excel spreadsheet to make bulk edits to that set of data. If this sounds like something you might want to do, then you should save the workbook, giving it a meaningful name. If this was a one-time import and you are happy with the results, you can discard the workbook. You can always create a work item query and open it in Excel if the need arises.

As a corollary, Microsoft makes it easy to create a query to return the items in the Product Backlog. From any of the backlogs, you can click the *Create Query* option and create a personal or shared query that will return the items in the backlog, including the exact columns that are currently being displayed. This query can then be opened in Excel for review, charting, or bulk-editing. Just remember not to futz with the *Backlog Priority* field, because the sparsifier will override those values at your most inconvenient opportunity.

Bulk-modifying Work Items

Excel is not your only choice for making bulk edits. Azure Boards provides a native bulk-modify experience as well, which you can use to quickly make the same change to a number of work items. With this feature, you can edit fields, add or remove tags, reassign work, or move work to a specific Sprint. You can also use bulk-modify to change the work item type or move work items to other projects.

From the Backlogs page or a query result, you can use the same multiselect approach that you use when selecting multiple items (such as files) in Windows. To select multiple work items, press and hold the *Shift* key and select the first and last work item at the ends of the range you want to select. Or, you can select multiple work items by holding down the *Ctrl* key as you click each work item until all are selected. You can also use a combination of the *Shift* and *Ctrl* key approaches.

With the work items selected, you can right-click the selection (or click the context menu of any of the rows) and select one of the bulk-edit options. You can directly assign the selected work items to a new team member, add a link to them, move them to a different Sprint, move them out of a Sprint (back to the backlog), or other operations. You can also click *Edit*, which allows you to bulk-modify any field or fields in those work items.

Editing a work item is different than opening it. When you *open* a work item, you can view and change all the fields for that one work item. When you *edit*, you can specify a value for one or more fields that will be applied to all of the selected work items, as I'm doing in Figure 5-10. You can also set a note for the history, which I recommend for increased transparency and traceability.

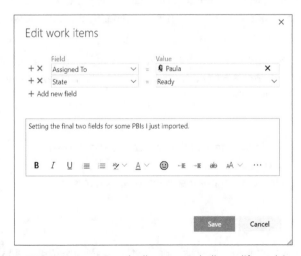

FIGURE 5-10 Azure Boards allows you to bulk-modify work items.

If you plan on importing items regularly and/or have a lot of complex data scrubbing to perform, you might consider building a custom solution. Azure DevOps has many wonderful, well-documented APIs. These APIs are based on REST, OAuth, JSON, and service hooks—all standard web technologies broadly supported in the industry.

Your custom solution could read the source data, perform any transformations that are required, and then use the Azure DevOps APIs to connect to the project and create or modify the work items. You can code against the APIs directly, or you can use one of these client libraries: .NET, Go, Node.js, Python, Swagger, or Web Extensions. For more information on extending Azure DevOps, visit *https://docs.microsoft.com/en-us/azure/devops/integrate*.

> **Tip** Fellow Professional Scrum Trainer Martin Hinshelwood created and manages the *Azure DevOps Migration Tools* project on GitHub. These tools allow you to migrate work items and other artifacts from one project to another, even between organizations. These tools do a lot more than just bulk editing, and there is some great functionality and sample code to reference. Visit *https://github.com/nkdAgility/azure-devops-migration-tools* to learn more.

Removing a Product Backlog Item

From time to time, you may want to remove a work item from the Product Backlog—or anywhere else in Azure Boards for that matter. The work item may have been created in error, may have been created in the wrong project, may have been a duplicate, or otherwise needs to go. You have a couple of choices when removing a work item. Which approach you take depends on how your organization feels about auditing work item changes and deletions.

Your first choice is to straight up *delete* the work item. Deleted work items won't appear in backlogs, boards, or queries. Deleted work items are moved to a Recycle bin on the Work Items page, from which you can recover them if needed. You can't open work items that have been moved to the Recycle bin. Work items in the Recycle bin can also be permanently deleted. Once permanently deleted, the work item is gone, with no record that it ever existed.

You can delete a work item from within the work item form or by right-clicking it in the Backlogs page; from query results; from the Boards page; or even from the Taskboard. You can also bulk-delete work items, using the previously mentioned bulk-edit approach.

> **Note** Azure Test Plans artifacts such as test plans, test suites, test cases, and so on are also types of work items. The method for deleting them, however, differs from deleting non-test work items. I will describe how to delete these test-related work items in Chapter 7, "Planning with Tests."

The other choice for removing a PBI work item is to simply change its state to *Removed*. By changing the state of a work item to Removed, you effectively remove it from all backlog and board views. The work item still exists, but Azure Boards knows not to display them except for in work item query results. To force removed items to *not* show up in query results, you must add a clause that filters on the *State* field. Removing work items—as opposed to deleting them—is generally preferred by organizations with strict audit requirements.

> **Tip** When changing a PBI work item's state to Removed, you should also add a comment to the Discussion section explaining why the item was removed. Adding a comment will help increased transparency and traceability.

Effective Product Backlog Creation

Writing a book about the fusion of Scrum, tools, and practices is difficult. I have to constantly balance any guidance I offer with that of the team's mandate to self-manage. I could just write "let the team decide" on every page of this book, but that would get very old. Most of my recommendations are for teams new to Scrum and Azure DevOps. I fully understand that these teams will develop their own behaviors over time, which is great. I only desire that these behaviors are healthy ones. To that end, Table 5-2 lists some of the proven practices that I recommend when creating a Product Backlog.

TABLE 5-2 Proven practices when creating a Product Backlog.

Tip	Reason
Keep titles short and to the point.	Sometimes the user will have only the title to go by and screen real estate is at a premium in Azure Boards.
Consider tagging bug PBIs with "Bug."	Using a query or a filter to identify which PBIs are bugs can be helpful.
Leave items in the root iteration.	Don't set the iteration path of a PBI until the Sprint in which they are fore-cast for development. It sets expectations that may not be met because things change. For release planning, consider using the Forecasting tool or a story mapping extension.
Don't create and link tasks or test cases.	You should wait until the Sprint in which the PBI is forecast for development to link tasks or test cases. Things change, and your planning effort could be wasted, or worse— an incorrect/outdated plan could be executed.
The Assigned To user should be the Prod-uct Owner or left blank.	Since the Product Owner is responsible for the Product Backlog, it makes sense that they are the person assigned.
Use the right tool for the job.	Use the Backlogs page for adding, ordering, bulk-modifying, refining, and forecasting. Use Excel for importing or (offline) bulk-editing.
Link to documents rather than attach them.	External documents can be linked to a work item, as well as discovered and manipulated independently. Attached documents are harder to find but may be more useful when archiving development efforts.

Reporting a Bug

In Scrum, no differentiation exists between a feature and a bug. Both represent something that must be developed in the software product. Both provide a value. Both have a cost. By default, Azure DevOps does differentiate. As I mentioned in Chapter 3, in Azure Boards the Bug work item type tracks additional information over a PBI work item, such as the steps to reproduce, severity, system informa-tion, and the build number that the bug was found and fixed in. Bug work items also don't have a *Business Value* field—although there can definitely be value in fixing a bug. Other than these differences, a Bug work item is treated just like a PBI work item in Azure DevOps. Insofar as Scrum is concerned, bugs are refined, sized, forecasted, and decomposed into a plan, and then developed according to a Definition of Done—just like any other PBI.

Fabrikam Fiber Case Study

Since the Scrum Team is using the custom Professional Scrum process, the Bug work item type has been disabled. Even if it hadn't, Paula had decided not to use the Bug work item type. This is because this work item type does not have a *Business Value* field and it also contains several extraneous fields. This is not to say that the Product Backlog won't contain bugs, but rather that the Fabrikam Scrum Team will use the PBI work item type to track them. They will tag the PBIs accordingly and put the repro steps and system information into the *Description* field.

Each Scrum Team may handle the discovery and classification of bugs in their own way. Before reporting a bug, someone should ensure that it is a valid bug. It could be that the odd behavior that someone experienced was by design, was a training issue, or was something that had already been reported or even fixed. This identifying and sorting process is known as bug *triage* or bug *repro*.

Triage also includes identifying the severity, frequency, risk, and other related factors. Triaging bugs can sometimes be a collaboration of the Scrum Team and stakeholders such as subject matter experts and business analysts. Their input can help elaborate on domain-specific issues and risks.

Smell Regardless of whether a team is using the Bug work item type or not, it's a smell when I *don't* see bugs in a Product Backlog. One concern is that the team isn't testing the product. It could also be that the stakeholders aren't reporting bugs. From a planning perspective, I'm more worried that bugs are being reported and tracked in another sys-tem. There may be another team (such as QA) or a sub-team that is handling testing and tracking with a separate tool.

Also, large organizations tend to have centralized trouble-ticket or issue-tracking systems. Software bugs found in production typically begin their life in these systems. They shouldn't end up there, though. Those bugs should be triaged and, if valid, added to the Product Backlog. Without having bugs in the Product Backlog alongside feature requests, the Product Owner won't be able to order the items effectively to maximize the work of the team and the value of the product. I would also caution against tracking bugs in both systems, which can be confusing and wasteful.

As with any PBI, anyone should be able to report a bug. Bugs, just like everything else in the Product Backlog, are validated, estimated (sized), ordered, and eventually forecast to be fixed. It's the Product Owner's prerogative to approve the bug fix, leave it in its current state (New), move it along on its journey to being ready, or remove it from the Product Backlog.

> ### Fabrikam Fiber Case Study
>
> Paula, exhibiting the Scrum Values of respect and openness, has invited any stakeholder, including users, to report an issue or bug directly. Currently these are emailed to Paula; she will triage them and, if valid, create a PBI with a "Bug" tag. Eventually she'd like the team to build feedback capability directly in the product—not just for sad things like issues, but for happy things like feature requests as well.

What Makes a Good Bug Report?

A bug report is just that—the reporting of a bug or other unwanted behavior in the software product. In order to write a good bug report, and thus create a good PBI, you must include enough information for the Scrum Team to understand it, gauge its impact on the business, and decide if it's worth fixing.

A good, clear title is a must. A team member should be able to grasp the essence of the bug from the title alone. If the Product Backlog contains many work items, having a clear title will help the Developers as they work the bug through refinement, forecasting, and development. It also saves a stakeholder from having to read the whole work item to grasp its context.

> **Tip** When I'm coaching a Scrum Team, I encourage them to brainstorm and then capture their definition of a bug—which is a lot harder than it sounds. It's important to have a definition and to have everyone—including stakeholders—understand it and abide by the definition before using the term in open conversation. With a transparent, generally understood definition of a bug, the usage of that potentially caustic moniker should drop, along with any unconscious perception of a low-quality product.

You should report only one bug per PBI. If you document more than one bug, some may be overlooked. Atomic bug tracking helps in the same way atomic testing does—it provides a precise understanding of what's working and what isn't.

Also, a picture is worth a thousand words. Sometimes words just can't demonstrate the issue that an annotated screenshot can. Team members will appreciate these extra efforts because they need to find the problem in the shortest amount of time. Any helpful documentation can also be linked or attached to the work item.

It is also a good idea to specify system information, including the build number that produced the failure. This build number can either be the one generated by Azure Pipelines or an assembly or product version number. Having a precise build number or version number will provide more information and help the team identify the exact problematic build or release. Otherwise, if the team uses a more current build or release version, they might search for a problem that was already fixed or otherwise mitigated. If they reference an older build or release, the problem code may not have been integrated yet. All of this information can be tracked in the *Description* field of the PBI work item.

Bug work items should always contain the observed as well as the *expected* results. This practice is useful because sometimes Developers don't think that a bug is really a bug, or they might claim that "it works on my machine"—not knowing what "works" actually means. The opposite can also be true, which is the expected result is working on their machines and the production deployment is different. Regardless, the variance between expected and observed should help prove the case. Also, generic descriptions like "This is a bug" or "It should work" are not helpful because the bug in question may not be immediately obvious to the others.

Note You should describe the observed results (the steps to reproduce the bug). This can be added to the *Description* field of the PBI work item. Also, you should consider using the *Acceptance Criteria* field to track the expected results (how it should work), as shown in Figure 5-11. Knowing the expected results can help the team create better tests. If you have additional acceptance criteria to list, you can do that in addition to the expected results. Be careful, though. You should avoid adding new "features" to a bug fix, unless the Product Owner is on board. If you have other unrelated improvements to make, consider creating a separate PBI work item to refine and plan. Gold plating (adding unrequested and thus wasteful features) should be avoided, even when fixing bugs.

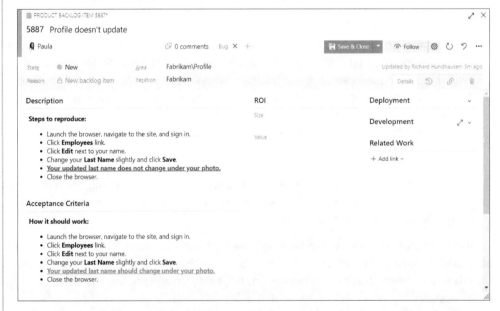

FIGURE 5-11 It's helpful to specify both the observed and expected results when reporting a bug.

When reporting a bug, be professional. Don't write titles like "Help!"; "It's broken"; "Got an error"; "What happened?"; or "Dude!" These kinds of titles are devoid of content at best and irritating at worst. Titles should be short and concise. Save the explanation (and commentary) for the *Description* and *Acceptance Criteria* fields. Your fellow team members can't read minds, so don't include notes like "Do you get what I mean?" Try your best to explain yourself. Don't assume the team members will just follow things that are written on the bug report. Also, don't play politics. Using a bug report to score political points is detrimental to the health of the team and the product. It also violates the Scrum Value of respect.

You've probably experienced firsthand that if you don't document the exact steps to reproduce an issue, you'll forget them very quickly. Be as specific as necessary, without being wasteful. Proofread the fields before saving. A proofread bug report has a much higher chance of being understood by others and of being fixed. Table 5-3 lists many proven practices when reporting bugs.

TABLE 5-3 Proven practices when reporting bugs.

Practice	Reason
Triage the bug.	To determine if it is a valid bug and not a feature, training issue, or by design
Keep titles concise but descriptive.	To save the user from having to read the entire PBI work item to get the context
Report only one bug per work item.	To estimate (size), order, and forecast each bug independently
Consider tagging bug PBIs with Bug.	To query or filter on which PBIs are bugs vs. feature requests/enhancements
Include screenshots with annotations.	To report a bug in an effective way
List repeatable steps to reproduce the bug.	To provide as much context to the team as possible in order to help them identify and fix the source of the failure; use the *Description* field
Provide expected results as well as observed results.	To show what success looks like as the team fixes the bug; use the *Acceptance Criteria* field
Include system information, the build number, and/or the version number.	To provide as much context to the team as possible in order to help them identify and fix the source of the failure; use the *Description* field
Use proper grammar, spelling, and tone.	To be professional

Where Do Bugs Come From?

Bugs can be introduced in a software product any number of ways. There is little point in attempting to name, blame, or shame any particular person, because there are so many sources that can cause a bug. Scrum Teams collectively own everything, including bugs. Besides, blaming and shaming have no place in Professional Scrum, because they violate the Scrum Value of respect.

Determining *why* something failed can sometimes take two or three times the amount of effort over just fixing it. Make sure the time spent analyzing the root cause adds value to your product or your process. The Sprint Retrospective is a good place to discuss any findings and potential corrections for the future. Whatever the finding, remember that the team collectively owns the quality (good or bad) of the product.

Here are a few reasons that bugs occur in software products:

- Feature creep
- Inadequate process
- Inadequate tooling
- Inexperienced developers
- Poor coding
- Poor test coverage
- Poorly understood requirements

Real bugs, as in the tiny crawly creatures, also have a well-defined lifecycle, according to their species. Their metamorphosis is a good metaphor for software bugs. It often starts with a simple observation of a probable source of customer dissatisfaction (the egg); proceeds to a well-defined report of the observed symptoms, steps to reproduce, and a technical investigation (the larva); proceeds to a fix for the problem (the chrysalis); moves to a working and tested build (the adult); and finally proceeds to a deployed fix (the death of the bug). Like real bugs, not all software bugs survive to adulthood. Of course, unlike real bugs, software bugs can retreat back to their larval state when the fix is unsuccessful. Those are called *reactivations*, which I will cover in a bit.

> **Tip** I often meet Developers who feel the need to trace a newly discovered bug back to the original PBI. Although this linking can be done easily in Azure Boards, I always ask, "Why?" If they want to see what the original acceptance criteria were, that's valid. They should consider copying and pasting the applicable criteria into the new PBI, making any tweaks. If, instead, they are looking for a reason *why* it broke, that can be better answered by looking at the code or tests. If they want to find out the size (such as story points) of the original PBI so that they can deduct it from their velocity, that's a dysfunction, not to mention a waste of time. Time spent dwelling on past mistakes is time that can't be used to work on the Sprint Backlog and achieve the Sprint Goal. Discuss your findings during the Sprint Retrospective. Remember, software development is complex. It is very hard and full of risk. We're not always going to get it right. Focus on improving in the next Sprint.

In-Sprint vs. Out-of-Sprint Bugs

Not all bugs are equal—so they shouldn't be treated equally. Bugs found in code running in production, or in a done Increment waiting to be released to production, are out of the scope of the Developers' forecast work for the Sprint. As such, those bugs should be handled the same as any feature request. The bug should be added to the Product Backlog, considered by the Product Owner, refined by the Scrum Team, and forecast in a future Sprint.

If the Product Owner deems the bug as critical and requiring an immediate fix, then the team should drop what they are doing and fix it. This interruption may or may not require their full capacity.

Regardless, everyone must realize that the Sprint's forecast may be missed and that achieving the Sprint Goal may be in jeopardy. The Product Owner weighs these risks and, after collaborating with the team and stakeholders, makes the decision.

> **Note** Unplanned work is just that. You can't plan it at the start of the Sprint. If the team's plan (such as tasks) consumes 100 percent of their available time, they will have no capacity left to handle unplanned work like an emergency bug fix. When the emergency does happen, it may cause the team to miss their forecast and therefore cause a drop in their velocity—if they are tracking velocity. Since velocity (or some similar measure) is an input to Sprint Planning, a decrease will provide slack time in the next Sprint for handling emergencies and other unplanned work. High-performance Scrum Teams watch their capacity while maximizing the value that their work produces.

I refer to bugs found in the code that the team is working on during the Sprint as *in-Sprint* bugs. These may not be bugs according to the team's official definition; rather, the code is just not quite there yet. Most teams' Definition of Done includes some form of "Code compiles," "There are no errors," or "All tests pass." In these cases, it's not really a bug—the team just isn't *done* yet.

> **Smell** It's a smell when bugs are created as a result of a failed test, such as an acceptance test. I often see this when there is another team, such as QA, that is doing the acceptance testing—which is not Scrum—and their preferred method of communication is not talking face to face but reporting a bug as a work item. Failed tests are not bugs. Failed tests simply indicate that the team is not done yet. At most, the team should update their plan (for example, add another task to the Sprint Backlog) to fix the code and pass the test.

An example of an in-Sprint bug would be when the team is developing new functionality, such as adding a new ticketing summary report, but gets blocked because of how they coded a controller in a previous Sprint. Fortunately, the controller issue doesn't affect any in-production functionality, but it does block them from completing the PBI in their current Sprint. Had they determined that the controller issue impacted functionality currently in production, then the Product Owner would've wanted them to fix it immediately. In this case, they have to fix it anyway in order to complete the new report.

The goal for in-Sprint bugs is to *fix* them, not manage them. Ideally, you want to fix all bugs discovered during a Sprint. If you don't, they could affect the Developers' ability to achieve their forecast or Sprint Goal. Here is the guidance I give to new Scrum Teams for handling in-Sprint bugs:

- If it's a small bug (< *n* hours to fix) and *won't* affect the Developers' ability to achieve the forecast or Sprint Goal, then just consider it part of the plan and fix it. The Developers decide what *n* equals and can adjust it during Sprint Retrospectives. A value of 2 feels right to me for smaller teams getting started.

- If it's a larger bug (> *n* hours to fix) and *won't* affect the Developers' ability to achieve the forecast or Sprint Goal, then create a PBI work item, add it to the forecast, associate a Task work item (if they are using tasks), and have a Developer address it during the Sprint. The PBI will help raise transparency and explain the reason behind any hiccup in progress.

- If it's a larger bug (> *n* hours to fix) and *will* affect the Developers' ability to achieve the forecast or Sprint Goal, then create a PBI work item and discuss it with the Product Owner. They may value the bug fix over another PBI in the forecast. If not, then the bug will be left in the code this Sprint and the bug PBI will be refined by the Scrum Team and forecast in a future Sprint.

But what about creating a bug PBI as a simple reminder? Suppose a team member finds a small bug and it's not possible to immediately collaborate on a fix. Rather than creating an additional PBI, I recommend simply updating the plan (such as creating another task associated with the PBI that the bug blocked). If the team is not using tasks, then adding another PBI to the Sprint forecast—to represent the bug—could be an option as well. Let the team decide. Regardless, an issue like this should be brought up with the other Developers at the earliest opportunity, even if collaborating to fix it isn't immediately possible. The Developers must decide if it can be fixed in the current Sprint or in a future one. If the bug is left in the code, the Product Owner should be consulted. Everyone will need to realize that the Sprint's forecast may be missed and that achieving the Sprint Goal might also be in jeopardy. What's worse, you have now added technical debt to your codebase and product.

Conversely, I refer to bugs found in production code, or in a done Increment waiting to be released to production, as *out-of-Sprint* bugs. Typically, these bugs don't affect the code associated with the forecast work that the team is currently working on. If code is affected, then treat them as *in-Sprint bugs*. Otherwise, consider the guidance I offer to Scrum Teams for handling out-of-Sprint bugs:

- If the Product Owner determines that the bug is critical, the team should do whatever needs to be done to create and release a fix. Everyone must realize that the Sprint's forecast may be missed and that achieving the Sprint Goal might not be possible as well.

- If the bug is *not* critical, then create a PBI work item. The Product Owner will decide if the bug should be refined and when it should be considered for Sprint Planning.

- If the number of critical bug occurrences is high and/or increases, consider adjusting the team's capacity or dedicating people to supporting issues like these. The Scrum Team should also look for the root cause of these critical bugs and experiment with solutions. Oftentimes, if quality is low, the Scrum Team should *slow down* and focus on getting done (according to their definition) and thus improving quality.

An example of an out-of-Sprint bug would be when the team is developing new functionality such as adding a new ticketing summary report and, while reviewing the controller code, sees that it's not working correctly. The team is not blocked from completing the PBI in their current Sprint, but they know that the bug has been deployed to production. Even though no users have reported it yet, they should inform the Product Owner, who may want them to immediately fix it. In that case, they may miss their forecast and possibly even their Sprint Goal.

Note If you suspect a bug exists, write a failing test to verify your suspicion. The test could be automated (such as a unit test) or manual (such as a Test Case work item). I know that I've said that unplanned items in the Product Backlog should not have any associated tasks or test cases. My reason for this guidance is to reduce waste, such as defining the how (through tasks and test cases) too early. An exception to this guidance would be a situation where a Test Case work item existed before the bug PBI work item. If this happens, don't discard the test case. It may still be of value in a future Sprint when the team works on the fix.

Bug Reactivations

Reactivations occur whenever a bug (previously thought Done) appears again. This could be caused by changes in the environment such as infrastructure upgrades, new deployments, or even having the closed the bug prematurely. People sometimes mark bugs as Done when the underlying problem has not been fixed. When this happens, it introduces waste into the process. A team member has to write and run a test and reopen the bug PBI. The original code may need to be refactored or scrapped and then retested. At a minimum, the reactivation doubles the number of context switches and usually more than doubles the total effort required to complete the corresponding work. Frequent reactivations are a smell of a deeper dysfunction.

Watching the rate at which reactivations occur is important. A tiny amount of waste might be acceptable, but a medium-to-high (or rising) rate of reactivations should be a smell, warning the team to diagnose the root cause and fix it. Although sloppy development practices are an obvious possibility, other potential causes include poor bug reporting, inadequate test management, and overly aggressive triage. The Sprint Retrospective is a great venue for such discussion.

Fabrikam Fiber Case Study

Paula has created an "Active Bugs" shared query, which returns all PBI work items with the tag Bug in the New, Ready, or Forecasted states—everything except for Removed and Done items. She also added a Query widget to the team's dashboard that shows a real-time count of Active Bugs. She configured the widget with conditional formatting so that when there are more than a specified number of bugs, it will change the background color of the widget to attract attention. You can see an example of this in Figure 5-12. Clicking the widget will open the query and show the active bug PBIs.

(Case Study continued on the next page.)

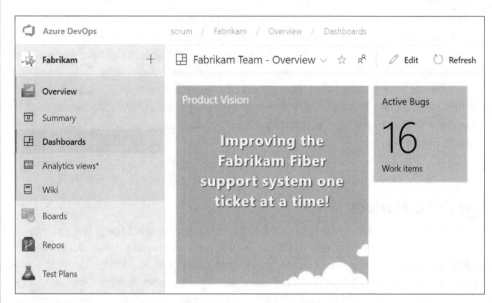

FIGURE 5-12 Query dashboard widgets show work item counts at a glance.

Refining the Product Backlog

Product Backlog refinement is an ongoing, part-time activity where the entire Scrum Team meets to better understand the upcoming items in the Product Backlog. When the time is right, the team will size the items to know if they are ready for Sprint Planning. If, when, and where the Scrum Team meets to refine the backlog is up to them. The *Scrum Guide* only recommends doing it and suggests that it take no more than 10 percent of the team's capacity during the Sprint. It is not an official event and is completely optional.

> **Tip** Although Product Backlog refinement is an optional activity, it is a critical one. It provides a regular opportunity to evolve the Product Backlog. I recommend that Scrum Teams establish a regular cadence of refinements. They can inspect and adapt the frequency as needed. Fellow Professional Scrum Trainer Simon Reindl suggests refining a PBI three times before considering it ready for forecasting and development.

Initially, a PBI work item need only have a title to be added to the Product Backlog. The values for the other fields will begin to emerge and continue until the time that it is forecast for development. It may still change after that. Prior to forecasting the PBI, the Scrum Team should have a solid understanding of the request, its value to the customer, what Done looks like, and a feel for its size—the level of effort required to develop it. This evolution can occur at Sprint Planning, at Sprint Review, or during a Product Backlog refinement session.

Tip Scrum Teams generally refine their Product Backlog from top to bottom, making sure that there is understanding and consensus around the highest-ordered items first. There are times, however, when the team may want to jump to a specific PBI that isn't near the top. To find a work item in a large Product Backlog, I recommend using the Azure DevOps search feature at the top. Searching will take you to the work items page and display the search results just as though you ran a query. As you can see in Figure 5-13, I am searching for the keyword *links* and have found the work item I'm interested in.

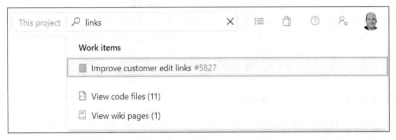

FIGURE 5-13 You can search your Azure DevOps project for artifacts in many ways.

As details emerge and consensus forms, you should edit the work item and make any updates. Updating can occur at any time during the Sprint, but especially during Product Backlog refinement. Here are some PBI changes that might be made as a result of refining the Product Backlog:

- **Add or improve description** A description that explains the who, what, and why is added.
- **Assign business value** The Product Owner assigns a business value.

Tip Business value is an abstract number. It doesn't necessarily relate to any direct value measure (sales, revenue, new customers, products sold, etc.). That said, just as a PBI's size/ effort can be helpful when comparing relatively to other PBIs, so can business value. For Product Owners learning to track business value, I coach them to base their number on a Fibonacci sequence, similar to how many teams base story points. The difference is that I multiply the story points by 100 to make them look more "value-y." The business value sequence I recommend to teams is 100, 200, 300, 500, 800, 1300, and 2100.

- **Link documents** The team links to or attaches any supporting documents.
- **Add acceptance criteria** After collaborating with the stakeholders (customers, users, and/or subject matter experts), you add acceptance criteria.

- **Size** After reaching a baseline understanding of what Done looks like, the team sizes the PBI relative to other items—already done or sized—in the Product Backlog.

- **Split** The Scrum Team decides that the PBI is too large to be completed in one Sprint.

- **Set state to Ready** The Scrum Team decides that this PBI is ready for Sprint Planning.

- **Set state to Removed** The Product Owner determines that the PBI is a duplicate or otherwise unnecessary.

I want to take a few pages to discuss a few of the specific actions taken while refining the Product Backlog: specifying acceptance criteria, sizing a PBI, and splitting a PBI. Although these are not new topics, and there are many good books out there, I want to discuss the Professional Scrum approach to these practices.

Specifying Acceptance Criteria

Professional Scrum Developers don't forecast any work until they know what Done looks like. For example, there's not a lot of detail to go on if a PBI is simply titled "Monthly ticket report." Even if it has an awesome description that reads, "As a technician, I want to see a monthly report of my support tickets so that I can better prepare for the future," a team would be able to build any of a hundred different reports that are fit for that vague purpose. Rather than commune this information on the fly, or build the wrong thing (and hope to be corrected by feedback), the team needs to take a more measured approach to knowing what to develop and what Done looks like.

Being more deliberate can be performed by collaborating with stakeholders and identifying acceptance criteria. This task can be performed by anyone on the Scrum Team, but often it is the responsibility of the Product Owner. A high-performance Scrum organization is one where the stakeholders themselves (such as the users) are able to craft their own PBIs, including acceptance criteria.

Using acceptance criteria is a lightweight, agile way of establishing requirements and defining success for a PBI. Think of it as the work item's specific Definition of Done. Acceptance criteria should define the *what* but not the *how*. The team should have free rein over how to build it, so long as it meets the acceptance criteria and the Definition of Done. Figure 5-14 shows a PBI with an emerging set of acceptance criteria.

Each individual acceptance criterion should be testable. In other words, you should be able to create and execute a manual or automated test to verify that each bullet is done. Sometimes it will take multiple tests to verify one criterion; other times one test can verify multiple criteria. I will dive deeper into this in Chapter 7.

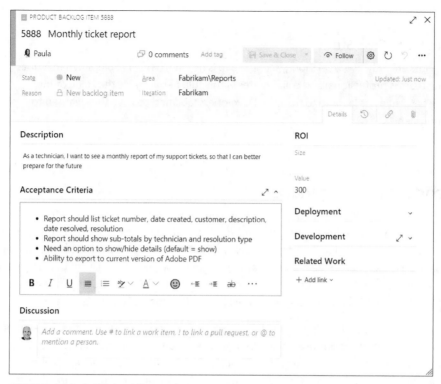

FIGURE 5-14 Each PBI should have adequate acceptance criteria.

Scope Creep

One of the agile values is "Responding to change over following a plan." So, what happens when the Product Owner needs to change the acceptance criteria? If the PBI work item hasn't been forecasted yet, there's no problem. Someone should update the criteria and the Developers would need to resize the item at the next refinement opportunity.

If the Developers *have* already forecast (and possibly started) the work for the current Sprint, the reflex is to resist the change. After all, don't the rules of Scrum "unionize" the team to protect against practices like *scope creep*? Some might think so, and they would be wrong. The Product Owner is responsible for maximizing the value of the product, as well as the work of the team. Sometimes these two responsibilities conflict, especially when business or market drivers demand an immediate change. This is not to say that there won't be waste.

The correct response is for the entire Scrum Team to collaborate on bringing value to the product with respect to this change. This is the N (negotiable) in the INVEST acronym I talked about in Chapter 1, "Professional Scrum." Yes, the forecast may slip. Yes, the Sprint Goal may not be achieved. Yes, there will probably be waste. But, in the opinion of the Product Owner, the immediate change to the product is worth these costs to ultimately avoid much greater risk and waste. The Sprint Retrospective should be used to better understand what happened and how to keep it from happening again in the future.

Sizing Product Backlog Items

Higher-ordered items in the Product Backlog are clearer and more detailed than lower-ordered ones. More accurate sizing is possible due to this increased clarity and detail. The lower in the order (the further down the backlog), the less detail exists. This is why a team should size the items toward the top of the Product Backlog rather than in the middle or at the bottom. It would be a waste of the team's time to size items that are too far down the backlog, especially ones that the Product Owner doesn't want in the first place. I discussed this, and the concept of the Product Backlog *iceberg*, in Chapter 1.

Each item in the Product Backlog is unique. It's difficult to size something unique that is being built for the first time, especially if it is unlike anything built before. Traditional estimation techniques do not work. As odd as it sounds, if you want to be more accurate, you should be less precise. It's okay to be precise on small things, like Sprint tasks ("create a web form," "update a stored procedure," "create a load test," etc.). Each of those tasks could be estimated in hours with some accuracy. For larger items, such as PBIs, you want to use a scale that is less precise, like T-shirt sizes (S, M, L, XL) or a Fibonacci sequence of numbers (1, 2, 3, 5, 8, 13, 21, etc.).

In Scrum, the team is responsible for all estimates. It follows that the people who will perform the work (analysts, designers, testers, coders, writers, etc.) make the final estimate. In smaller teams, where the Product Owner and/or Scrum Master is also a team member, they will have input on the estimations as well. Never underestimate how long good estimation takes, but also realize that too much analysis and estimation can have a diminished return on the time invested.

> **Tip** Scrum offers three formal opportunities to refine the Product Backlog: during Sprint Planning, during Product Backlog refinement, and during Sprint Review. It's best to do the bulk of your estimation during Product Backlog refinement. This way, the Developers can spend less time sizing and forecasting work at the next Sprint Planning.

Sizing should be performed as late as possible. Early estimates are less accurate than ones made later. You always know more today than you did yesterday. You also don't want to waste the team's time estimating items that are way down the list. Proper ordering of the Product Backlog can reduce waste when estimating. The Product Owner should know what's coming up next or soon thereafter, and focus on those items. The Product Owner should wait until obviously large (epic) PBIs are split.

Sometimes it's a chicken-and-egg problem, though. The Product Owner needs an idea of the cost (size/effort) of a PBI before they can order it. A solution for adding value to the estimation process is for the team (or a proxy) to provide the Product Owner a rough order of magnitude estimate, such as a T-shirt size (S, M, L, XL). This approach can give the Product Owner enough insight to be able to order the item effectively, or even decide that they don't want it at all. A more thorough estimate, provided by the entire team and using a more precise technique, can be performed at a future refinement session.

The *Scrum Guide* does not prescribe any particular size technique or unit of measure. Teams can use whatever practice and values they wish. Planning Poker is a very popular method. Story points are popular too—although some teams prefer to call them "complexity points" or something very abstract like "acorns." I once coached a team building pharmaceutical software that went so far as to use the term "Vicodins." (Vicodin is a prescription pain medication, which is fitting given some of their painful user stories.) Whatever term you decide on, using an abstract measure like a story point is preferred for sizing complex work. Abstract values have an advantage over temporal values (hours, days, ideal-days, weeks) because their usage doesn't imply a commitment or a plan or anything that smells like a schedule. For example, if you were estimating in days, a stakeholder might conjure up a specific expectation—and then write it on the calendar!

Note An agile estimation technique is not a silver bullet. In fact, it's the worst form of estimation except for all the others that have been tried. Agile estimation techniques won't remove uncertainty from early estimates, but they also won't waste unnecessary time. Estimates will become more accurate over time. This is due to the empirical nature of agile estimation techniques, where actual work is taken into account. The most important outcome of any agile estimation practice is the conversation and shared understanding—not the number. After shared understanding is established—as manifested by a consensus of the PBI's size—refinement of that PBI can be considered complete.

By default, the PBI work item has an *Effort* field to hold the estimate. In the customized Professional Scrum process, I kept this field and simply changed its label to *Size*. This numeric field indicates the relative rating for the amount of work (or complexity) that the PBI implementation will require. Larger numbers indicate more effort/complexity than smaller numbers.

Tip I'm often asked if *size* is the same as *effort*. My answer is yes, as long as the unit of measure is abstract, is not a measure of time, and is used relative to other PBIs. I'm also asked if size is the same as *complexity*. My answer again is yes, as long as the unit of measure is abstract, is not a computer science measure of complexity (such as function points), and is used relative to other PBIs. In other words, if the unit of measure is abstract, size can mean effort or complexity. Since it's confusing, Professional Scrum Teams simply use the generic term *size*. Regardless of what you call it, the *Effort* field in the PBI work item will hold the value.

Planning Poker™

Planning Poker is a tool for estimating complex work, such as software development. It is a technique where each team member selects an estimate card such that it cannot be seen by the other players. After everyone has selected a card, all cards are exposed at once. The Product Owner and Scrum Master don't participate in the estimation game unless they are also a team member. If necessary, the Scrum Master facilitates the process. The Product Owner should offer support where needed, such as answering questions. Other stakeholders or domain experts may attend the refinement session to offer support but not to participate in estimation.

The units represented are typically story points in a limited Fibonacci sequence (0, 1, 2, 3, 5, 8, 13, 21, etc.). The cards are numbered in this sequence to account for the fact that the larger an estimate is, the more uncertainty or risk it contains. The 0 card, though rarely used in my experience, is a way of letting the others know that the PBI is invalid or has already been implemented in some form. Some decks may also contain larger numbers, question marks, infinity symbols, or coffee break cards. Remember, it's about the conversation, learning, and shared understanding—not the number on the cards.

Here's how I facilitate a Planning Poker session:

- **Select a baseline PBI** The Developers select and reference a medium-sized PBI that they have worked on recently. This PBI will be arbitrarily assigned a size of 5 and will be used as a baseline. The referenced item doesn't have to be similar to the one being estimated, but it helps. Over time, as the team collaborates and learns more about the domain and its technology, it can select a new baseline PBI and then resize items in the Product Backlog.

- **Read the PBI** Typically, the most knowledgeable Developer or domain expert for the given PBI provides a short overview. This could also be the Product Owner.

- **Discuss the PBI** The Developers are given a timeboxed opportunity to ask questions and to discuss and clarify assumptions and risks. A summary of the discussion can be recorded.

- **Size the PBI** The Developers consider the size of this PBI as compared to the baseline PBI, selecting an initial size (1, 3, 5, 8, 13, or 21). Everyone reveals all cards at the same time—to avoid some Developers influencing (*anchoring*) other team members' opinions. If, for example, a team member initially thinks a PBI's size is 3 (as compared with the baseline PBI), they should also consider the perceived uncertainty, technical debt, risk of using new technology, lack of tests, and so forth and consider increasing the size, possibly to 5.

> **Note** *Anchoring* occurs when a team member discusses or hints at their estimate prior to the revealing of their card. A cross-functional team normally has a mix of conservative and impulsive estimators. Some Developers may have an agenda too, like wanting to maximize the time they can have to work on an item. Conversely, the Product Owner—who is in the room during refinement—is likely to want something done as quickly as possible. Compromise through collaboration becomes important at this juncture.

The estimate becomes anchored when the Product Owner, or one of the more experienced Developers, says something like, "This should be easy" or "I could do that in a day." Anchoring can also go the other way, when someone says something foreboding like, "Isn't that the component riddled with technical debt?" or "Wait, that code doesn't have any unit tests." Whoever starts the conversation with the statement, "That'll take the entire Sprint!" immediately has an impact on the thinking of the other Developers. Their estimates have now been anchored by that statement, even subconsciously. They are now likely to make at least a subliminal reference to that opinion. For example, those who were thinking 5 points are likely to increase their size estimate.

Anchoring becomes a particular problem if an influential team member makes the original statement that described the PBI. Because the remainder of the team has been anchored, they may consciously or otherwise fail to express their original unity. In fact, they may fail to even discover that they were thinking the same thing. This can be dangerous, resulting in sizes that are influenced by agendas, attitudes, alphas, or opinions that are not focused on developing done, working product.

- **Discuss the outliers**   The team member(s) with the highest and lowest estimates are given an opportunity to explain why they picked such an outlying number. If no outliers exist, and the team is generally in consensus on a size, then you can record it and move on to the next PBI.

- **Repeat until consensus**   The team should repeat the estimate process until a consensus is reached. The size is then recorded in the PBI work item. If consensus is not reached after a few rounds, estimation should be tabled until the next refinement session since a compromise can't be attained at this time. More will be known later. For situations where consensus is far off, I coach the outliers to do "homework" to verify/prove that the work is so hard or so easy. This homework will be "due" at the next refinement session.

Although several online estimation and sizing tools, including some that integrate with Azure DevOps, are available, I prefer in-person, physical cards that support high-fidelity conversations and learnings. For distributed teams or remote Developers, tools may be necessary.

Wall Estimation

A Scrum Team can employ many practices to estimate the size of a PBI. Though Planning Poker is the most popular, several of my fellow Professional Scrum Trainers use an affinity estimation practice called *wall estimation,* also known as *white elephant sizing.* It is loosely based on the idea and workflow of a white elephant gift exchange. If you've never heard of it, feel free to look it up. Here's how I facilitate a wall estimation session:

- **Create a board**   Start with the team standing in a half circle, facing a whiteboard with seven columns—each representing a Fibonacci number (1, 2, 3, 5, 8, 13, and 21). On a nearby table is a timer and a stack of PBIs to be sized.

- **Select a baseline PBI**   The Developers select and reference a medium-size PBI that they have worked on recently. This PBI will be assigned a size of 5 and be placed above the 5 column. It will be referenced as the baseline. Over time, as the team collaborates and learns more about the domain and technology, it can select a new baseline PBI and then resize items in the Product Backlog.

- **Prepare the PBIs**   The PBI cards can be index cards or sticky notes. The cards contain the titles and short descriptions of the items that you are about to estimate.

- **Size a PBI**   The first team member will start the timer, pick the card off the top of the deck, read it out loud, stick it in one of the columns, provide a reason for their decision, and then stop the timer. I leave the timer duration up to the team. Some teams practice a faster, "silent" version, where they skip the explanation.

- **Change the size**   The next team member can choose to size a new PBI just like above, or change (steal) the size of a previously sized PBI—just like the classic white elephant exchange. If changing a size, the person needs to move the PBI to a new column and provide a reason for the change.

- **Repeat until done**   The next team member repeats the above, sizing a new PBI or changing the size of a PBI already on the board. This continues until all of the PBIs are sized or the timebox expires. The sizes are then recorded in the PBI work items. Once all new PBIs have been sized—and only disagreements with existing sizes remain—Developers can start skipping their turn. Skipping a turn suggests that team member agrees with the estimated sizes on the board.

Everyone should have laser focus and listen actively to the estimator. Everyone else should remain silent. There should be no discussion, judgment, or sighing. If a team member does not size the PBI within the time limit, the PBI card is returned to the bottom of the pile. If consensus is not reached and a compromise can't be made for a specific PBI, estimation should be tabled until the next refinement session when more is known. For situations where consensus is far off, I coach the outliers to do homework to verify/prove that the work is so hard or so easy. This homework will be due at the next refinement session.

Fabrikam Fiber Case Study

The team has decided to use Planning Poker to estimate the size of the PBIs in the Product Backlog. This occurs primarily at the Product Backlog refinement sessions on Friday mornings. Since they have been working together for some time on the same domain, using familiar tools and technologies, the team's baseline is well established and understood. Estimation sessions go smoothly and consensus is usually achieved quickly.

Splitting Product Backlog Items

Earlier in this chapter, I discussed splitting epic PBIs into two or more PBIs. Now that I am talking about refining the Product Backlog, let's review this guidance. When a PBI is too large for the team to complete in one Sprint, it must be split. For some teams, even that is too risky and their working agreement states that a PBI cannot be larger than what can be done in a few days' work.

Regardless of what the team defines as "too large," there will be times that a PBI needs to be split. For example, if a team tracks velocity and computes that the next Sprint's velocity will be between 16 and 20, then any PBI that is larger than 13 points should be split. If the team wants to further reduce risk, it may have a working agreement for one-half of that so that any PBI larger than 8 points should be split.

There are several recommended ways to split a PBI, and they all have one thing in common—they split the PBI *vertically*. That is to say, there will be a vertical slice of functionality containing a small piece of each architectural/component layer in the solution. For example, if our application has a model, view, and controller (MVC) architecture, each PBI's implementation would include a piece of work in each MVC component, as well as the persistence store (such as a database) and the user interface (UI). This way, each PBI could be inspected, because it has a UI and is functional all the way down to persistence. You can see an example of how I might split a large PBI by acceptance criteria in Figure 5-15.

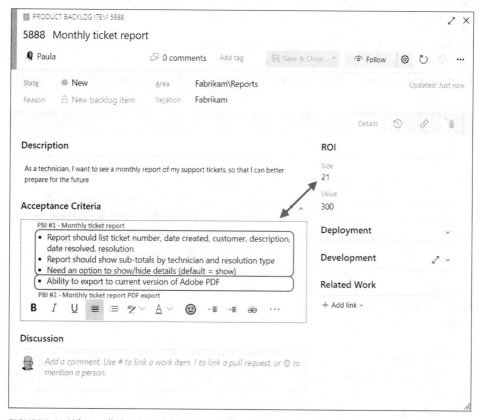

FIGURE 5-15 When splitting a PBI, consider splitting by acceptance criteria.

When splitting a PBI, the team should resist the urge to split it *horizontally*—which would be by architectural or component layers (data, service, controller, UI, etc.). This may seem a logical way to go, but it will result in undone work at the end of a Sprint. Even though the PBI will build and pass tests, there is nothing that is potentially releasable or even inspectable at the Sprint Review. Stakeholders don't want to inspect mock-ups, wireframes, code, or passing tests. In Scrum, especially Professional Scrum, every PBI should offer an independent slice of value when released and that value should be able to be inspected during Sprint Review.

Definition of Ready

When a PBI work item is initially created, it is in the *New* state. Once the item is ready for Sprint Planning, its state should be changed to *Ready*. When the Developers forecast to deliver the item in the current Sprint, its state should be changed to *Forecasted*. Finally, when the PBI work item is done according to the Definition of Done, the state should be changed to *Done*. Unfortunately, there can be many states in-between New and Ready that Azure Boards does not support by default.

More and more teams are adopting a definition of ready as a way to avoid starting work on PBIs that are ill-defined, such as those that do not have clearly defined acceptance criteria. Although some teams can do this implicitly, others desire a formal definition to explicitly communicate this working agreement with their Product Owner and stakeholders. Some teams even want to make this definition actionable and visual, such as using a board to move a PBI through its workflow on its journey to ready.

During Product Backlog refinement, detail, order, and estimates are added or improved until each PBI is ready. In effect, Product Backlog refinement helps "de-risk" Sprint Planning as well as development. By observing a definition of ready, a Scrum Team reduces the risk of the team being caught flat-footed in Sprint Planning and during development. With this agreement, the Developers won't begin working on items (or even consider forecasting them) that are not sufficiently well described and understood.

Smell It's a smell when I see a definition of ready used as a gate that prevents a team from considering more valuable, late-breaking items, in favor of better understood, ready items. I have no objection against Developers deciding that a PBI is too large to forecast, but using the definition of ready as leverage against the Product Owner is a dysfunction. Remember, agility comes from the ability to "change on a dime, for a dime" and a definition of ready used in a dysfunctional way prevents that. If the Product Owner wants to bring brand-new PBIs into Sprint Planning—even if they are not ready—they should be allowed to.

The "definition of ready" is not an official Scrum artifact or practice. It's not even mentioned, per se, in the *Scrum Guide*. In other words, it doesn't have "teeth" like the Definition of Done does. This is not to say that it's not useful. It can be very useful, if not abused. Scrum is a great framework for experimenting with practices, such as the definition of ready.

The creation and use of a definition of ready should be a team decision—to include the Product Owner. It can be codified into a working agreement. This working agreement will apply to Product Backlog refinement, or any other time the Scrum Team works on making a PBI ready for Sprint Planning.

A definition of ready is often loosely based on the INVEST mnemonic covered in Chapter 1. The INVEST mnemonic was created by Bill Wake in 2003 as a reminder to the agile community of the characteristics of a good quality user story. Here is a review of INVEST:

- **I (Independent)**   A PBI should be self-contained. It should be possible to bring it into a Sprint without a dependency on another PBI or an external resource.

- **N (Negotiable)**   A PBI should leave room for discussion about its optimal implementation.

- **V (Valuable)**   A PBI should deliver a measure of value to the stakeholders.

- **E (Estimable)**   A PBI must have a size, ideally relative to other PBIs.

- **S (Small)**   A PBI should be small enough to be forecasted and developed in a Sprint.

- **T (Testable)**   A PBI should have clear, testable acceptance criteria that define Done for this item.

I recommend teams not use INVEST directly as their definition of ready but rewrite it to make it more emergent and Scrum-friendly. Also, additional steps can be added that are specific to their organization, team, and culture. These steps can also help visualize the PBI's evolution to ready. Here is a sample definition of ready workflow loosely based on INVEST:

- **New**   The PBI was just added to the Product Backlog. Maybe only a title exists.

- **Interested**   The PBI is of interest to the Product Owner.

- **Description**   The PBI contains a short, meaningful description. The user-story description format ("As a...," "I want...," "So that...") can work well.

- **Value**   The PBI has been assigned a measure of value—possibly relative to other PBIs.

- **Acceptance Criteria**   The PBI's acceptance criteria are identified, describing what done looks like for this PBI.

- **Dependencies**   The PBI's dependencies (technical, domain, people) are identified.

- **Sized/Estimated**   The PBI's size—relative to other PBIs—is determined.

- **Ready**   If the PBI is understood and small enough, then it is ready for Sprint Planning and development.

Implementing a Definition of Ready

For Scrum Teams wanting to create a visual—and actionable—definition of ready, I recommend using the Kanban board. Rather than further modify the already customized Professional Scrum process and add extra workflow states, it's much easier to customize the columns on the Kanban board. Using the Kanban board, a Scrum Team can enact their definition of ready by adding new columns in-between the New and Ready states defined in the PBI work item type.

By default, the Board's columns match the workflow states of the PBI work item type. In our case, using the Professional Scrum process, those columns are *New*, *Ready*, *Forecasted*, and *Done*. I will add additional columns in-between *New* and *Ready* to represent my definition of ready. To do this, I open *Team Settings* on the Boards page, go to *Board Columns*, and add the new columns, taking care to map the new columns to the New workflow state. I also set all the Work in Progress (WIP) limits to 0, since the Scrum Team won't actually be practicing Kanban as they refine their Product Backlog. Before leaving *Team Settings*, I also added the *Business Value* field to the cards. You can see the final result, split into two screenshots to represent the left and right sides of the board respectively, in Figure 5-16.

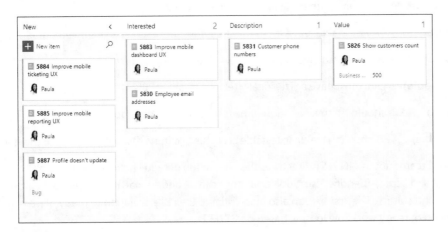

FIGURE 5-16 The Kanban board (left and right sides) can be used to model a team's definition of ready.

Notice in Figure 5-16 that the PBI cards contain more and more information as they progress through the columns from left to right. For example, cards don't enter the *Value* column unless they have a *Business Value*. Cards don't enter the *Sized* column unless they have an *Effort* (size). Cards that are too large (for example, 21 points) don't make it into the *Ready* column.

If I had added more fields to the cards (which can make for a messy board), you'd see *Description* and *Acceptance Criteria* also emerging in their respective columns. What you wouldn't see is the *State* changing. Cards remain in the New state all the way up through *Sized*. On this journey, only their board *column* changes. Once a card moves into the *Ready* column, the underlying workflow State changes to Ready. If you return to the Stories backlog and add *Board Column*, then you can see this *State* versus *Board Column* relationship, as shown in Figure 5-17.

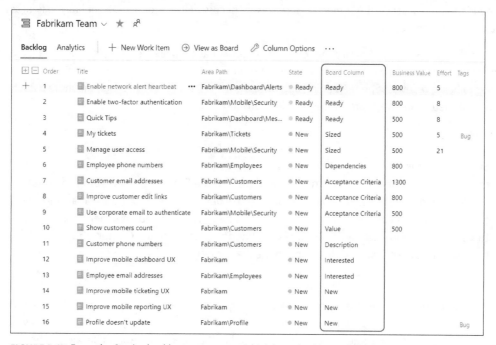

FIGURE 5-17 From the Stories backlog, you can see which board column a PBI is in.

 Tip Even though I have disabled the Bug work item type in my custom Professional Scrum process, some pages in Azure Boards might still have UI elements pertaining to bugs. For an improved experience, you should select the *Bugs are not managed on backlogs and boards* option in the *Working with bugs* section of Team Settings, as I've done in Figure 5-18.

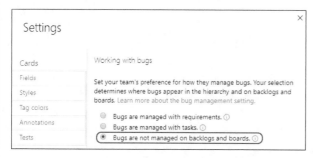

FIGURE 5-18 Configure Azure Boards to not manage bugs on the backlogs and boards.

Ordering the Product Backlog

The Product Backlog should be ordered by the Product Owner to maximize the value of the product being developed. This order can be influenced in many ways: ROI, value, risk, effort, business priority, dependency, technical debt, learning value, votes, or necessity. The Product Owner has the final say in the ordering of the items.

Items at a higher order (toward the top of the Product Backlog) are clearer and more detailed than lower-ordered ones (toward the bottom of the Product Backlog). The items toward the top are also closer to being ready for Sprint Planning and development. The higher the order, the more a PBI has been considered, and the more consensus exists around its size.

> **Smell** It's a smell when I see Developers ordering the Product Backlog. This is the responsibility of the Product Owner. It could be that the Developers are doing so at the Product Owner's request. It could also be that the pertinent items need to be arranged to minimize technical dependencies. This is fine, as long as the Product Owner is okay with the order. The worry is that the Scrum Team has an absent Product Owner who defers both the "what" and "when" decisions to the Developers.

In Azure Boards, you can order the Product Backlog by dragging items up or down. If you click and hold on an item, you can drag it above or below another item and then drop it. You can see this in Figure 5-19, as the Monthly sales report PBI is dragged to the top of the Product Backlog. When it is dropped, it will be the highest-ordered item in the Product Backlog. Reordering like this assumes you have at least a Basic access license. Stakeholder access doesn't support dragging and dropping in the Product Backlog.

Fabrikam Team
Backlog
Analytics
New Work Item
View as Board
Column Options
Stories
Order
Title
Area Path
State
Board Column
Business Value
Effort
Tags
1 Enable network alert heartbeat Fabrikam\Dashboard\Alerts Ready Ready 800 5
2 Enable two-factor authentication Fabrikam\Mobile\Security Ready Ready 800 8
My tickets
3 Quick Tips Fabrikam\Dashboard\Mes... Ready Ready 500 8
4 My tickets Fabrikam\Tickets Ready Ready 500 5 Bug
5 Manage user access Fabrikam\Mobile\Security New Sized 500 21
6 Employee phone numbers Fabrikam\Employees New Dependencies 800
7 Customer email addresses Fabrikam\Customers New Acceptance Criteria 1300
8 Improve customer edit links Fabrikam\Customers New Acceptance Criteria 800
9 Use corporate email to authenticate Fabrikam\Mobile\Security New Acceptance Criteria 500
10 Show customers count Fabrikam\Customers New Value 500

FIGURE 5-19 A Product Owner can order the Product Backlog by dragging and dropping PBI work items.

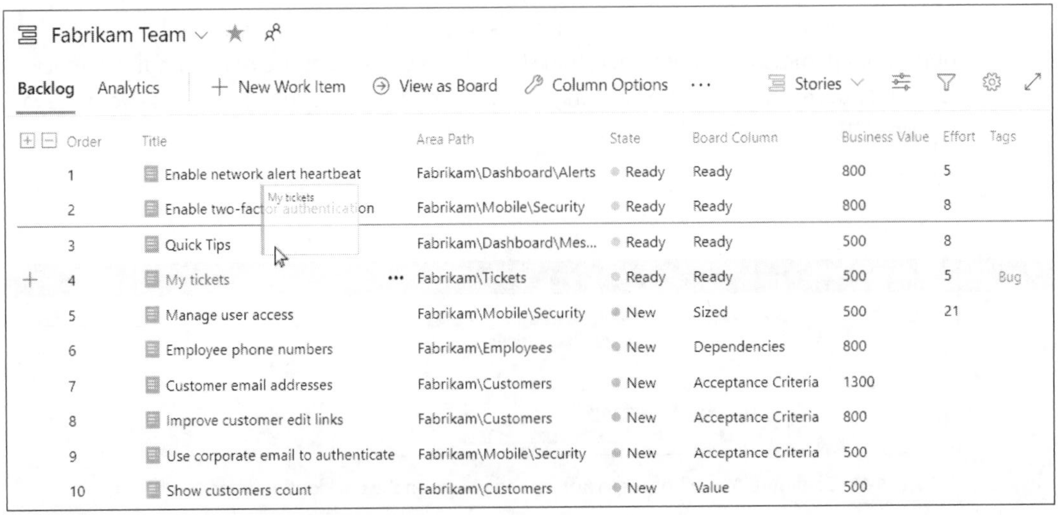

As your Product Backlog grows, you could potentially find yourself looking at hundreds of items. Although each backlog can display up to 10,000 work items in Azure Boards, if you get more than a few hundred, dragging, scrolling, and dropping can become a nightmare. That is why a professional, engaged Product Owner should be constantly paying attention to the Product Backlog, keeping it in good shape and ordering it. The more important and well-known items should be at the top, with "ready" items at the very top.

Ordering a large Product Backlog can be very tedious—you may have to drag up or down several "screens" of items. If this is your situation, your Product Owner my find it faster to right-click and select *Move to position*. Having to use this feature often, however, can be a smell of having an out-of-control Product Backlog.

> **Tip** It's fine to remove or delete old, stale PBIs from a Product Backlog. These might be the PBIs that always sink to the bottom, never getting implemented. This decision should lie with the Product Owner and may be influenced by the organization's policies and guidance. If a deleted feature is really valuable, it will pop up again at some point in time.

Planning a Release

Every organization will do some level of release planning. A Professional Scrum Product Owner is able to forecast and plan releases at any time—with varying degrees of confidence. Release planning depends on the Product Backlog being refined and correctly ordered. It also requires the Developers to know what they are historically capable of delivering each Sprint. This is typically quantified in either a velocity or Throughput metric. You will learn more about flow metrics, such as Throughput, in Chapter 9.

In Scrum, the Product Owner is responsible for managing the expectations of the stakeholders. This is done through Product Backlog management (refinement and ordering) and then forecasting when an item or items will be released—or which items will be released by a specific date. The release plan is often a separate visualization, such as a query result or report, but it can also be inferred by observing the Product Backlog.

A Product Owner, Developer, or stakeholder can use the Forecasting tool to help visualize their release plan in real time, directly on the refined Product Backlog. By specifying a value for the Developers' velocity, the Forecasting tool will show which items in the backlog can be completed within future Sprints, as you can see in Figure 5-20. As the Product Backlog is refined and reordered and velocity changes/stabilizes, the Forecasting tool can be run again.

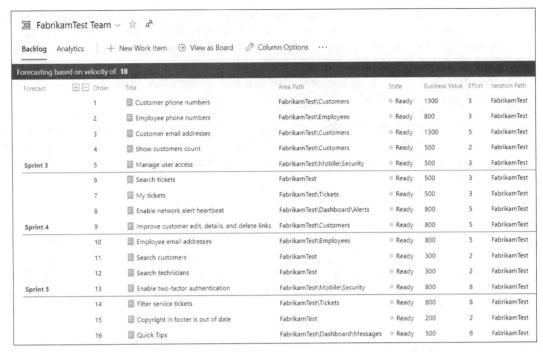

FIGURE 5-20 A team can use the Forecasting tool to see which PBIs might be completed in upcoming Sprints.

A risk of generating a separate release plan, outside the context of the Product Backlog, is that stakeholders may infer an expectation of scope, schedule, or cost. Any release plan is just a forecast, and the stakeholders need to be made aware of that. As with all types of work in the complex space, especially the *invisible* IT work of software development, things can change, especially plans.

Note Release planning is not officially part of Scrum, although it is a popular complementary practice. Keeping the Product Backlog healthy (refined and ordered) is the best input for release planning, regardless of the type or scope of release planning.

A release plan establishes the goal of the release and includes the highest-ordered PBIs, the major risks, and the overall features and functionality that the release will contain. It also establishes a probable delivery date and/or feature set, and a cost that should hold if nothing changes. The organization can then inspect progress and make changes to this release plan on a Sprint-by-Sprint basis.

The release plan will probably start with a large margin of error unless the Developers have a decent amount of empirical evidence of executing, sizing PBIs, failing, learning, and succeeding as a team. Over time, this mounting empirical data will become apparent as a stable velocity or Throughput. The release plan will become increasingly refined (and accurate) as development progresses and empirical data is gathered. Again, the key to release planning is to have the Product Backlog in good shape, which is the result of an engaged Product Owner and constant refining by the whole Scrum Team.

In traditionally managed organizations, their planning process is done at the beginning of the effort and largely left unchanged as time passes. In an organization practicing Professional Scrum—where inspection, adaptation, and transparency are sacrosanct—the Product Owner defines the overall goal with stakeholders and works with the Developers to define probable outcomes. This style of planning requires substantially less time than to build a traditional release plan.

Traditional release planning efforts are up-front "guesstimates" that seldom prove true. Scrum's just-in-time planning is ongoing during all of its events. Because of that, agile release planning practices consume more effort than traditional release planning practices. Empirical methods usually do take longer than guessing. Scrum's approach, however, adds more value and probability of success because dynamic planning is more valuable than a static plan.

Release planning in Scrum supports three types of release models:

- **By date**   A date-target planning model is one where the Increment must be released by a specific date. The scope (features and bug fixes) that will be in that release must be negotiable. Having a refined Product Backlog will help the Scrum Team and stakeholders identify that scope.

- **By feature**   A feature-target planning model is one where the Increment is released once it contains a minimum set of features and bug fixes (to be determined by the Product Owner). The release date must be negotiable. Having a refined Product Backlog will help the Scrum Team and stakeholders identify that date.

- **Continuously**   A Continuous Delivery ("on-demand delivery") release model is one where the Increment can be released to production as each PBI (feature or bug fix) is done. Release can occur at any time during the Sprint, assuming the Definition of Done is met. This model is popular for server/cloud-hosted software. Having a refined Product Backlog will help the Scrum Team and stakeholders know what will be released next.

Release planning occurs all the way through a product's lifecycle, not just at the beginning. Release planning typically corresponds to the version increment of the software product. As the Developers progress through the Sprints, their empirical data (such as velocity) can be applied to the Product Backlog to assess how the release plan compares to reality. If a release burndown chart is being maintained, it will include data from past Sprints and can provide a view of the progress. This information is an important input to release planning.

Fabrikam Fiber Case Study

Originally, the Fabrikam Fiber website was released to production quarterly. After the adoption of Scrum and Azure DevOps, this improved to monthly, and then to every two weeks. The plan is to move to Continuous Delivery (CD) once more automated testing is achieved.

Story Mapping

It is a fact of life that developing the desired functionality in a complex product will require more than the available time and budget. Based on its assessment of progress, a Scrum Team can make changes to its release plan without compromising quality. The most effective variation to make is to the scope. This involves deferring some PBIs until later, thus reducing the scope for the current release.

That said, when stakeholders become dismayed when their desirements are removed from a Sprint, it's not productive to sit back and quote the *Scrum Guide* to them or even hand them a copy of this book—although I wouldn't mind if you did. Yes, there are rules in place to minimize scope creep from occurring during the Sprint and the release. Unfortunately, the Scrum Values, as well as the Agile Manifesto's values, outweigh them. In other words, it is important that PBIs—even up to and including those in the Sprint—be "negotiable" so that their desired outcome can be achieved by pragmatically adjusting the sophistication of the product's implementation. In other words, it's more important to deliver business value in the form of working software than to follow a plan. I do believe I've read that somewhere.

Enter story maps. Where a Product Backlog is one-dimensional, easily showing the ordering of the items, story maps are two-dimensional. Working from a large *flat* list like a Product Backlog can be difficult to navigate and stakeholders can become lost, not knowing what should come next. Story maps can be arranged in many ways to visualize a release plan and even the Product Goal or vision. The goal of a story map is to keep the focus on the users and their experience, resulting in a better conversation and shared understanding.

Story *mapping* is the process of creating a story map. Story mapping begins by taking a selection of PBIs (typically user stories) and organizing them on a board. This organization will represent the order in which new functionality is delivered and can be used to plan future Sprints. By visually arranging the PBIs on a story map, your team is breaking the user's journey down into understandable parts. Having the stakeholders involved will ensure that desires are represented correctly and nothing gets missed. It also helps them understand the realities and trade-offs of release planning. If a stakeholder wants a medium-sized PBI added to the release, then they will first have to pull a medium-sized PBI out of the release. Story maps make this very obvious.

Story mapping consists of ordering PBIs along two dimensions. The horizontal axis is for arranging user activities and functionality in an order that describes the behavior of the product, possibly as a user's journey. The vertical axis is for planning the increasing sophistication of the product over time. Figure 5-21 shows a sketch of a story map and these two axes.

Even with only two axes, story maps can be created in many different configurations. PBIs can be grouped in many ways on the horizontal axis (by feature, by epic, by journey, by persona, etc.). PBIs can also be grouped in many ways on the vertical axis (by Sprint, release, or date).

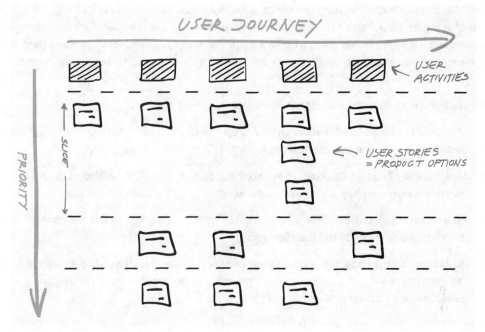

FIGURE 5-21 Story maps keep the focus on the users and their experience, resulting in a better conversation and shared understanding.

Image provided by SpecFlow

 Tip Because the story map involves the entire product, the entire Scrum Team and stakeholders should be involved in its creation and evolution. Perhaps these updates are done during Product Backlog refinement. Perhaps it's done at Sprint Planning, or even Sprint Review. Let the team decide when and where. Having a physical story map posted in a public area where it is visible to the stakeholders can be a powerful information radiator.

Traditionally, story maps are created using sticky notes that are placed on walls or whiteboards. These physical maps are not without their disadvantages. Walls are not transportable, and unfortunately, that means those story maps are only temporary. It is also difficult for remote stakeholders to view these boards in real time. Also, let's not forget the cleaning person at night who can wreak havoc on physical boards and maps. I'll explore an electronic tool alternative in the next section.

SpecMap

SpecMap is an Azure DevOps extension by Tricentis that provides story mapping capability for Azure Boards. SpecMap enables a Scrum Team to build story maps from work items, providing a visualization and basis for discussing the needs of your stakeholders. These conversations can assist a Product Owner in ordering the Product Backlog to deliver the biggest impact.

SpecMap imitates traditional story maps that use sticky notes. A team can depict the user journey as a series of activities, place PBIs on the map, and easily create slices. These artifacts are represented by work items and even Sprints. Assigning a PBI to a particular activity automatically creates a parent-child link between the two. This means that creating a story map in SpecMap can assist a team in Product Backlog refinement and even Sprint Planning.

The key features of SpecMap include the following:

- **Map user activities to PBIs via drag and drop** Dragging PBI work items from the Product Backlog to a story map can create a mapping to a Feature or Epic work item.

- **Structure the Product Backlog** New work items are automatically added to the Azure Boards backlog hierarchy with the corresponding feature as the work items' parents.

- **Plan releases with slices** Slice a story map to represent Sprints and releases. PBI work items don't have to be assigned to future Sprints.

- **Add tasks to PBIs and keep track of your progress** Use SpecMap during a Sprint, adding tasks directly to a PBI card. An overview of the completion status of those child tasks is displayed, allowing a team to keep track of its progress.

After you install the SpecMap extension, it will become available in the Boards hub. First, you will need to add a new map to the project and give it a name. Next, you'll map out the user's journey (or sets of features) from left to right as a series of high-level activities that your product has an impact on. These activities can optionally be linked to a work item, such as a Feature or an Epic. You can also track the outcome expected when this activity has been completed.

Next, you can assign related PBIs to these activities by dragging and dropping from the appropriate backlog on the left. You can drag and drop these PBIs up and down to change their priority, and you can drag the PBIs horizontally to position them under a different user journey or activity. A Product Owner can do this before a Sprint Review or other stakeholder meeting, or during any Product Backlog refinement workshop.

Next, you can slice groups of PBIs horizontally to plan Sprints or releases. You can drag PBIs up and down to put them into these different slices. These slices may be linked to iteration paths (Sprints) to automatically assign the PBIs when they are added or dragged to that slice. You can see a SpecMap story map being constructed in Figure 5-22.

> **Tip** Professional Scrum Teams don't preassign, or *pre-bucket*, PBIs to future Sprints. In addition, undone PBIs that aren't in the current Sprint should have their iteration set to the root, indicating that they are in the Product Backlog. For this reason, it is not recommended to link SpecMap slices to iteration paths, because you'd be pre-bucketing work into these future Sprints. The good news is that you don't have to do this. SpecMap maintains its slice state in the map, and not in the work items.

Unlike SpecMap, Microsoft's *Delivery Plans* requires work items to have their iteration set in order to arrange them in the release. This pre-bucketing of work is wasteful because things can change. New PBIs may be introduced or existing plans may change, requiring someone to go back and change items that were set in future iterations. What's riskier is that stakeholders, or others who don't understand that this is just a forecast, might infer scope, schedule, or cost by observing the work items directly (and seeing that they have been assigned to a future Sprint).

FIGURE 5-22 SpecMap is an Azure DevOps extension that lets you easily create story maps.

Product Backlog Checklist

Each Scrum Team and Product Owner will create and manage their Product Backlog differently. Those just getting started with Scrum and/or Azure DevOps may have more work to do to get their Product Backlog imported and refined. Other teams and Product Owners may have a fairly light flow of new PBIs, resulting in a Product Backlog that stays in good shape. In either case, it's safe to bet that your Product Backlogs will go through seasons of churn. For example, right after a large release there may be lots of feedback that needs to be captured in the Product Backlog.

Regardless of where you and your team are, this checklist can be helpful in preparing your Product Backlog and keeping it healthy and refined.

Scrum

❏ Talk to stakeholders to see if there is interest in having an Epic or Features backlog.

❏ Identify existing lists in your organization that would constitute a Product Backlog.

❏ Decide which items would go on the Epics, Features, or Stories backlog.

Azure DevOps

- ❐ Import the PBI, Feature, and Epic work items from existing lists.
- ❐ Ensure that PBIs all have short, meaningful titles.
- ❐ Tag those PBIs appropriately (Bug, High Risk, etc.).
- ❐ Assign PBIs to the Product Owner (or leave them blank).
- ❐ Set the PBI States appropriately (New or Ready).
- ❐ Ensure all PBIs are set to the appropriate area.
- ❐ Ensure all PBIs—not in the current Sprint—are set to the root iteration.
- ❐ Ensure all PBIs have a short, meaningful description.
- ❐ Ensure all PBIs have adequate acceptance criteria.
- ❐ Ensure all PBIs have been assigned a value.
- ❐ Ensure all PBIs have been sized.
- ❐ Promote large PBIs to Features.
- ❐ Promote large Features to Epics.
- ❐ Map PBI work items to Features and Features to Epics accordingly.
- ❐ Order the Product Backlog with "ready" PBIs at the top.
- ❐ Consider ordering by ROI (value/size) in lieu of another approach.
- ❐ Avoid linking PBI work items to other PBI work items.
- ❐ Avoid assigning PBIs to future Sprints.
- ❐ Avoid creating associated tasks prior to Sprint Planning.
- ❐ Avoid creating associated test cases prior to Sprint Planning.

Chapter Retrospective

Here are the key concepts I covered in this chapter:

- **The Backlogs page** This page is used to create and manage the Product Backlog in a flat, one-dimensional format.

- **Importing work items** You can natively import work items from CSV files in Azure Boards.

- **Microsoft Excel** Excel is used to query and bulk-update multiple fields with different values.

- **Bulk-modifying**   Azure Boards offers the ability to make bulk edits to work items. This can be done from query results or the Backlogs page.

- **Epic**   An Epic is any PBI that is too large to be developed in a single Sprint. Epics must be split. Not to be confused with the Epics backlog.

- **Features backlog**   This is a hierarchical backlog of work items that sits above the Stories Backlog in Azure Boards. A Feature work item can be the parent of a PBI work item.

- **Epics backlog**   This is a hierarchical backlog of work items that sits above the Features backlog in Azure Boards. An Epic work item can be the parent of a Feature work item.

- **Removing vs. Deleting**   When you set a work item's state to Removed, it will disappear from the various backlogs and boards but can be discovered by running a query. Deleted work items are sent to the Recycle bin, where they can be recovered or permanently deleted.

- **Reporting a bug**   Bugs found during a Sprint should just be fixed; otherwise, create a PBI work item to report the bug. These PBIs can be optionally tagged as "Bug." Annotated screen-shots and listing expected results are good ideas. Failed tests are not bugs.

- **Refinement**   This is an optional activity in which the Scrum Team works to understand and estimate upcoming items in the Product Backlog in order to get them ready for Sprint Planning.

- **Sizing**   Use empiricism and low-precision practices, such as T-shirt sizing, to size PBIs and develop a shared understanding of its complexity. There are many agile estimation techniques available, such as Planning Poker.

- **Definition of ready**   This is an optional practice a Scrum Team can use to visualize the work-flow for getting a PBI ready for Sprint Planning. The Boards page can be used to help track PBIs on their journey to becoming ready.

- **Ordering the Product Backlog**   Use drag-and-drop in the Backlogs page to order the Product Backlog. There are many ways to order the Product Backlog—by ROI, risk, dependency, learning, or by the whim of the Product Owner.

- **Velocity**   Use the built-in analytics to see (in real time) the Developers' velocity in order to use it for forecasting.

- **Forecasting**   Use the Forecasting tool and the Developers' velocity to view which PBIs might potentially be developed in the upcoming Sprints.

- **Release planning**   Use a refined Product Backlog to forecast *what* will be released by a specific date or *when* a specific set of functionality might be released. Professional Scrum Teams value planning over having plans.

- **Story maps**   A story map is a two-dimensional visual representation of the Product Backlog that can be used to develop a shared understanding with stakeholders. SpecMap is an Azure DevOps extension that supports the creation of story maps.

CHAPTER 6

The Sprint

The heart of Scrum is the Sprint—a timebox of one month or less during which a done, usable, and releasable product Increment is created. Sprints have consistent durations throughout a development effort. A new Sprint starts immediately after the conclusion of the previous Sprint. Sprints consist of Sprint Planning, Daily Scrums, the development work, Product Backlog refinement, the Sprint Review, and the Sprint Retrospective.

Sprints are limited to one calendar month. When a Sprint's horizon is too long, the definition of what is being built may change, complexity may rise, and risk may increase. Sprints enable predictability by ensuring inspection and adaptation of progress toward a Sprint Goal at least every calendar month. Sprints also limit risk to one calendar month of cost. The frequency of feedback, experience and technical excellence of the Developers, and the organization and Product Owner's need for agility are key factors in determining the length of a Sprint.

The first event within a Sprint is Sprint Planning. The entire Scrum Team attends and participates in Sprint Planning. Stakeholders may attend—if their presence adds value to the conversation by helping clarify details about the PBIs in question. The first part of this chapter will cover Sprint Planning—its inputs and outputs, and how Azure DevOps supports Sprint Planning to capture the Sprint Goal and the Sprint Backlog.

Each day of the Sprint, the Developers meet for the Daily Scrum. This is a timeboxed meeting, lasting no longer than 15 minutes. The Developers use this opportunity to synchronize with one another and to develop a plan for the next 24 hours. The Developers should update their remaining work, at least daily, so that they can inspect the progress of their forecasted work. Beyond that, what the Developers do during the course of the day depends on what is required to get the forecasted work done and achieve the Sprint Goal. Professional Scrum Developers make sure that the work performed is always of value to the organization. The second part of this chapter will be spent on those daily activities that the Developers perform as they pertain to Azure Boards.

In summary, this chapter focuses on how to use the various tools in Azure DevOps, especially Azure Boards, to plan and manage the work in the Sprint Backlog. If you are more interested in the concept of the Sprint and the Sprint Backlog, and less on how to manage them using Azure Boards, I refer you back to Chapter 1, "Professional Scrum."

Note In this chapter, I am using the customized Professional Scrum process, not the out-of-the-box Scrum process. Please refer to Chapter 3, "Azure Boards," for information on this custom process and how to create it for yourself.

Sprint Planning

At the beginning of the Sprint, the Scrum Team attends Sprint Planning, where the work to be performed in the Sprint is planned. This meeting is timeboxed to a maximum of eight hours for a one-month Sprint. For shorter Sprints, the event is usually shorter. For teams practicing a two-week Sprint, I suggest a timebox of four hours or less. Initially, the Scrum Team may take the entire timebox, but after a few Sprints, they should start getting the hang of it and take less time. Also, by regularly refining the Product Backlog, time spent in Sprint Planning should decrease.

Sprint Planning can be conceptually broken down into three topics: the *what,* the *how,* and the *why.* In the first topic, the Developers work to forecast *what* functionality (PBIs) will be developed during the Sprint. The Product Owner discusses the objective that the Sprint should achieve and the PBIs that, if completed, would achieve this goal. The Sprint Goal provides direction as the Developers create the forecast. The entire Scrum Team collaborates on understanding the work of the Sprint.

In the second topic, the Developers decide *how* they will build this functionality into a done product Increment during the Sprint. This is called the plan, and it can be represented in many ways. The forecasted PBIs selected for this Sprint, along with the plan for delivering them and the Sprint Goal, make up the Sprint Backlog.

Tip It is not expected that the Developers will completely generate the Sprint Backlog during Sprint Planning. Some work may be stubbed out for later discovery. It's important for work to be planned for at least the first few days of the Sprint. Also, Professional Scrum Developers don't fill their capacity to 100 percent during Sprint Planning—they leave slack time for other activities.

What happens during Sprint Planning in these topics is completely up to the Scrum Team. In early Sprints, they may spend 75 percent of their time forecasting and 25 percent of their time creating a partial plan. Later, as they have more experience practicing Scrum, they may balance out the topics. After regularly refining the Product Backlog for a period of time, Sprint Planning may look more like 25 percent forecasting and 75 percent creating the plan, with a much shorter meeting.

Sprint Planning has the following inputs:

- **Product Backlog** The refined Product Backlog that, hopefully, contains at least a few ready PBIs toward the top.

- **Objective** What the Sprint should achieve. Provided by the Product Owner. This may end up being the Sprint Goal, or at least contribute to the Sprint Goal.

- **Increment** Knowledge of the Increment's content and health. If the product is software, then this would include the quality of the code base, technical debt, lack of tests, etc.

- **Definition of Done** A shared understanding of what it means for work to be complete in order to ensure transparency. Generally speaking, if the Definition of Done is stringent, then less work would be forecasted than with a simpler Definition of Done.

- **Past performance** The Developers' performance in recent Sprints. This could be velocity or even a flow metric like Throughput. I will cover flow metrics in Chapter 9, "Improving Flow."

- **Availability** A shared understanding of the Developers' general availability/capacity during the Sprint. This takes into consideration training, travel, vacations, holidays, etc. This does not need to be detailed or the result of traditional capacity planning. In fact, using the term "availability" instead of "capacity" helps the Developers distance themselves from this command-and-control (C&C) waterfall practice.

- **Retrospective commitment(s)** Implementing one or more process improvements— as identified in the previous Sprint Retrospective(s). This may take time away from the development.

During Sprint Planning the Scrum Team also crafts a Sprint Goal—to represent the *why*. The Sprint Goal is an objective that will be met within the Sprint through the implementation of the Product Backlog, and it provides guidance to the Developers on why it is building the Increment. The Product Goal and the Product Owner's objective may influence the Sprint Goal, or it may simply become the Sprint Goal.

Sprint Planning has the following outputs:

- **Forecast** The selected PBIs that the Developers feel that they can accomplish during the Sprint. The Sprint Backlog contains the forecast.

- **Plan** How the Developers will develop and deliver the forecasted PBIs. The plan can be represented in many different ways, including tasks, tests, and diagrams. The Sprint Backlog contains the plan.

- **Sprint Goal** An objective that will be met within the Sprint through the implementation of the forecasted PBIs. The Sprint Goal provides guidance to the Developers during the Sprint. The Sprint Backlog contains the Sprint Goal.

By the end of Sprint Planning, the Developers should be able to explain to the Product Owner and Scrum Master how they intend to work in a self-managing way to accomplish the Sprint Goal and create the anticipated Increment.

Sprinting in Azure Boards

Using Azure Boards, a Scrum Team can plan a Sprint, create the Sprint Backlog, and manage their daily work. Developers can also inspect their progress toward the Sprint Goal by tracking that work on the Kanban board or the Taskboard, or by referencing the built-in analytics.

Azure Boards enables a Scrum Team to perform the following Sprint and Sprint Backlog–related activities:

- Capture the Sprint Goal.

- Plan a Sprint by creating the forecast and the plan.

- Visualize and update the Sprint plan (using Task work items).

- Take ownership of tasks.

- Track progress using the Taskboard.

- Visualize blocked work.

- Track progress using a burndown chart or the Taskboard itself.

- Move incomplete PBIs back to the Product Backlog.

Creating the Sprint Backlog

The Sprint Backlog is a Scrum artifact. It contains the forecasted PBIs along with the plan for delivering them. It is created during Sprint Planning, but the plan will continue to emerge throughout the Sprint. In Azure Boards lingo, the Sprint Backlog contains those PBI work items that have their Iteration set to the current Sprint and any related Task work items that represent the plan.

 Note There are many ways Developers can formulate their Sprint plan. The *Scrum Guide* is silent on this. Tasks are a very popular way to express a plan, but so are failing acceptance tests, workflow states modeled on a Kanban board, diagrams, and even conversations. In this chapter, I will show you how to use tasks (Task work items) as a way to express a Sprint plan. In Chapter 7, "Planning with Tests," I'll explain how and why Developers may want to graduate to expressing their plan as failing acceptance tests (using Test Case work items).

In this section, I will show you how to use the various tools in Azure Boards to create the Sprint Backlog.

Creating the Forecast

Assuming that the Developers have already come to a consensus on what feels like the right amount of work, the act of forecasting the PBIs in Azure Boards is straightforward. In fact, there are only two options you have to set to make in order to forecast a PBI work item:

- Set *Iteration* to the current Sprint.

- Set *State* to *Forecasted*

The easiest way to set Iteration for a PBI work item is to drag it from the Product Backlog to the current Sprint in the Planning pane. You can see this in Figure 6-1. After you drop the item, *Iteration* will be updated. You can also set *Iteration* manually and get the same result—which will help users with only a Stakeholder access license.

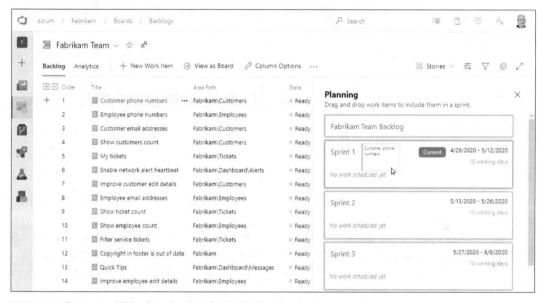

FIGURE 6-1 Forecast a PBI by dragging it to Sprint 1 in the Planning pane.

> **Note** If the current Sprint is not visible in the Planning pane, it's because it hasn't been created yet or it hasn't been selected by the team. Rather than go to Project settings, you can use the shortcut in the Planning pane to add a new Sprint (or select an existing one). Whether or not you use the shortcut, being able to select just the Sprints a team wants is a nice feature that keeps the list of Sprints short and manageable in the Planning pane.

Once a PBI work item's Iteration is set to a Sprint, it will become visible in that Sprint Backlog. Further, when the State of the PBI work item is changed to Forecasted, the work item may disappear from the Product Backlog page. This behavior can be controlled by setting the *In Progress Items* option, as you can see in Figure 6-2. I don't like seeing In Progress Items in the Product Backlog, so I keep this option cleared. You may have a different preference.

FIGURE 6-2 We are disabling the In Progress Items option in the Product Backlog.

> **Tip** To remove a PBI work item from the forecast (also known as *deselect* or *descope*), just drag that work item to the top node in the Planning pane. This node represents your team's Product Backlog (for example, Fabrikam Team Backlog) and will reset *Iteration* back to the root value, indicating that the item is no longer forecasted. You can also reset the *Iteration* value manually. You will also need to reset the work item's *State* field. Although you can remove a PBI from the Sprint at any time, even after Sprint Planning, ideally all forecasted PBIs will be completed, according to the Definition of Done, and never return to the Product Backlog—because they will have their *State* fields set to *Done* at that point.

The other step in forecasting a PBI work item is to set its *State* to *Forecasted*. This signals Azure Boards that the PBI is now planned—for the sake of analytics. Simply dragging to the current Sprint does not do this for you automatically—which is actually a good thing. Sometimes the Developers will bring many (maybe too many) PBIs into the Sprint, have a conversation about which items to forecast, and then set those to the *Forecasted* state. The others would be deselected for that Sprint. You will need to update the PBIs manually by opening each work item one at a time and setting *State* to *Forecasted*. For a large number of items, you can use the bulk-modify feature to multi-select and change their *State* fields in a single operation.

From the perspective of Azure Boards, once the *Iteration* and *State* fields are set, those PBI work items are forecasted for that Sprint. Your primary tool for viewing and managing them is now the Sprint Backlog.

Fabrikam Fiber Case Study

Typically, Paula (the Product Owner) is the user who edits the items in the Product Backlog, setting the *Iteration* and *State* fields accordingly. Other Scrum Team members have done this in the past as well, and they are all capable of doing it. These changes are done during Sprint Planning.

Forecasting Based on Past Performance

If the Developers have completed multiple Sprints, they can start to use their empirical data to improve forecasting. In other words, the Developers can review how many PBIs they have completed in previous Sprints—according to their Definition of Done. This measure is known as *velocity*, and it can be used as an input to Sprint Planning.

Velocity is a complementary practice and is the most popular way Scrum Teams measure their past performance. It relates to how much PBI effort/size the Developers can develop and deliver in a single Sprint. Velocity is the sum of those effort/size points for PBIs done—according to the Definition of Done—for a Sprint or series of Sprints. Once established, the Developers' velocity can be used to forecast Sprints and plan releases. Velocity is most accurate when the team composition, Sprint duration, Definition of Done, domain, and estimation practices remain constant.

Note I'm often asked how velocity can be meaningful when PBIs are estimated using such crude precision and in such abstract values as story points. Fortunately, it is. Thanks to the wisdom of the crowd, the law of large numbers, and relative sizing, the average of the results obtained from a large number of trials (Sprints) should be close to the expected value and will tend to become closer as more trials are performed. Remember, though, the goal of estimating is not the number—it's the conversation and learning that accompany it.

A key piece to maintaining an honest and consistent velocity is having a stable Definition of Done and sticking to it. It may seem like a small thing, but a Definition of Done is used to establish transparency as to when work is complete—and, as a corollary, when it is not. Without a consistent meaning of *done*, velocity cannot be measured.

A Definition of Done also helps ensure that the Increment is of high quality, with minimal defects. Professional Scrum Teams will expand their Definition of Done over time. They know that these changes may affect their velocity for a short period of time, but the quality improvements are worth it. Depending on the types of changes made to the Definition of Done (such as adding automated testing), productivity—and, as a corollary, velocity—should improve as well. This is a part of continuous improvement, which I will discuss in Chapter 10, "Continuous Improvement."

 Smell It's a smell when I see Developers *only* use velocity to forecast their work. It's also a smell when I see Developers only use availability/capacity to forecast their work. Professional Scrum Developers use all Sprint Planning inputs as data points for a rich conversation on what feels like the "right" amount of work to forecast.

There are several practices that, if adopted, can improve the Developers' productivity and, by extension, their velocity. Here is a partial list:

- Establish a right-sized, cross-functional team of Developers.
- Keep team membership consistent.
- Keep Sprint length consistent.
- Remove identified impediments.
- Keep a refined Product Backlog.
- Execute as a team (swarming, pairing, mobbing, etc.).
- Create and run tests early (also known as *shift left*).
- Build, deploy, and test continuously.
- Avoid generating technical debt.
- Avoid meetings beyond those defined by Scrum.
- Avoid starting work on a new PBI before finishing work on a previous one.
- Avoid multitasking and interruptions.
- Inspect, adapt, and improve continuously.
- Learn from your mistakes.

For Developers new to Scrum, after the team has completed a few Sprints, average velocity should begin to stabilize and become more valuable. In Azure Boards, you can find the team's velocity on the Analytics page. Figure 6-3 shows a Velocity chart generated from the prior nine Sprints. The Velocity chart can be configured to show a count of PBI work items, a sum of Business Value, a sum of Size/Effort, or a sum of Remaining Work.

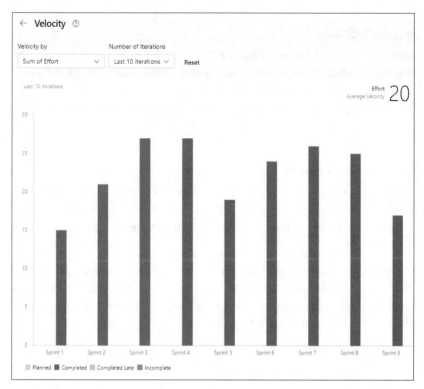

FIGURE 6-3 Azure Boards provides informative analytics, such as the Velocity chart.

The color of the vertical bars on the Velocity chart indicates the type of work in the Sprint:

- **Dark green** Represents completed work (PBI work items set to Done prior to the Sprint end date).

- **Light green** Represents completed work (PBI work items set to Done *after* the Sprint end date). This is a smell that should be discussed by the team.

- **Dark blue** Represents incomplete work (PBI work items still set to Forecasted). Ideally a Professional Scrum Team wouldn't show any Incomplete work because, after the Sprint, they would move incomplete PBIs back to the Product Backlog by setting the Iteration back to the root value.

- **Light blue** Represents planned work (PBI work items assigned to a Sprint prior to the Sprint begin date). This includes work that was moved to a different Sprint after the start of the Sprint. Ideally, a Professional Scrum Team wouldn't show any Planned work because they don't pre-bucket work into future Sprints.

Tip The Velocity chart provides an average velocity based on a simple average of the Sprints being displayed. Your team may want to compute this differently, based on fewer Sprints, more Sprints, tossing out highs/lows, and so forth. In other words, don't assume you have to use Azure Boards' average.

Fabrikam Fiber Case Study

The team's velocity seems to have normalized around 21. This has been the average of their last few Sprints. They will use a range of 17–25 as they have probabilistic conversations with stakeholders. They are also just starting to play with Kanban as a complementary practice. Down the road, they will consider using flow analytics, such as Throughput, to help forecast their work.

Forecasting Using the Forecasting Tool

As I explained in the last chapter, the Forecasting tool helps visualize a release plan in real time, directly on the Product Backlog. The same tool can be used to assist in Sprint Planning as well. By specifying a value for the team's velocity, the Forecasting tool shows which items in the backlog can be completed within that Sprint.

You can enable the Forecasting tool on the View options pop-up. After enabling it, you'll need to provide a velocity number. As you can see in Figure 6-4, I've entered a velocity of 21. Azure Boards then adds a *Forecast* column and displays horizontal lines to the backlog. The Forecast column shows the Sprint that the tool predicts the PBI work item will be developed in.

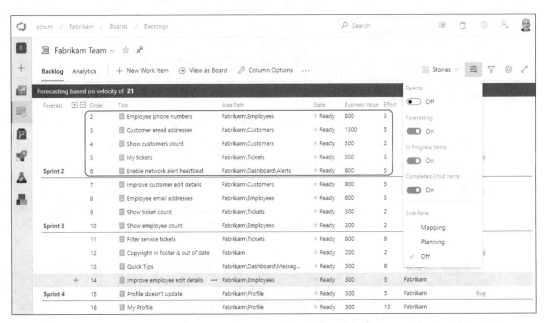

FIGURE 6-4 The Forecasting tool can help identify which PBIs might be selected for the current and future Sprints.

The Forecasting tool shows only future Sprints. There will typically be some PBI work items at the top of the backlog that don't appear to be in a Sprint. This is because the tool considers them candidates for the current Sprint. Items that fall between two forecast lines are considered candidates for the Sprint named on the top line. Also, the Forecasting tool ignores PBI work items set to Forecasted. To minimize confusion, you should hide the In Progress Items when using the Forecasting tool.

The Forecasting tool is not based on magic. In fact, its algorithm is quite simple. It walks down the ordered backlog, adding up the size/effort points of each item. Once the sum is greater than the provided velocity, it increments the Sprint number on the horizontal line. When it's out of work items or Sprints selected by the team, it stops.

Smell It's a smell when I see a team blindly follow what the Forecasting tool suggests. I've even seen feature requests on Microsoft's Developer Community asking for the Forecasting tool to auto-assign those PBIs into respective Sprints. Please, no! Remember, velocity is just one of many inputs to Sprint Planning.

You may encounter one issue with the Forecasting tool: if it finds a PBI with a size that is greater than the provided velocity, rather than skip it, flag it, or convert it to a Feature work item, the Forecasting tool assumes that the Developers will start on it in one Sprint and then carry it over and finish it in the next Sprint. For example, with a velocity of 15, if a PBI has a size of 21, the tool will carry over 6 points of effort/size to the next Sprint. Professional Scrum Developers don't start work that they know they cannot finish. Those large PBIs are not ready and would need to be split prior to being forecasted. This would most likely occur during Product Backlog refinement.

Tip Run your low-, medium-, and high-velocity numbers through the Forecasting tool to get an idea of the range of PBIs that could be forecasted for the Sprint. For our Fabrikam Fiber case study, this means running it for a velocity of 17, 21, and finally 25. The difference between low and high can result in a substantial variation in the number of PBIs to consider. All of these PBIs should be discussed during Sprint Planning.

Fabrikam Fiber Case Study

When the team was first learning Scrum, velocity meant everything. It was their only metric and they idolized it. They would try to improve it each Sprint. As the team improved, they learned that velocity is a measure of output, not outcome. Delivering business value in the form of working product is the most important metric. The team still uses velocity but only as *one* of several inputs into Sprint Planning. They now just forecast the number of PBIs that feels like the right amount of work, given all of those data points. Paula still uses velocity and the Forecasting tool for release planning.

Forecasting Using the Capacity Tool

One of the inputs to Sprint Planning is the expected availability of the Developers during the Sprint. This will vary from Sprint to Sprint, based on factors such as training, travel, vacations, and holidays. Availability is just another data point the Developers should consider when determining how many PBIs to forecast. In other words, if availability is low, the Developers should forecast fewer PBIs than normal, and vice versa.

As much as I want Developers to be able to have a collective *feeling* of their availability, neophyte teams may need some tooling to help visualize this. For those teams, they might consider the Capacity page in Azure Boards to better determine their availability. The Capacity page allows individual Developers to track their days off as well as their daily capacity. Team days off can also be specified, and they will apply to all Developers—as you can see at the bottom of Figure 6-5.

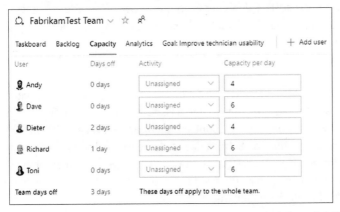

FIGURE 6-5 The Capacity page allows a team to track their daily availability and days off.

> **Tip** If your team decides to use this tool, don't waste time setting the Activity type. Rather than specify Deployment, Design, Development, Documentation, Requirements, or Testing, just leave it set to *Unassigned*. Doing so will save time. It also stops reinforcing that certain people only do certain work. Professional Scrum Teams strive for all Developers to be T-shaped—which are individuals who do more than just their specialty. I'll talk more about the concept of T-shaped in Chapter 8, "Effective Collaboration."

Once availability and capacity have been entered, the Work details pane can be displayed as the Sprint Backlog is created. The Work details pane features visualizations that show the total hours of planned work compared to the capacity of the team and the individual Developers. Work details can be displayed for the overall team, by activity, or by individual. Since Professional Scrum Teams don't track capacity by activity or preassign tasks to Developers, only the top, team-level visualization might be of value. The other sections can be ignored and even collapsed, as I've done in Figure 6-6.

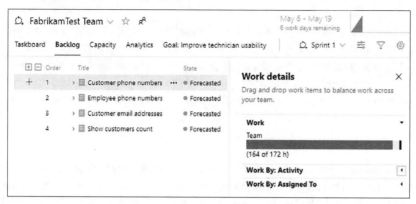

FIGURE 6-6 The Work details pane shows hours planned vs. capacity.

Smell It's a smell if I see a team using capacity planning tools of any kind. Though it might be helpful for teams new to Scrum to avoid forecasting too much work, Professional Scrum Teams recognize traditional capacity planning as a source of waste and potential dysfunction. They are already aware of their daily availability and upcoming days off, and they will use that as an input when forecasting a comfortable amount of work for the Sprint. Planning capacity by individual or activity type is counter to the self-managing attributes of the team. Let the Developers decide what seems like the right amount of work, what to work on next, and who should do what type of work. Let them do this as late as is responsible during the Sprint.

Capturing the Sprint Goal

One of the outputs of Sprint Planning is the Sprint Goal. It may have been an input as well, but assuming it wasn't, the Scrum Team will need to collaborate and craft one. The Sprint Goal is an objective set for the Sprint that can be met through the implementation of the forecasted PBIs. It provides guidance to the Developers on *why* they are building the Increment and gives some flexibility regarding the functionality implemented within the Sprint. The selected PBIs deliver one coherent function, which can be the Sprint Goal. The Sprint Goal can be any coherence that causes the team to work together rather than on separate initiatives.

Unfortunately, there is no "first-class" support for capturing a Sprint Goal out-of-the-box in Azure DevOps. Previous versions of the Scrum process had a Sprint work item type that included *Sprint Goal* and *Sprint Retrospective* fields, but this work item type is no longer part of the Scrum process. The wiki is an obvious place to capture a Sprint Goal. A Scrum Team could create a dedicated page for all its Sprint Goals or separate pages for each Sprint. The problem with using the wiki is that the Scrum Goal is a long way away from the Sprint Backlog and Taskboard in the Boards hub.

Smell It's a smell when a Scrum Team does not have a Sprint Goal. Yes, it's difficult to craft a good goal, especially when the forecasted PBIs span many areas and capabilities and might even contain a bug fix or two. Having a goal, however, gives the Developers something to focus on and commit to. It's also good for people to have goals. Professional Scrum Teams understand the psychological value in having a Sprint Goal, working toward it, and achieving it.

A Scrum Team can also install and use the Sprint Goal extension by Kees Schollaart. This extension allows a Scrum Team to capture the Sprint Goal and keep it visible within Azure Boards as a way for the team to keep focus on the goal. The Sprint Goal extension can be configured each Sprint with that Sprint's goal and display it as the Goal tab's label, as you can see in Figure 6-7. As the Developers interact with the Sprint Backlog and Taskboard, the Sprint Goal is visible at all times. This extension can be used in addition to the wiki.

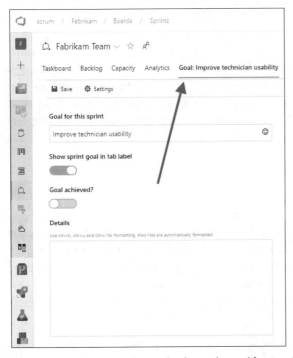

FIGURE 6-7 The Sprint Goal extension keeps the goal front and center during the Sprint.

Note The Sprint Goal doesn't have to be captured electronically—it can be written on a whiteboard in the team area or another public place. The advantage of using an electronic format is that it can be shared with team members and stakeholders outside the office. A list of past Sprint Goals can be maintained as well, for reference, in the wiki.

Creating the Plan

The forecast and Sprint Goal are only two of the three outputs of Sprint Planning. The Developers must also identify the *plan* for implementing the forecasted items. Although the plan should be created during Sprint Planning, it will also continue to emerge throughout the Sprint.

For some teams, the plan may have begun to emerge in the weeks prior to Sprint Planning, during Product Backlog refinement. These ideas should be captured in some easy, low-fidelity format such as notes or whiteboard photos, rather than as any kind of work items. Plans change and the latest responsible moment for creating an actionable plan is during Sprint Planning. Not only does this allow the Developers to create a plan with the latest information available, but it also reduces waste.

In Azure Boards, the plan is typically represented by Task work items. This emerging hierarchy of work items can be seen and managed in the Sprint Backlog and Taskboard. From the Sprint Backlog, you can create and associate Task work items for each PBI work item, as you can see in Figure 6-8. The new Task work item will default its *Area* and *Iteration* to that of its parent PBI. The Developers should also provide a short, meaningful *Title* and, optionally, *Remaining Work* (in hours) at a minimum. Additional details can be entered in the *Description* field.

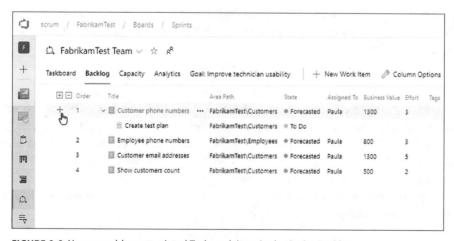

FIGURE 6-8 You can add an associated Task work item in the Sprint Backlog.

Tip Estimating the duration of tasks is optional. If the Developers decide to do it, they should estimate as a team. As estimates are made, it should not be assumed that any particular Developer will be doing any particular piece of work. In other words, don't let the Developer with the best skills for completing a specific task provide the estimate for that task. During development, it might be the case that someone else ends up doing the work. Once a Developer owns the task, they should reestimate the remaining work regularly until it is done. The estimate variability of having specialists and generalists work on tasks should be a wash because some team estimates will be lower and others will be higher.

As the Developers formulate their plan, it's best to leave the *Assigned To* field blank. Since all the work will be done *after* Sprint Planning, knowing who will be performing each task is difficult. Some teams may prefer to leave Sprint Planning with each Developer having at least one task to get started on. I will cover this in detail in Chapter 8.

Tip Consider *not* having Azure DevOps up and running during Sprint Planning. The absence of any electronics and tools will increase focus. Keystrokes and mouse clicks can interrupt a productive conversation. Sticky notes or whiteboard notes can be converted to work items after Sprint Planning.

Fabrikam Fiber Case Study

The team likes to work on sticky notes or a whiteboard during Sprint Planning. This way, changes to the plan can be made quickly and easily. Toward the end of the meeting, those notes are converted to Task work items.

Work planned for the first days of the Sprint by the Developers should be defined by the end Sprint Planning, often to units of one day or less. It is common for only 50 to 80 percent of the total plan to be identified during Sprint Planning. The rest should be at least "stubbed out" for later detailing or given rougher estimates that will be decomposed later in the Sprint. The act of decomposing work implies that the Developers have a consensus on how the work will be accomplished. Professional Scrum Developers are good at reaching consensus because they operate by the Scrum Values, especially courage and openness in these situations.

Note I get pushback from some Developers who think that creating Task work items is waste. Before I share my opinion with them, I like to find out more about their team. If they are a high-performance team—with PBIs that take only a day or two to complete—then perhaps they don't need to break work down and track individual tasks. They can decide—in real time—how to decompose, execute, track work, and assess progress.

I explain that there are broader benefits to creating and tracking work items. As the Developers work, their code and other artifact commits can be associated with these work items to enable traceability all the way through to a build and even a release. This is important to the Product Owner and stakeholders, as well as your fellow Developers, who want to trace which work was implemented by which commit, included in which build, and deployed with which release. Traceability in Azure DevOps is bidirectional so that a change in a release can be traced back to the work item(s) to explain why the change was made. Developers new to Scrum can also use this information to learn what they are actually capable of and to improve their confidence for the next Sprint.

When creating a plan for a PBI, do what works for the Developers: consider decomposing along design seams. The goal is to achieve independence from other tasks as much as possible. In other words, you should try to avoid interdependencies, and thus blocking tasks. Initial blocking tasks are those that have to be done first, usually by a single Developer (or a pair), and they prevent a majority of work from moving forward until they're complete. Ideally, each Developer (or pair) can work on separate tasks in parallel and make meaningful progress when those tasks are done.

Here are some questions to consider as the Developers create their plan:

- Does this task provide a meaningful step toward the completion of the PBI?

- Does this task have criteria for being done (either explicit or implicit)?

- Can this task be worked on by a pair of Developers effectively?

- Can this task be worked on by a mob (all Developers) effectively?

- Does this task depend on any other tasks being done first?

- Do other tasks depend on this task being done first?

- Can this task be completed in one day or less?

- Has anyone else created this task already?

- Is this task done already?

Remember, tasks are what you actually must do during the Sprint. They include work across all disciplines: analysis, design, coding, database, testing, documentation, deployment, security, operations, and everything else. The tasks must bring together everything required from the perspective of completing the PBI (according to the Definition of Done) as well as the product's quality needs and the practices of the people doing the work. In other words, completion of all tasks should result in the Definition of Done being met for that PBI. You can see an example of just such a plan starting to emerge in Figure 6-9.

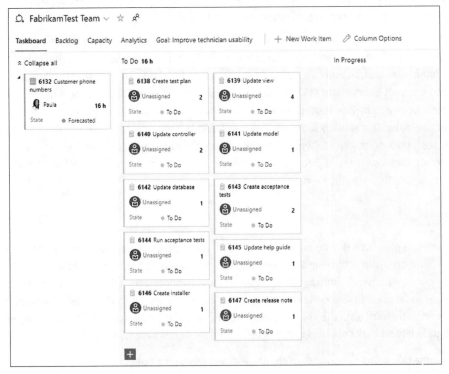

FIGURE 6-9 This example plan is represented by tasks in the Taskboard view of the Sprint Backlog.

Fabrikam Fiber Case Study

The Developers have become quite good at envisioning the plan at the Product Backlog refinement sessions, finalizing it during Sprint Planning, and then creating a representation of it using Task work items. New Task work items are created during the Sprint as the plan evolves and changes.

Sprint Activities

What the Developers should be doing in-between Sprint Planning and Sprint Review is deliberately vague in the *Scrum Guide*. It doesn't even have an official name—it's simply called "development." This ambiguity is by design. Scrum is just a framework. Besides meeting for the Daily Scrum and ensuring that progress is inspected, the Developers are on their own to manage and execute their own work throughout the day—which is how self-management should be.

Once the Sprint has been planned, the Developers should begin executing against their plan. If the plan was expressed using Task work items, then the Developers should begin working on these tasks. It is important to make progress toward the Sprint Goal on a daily basis so that, by the end of the Sprint, the Developers will have accomplished what they set out to do. Azure Boards includes a number of tools that can help Developers manage their Sprint and track its progress.

Developers can use the Taskboard to help visualize the work in progress, what work has been done, and what work remains. The Developers can use the Taskboard collectively or individually, as each person tracks their work and the overall work for the team. The Developers can also use the burndown and other charts to review the state and rate of progress. All of these analytics are calculated automatically by Azure DevOps. As the Sprint progresses, these calculations and tools can be used to decide whether to adjust plans, add more work, or make changes to help deliver a Done Increment of the product.

The rest of this chapter includes those Scrum-related activities that the Developers need to perform throughout the development portion of the Sprint. I'll relate those activities back to the respective tools in Azure DevOps.

The Daily Scrum

As I mentioned in Chapter 1, the Daily Scrum is a 15-minute, timeboxed meeting for the Developers to synchronize their activities and create a plan for the next 24 hours. It allows Developers to listen to what other Developers have done and are about to do. This meeting leads to increased awareness, collaboration, and even accountability. Developers make plans at this meeting, and these plans and their outcomes will be inspected 24 hours from now.

The Developers should use the conversations during the Daily Scrum to assess their progress. By hearing what is or isn't being accomplished each day, the Developers can determine whether they are on their way to achieving the Sprint Goal. As Developers improve in their collaboration, this vibe will become more noticeable—even outside the Daily Scrum. High-performance teams may even outgrow the need for a tool, such as a burndown chart, to help them assess progress. A team's conversations and intuition might suffice instead. Any method they choose should be transparent to the team.

Handling Impediments

During the Daily Scrum, someone might raise an impediment. An *impediment* is anything that blocks or slows down the progress of the team toward achieving the Sprint Goal—*that cannot be easily removed by the team*. Impediments are different than challenges, which tend to be smaller and able to be resolved quickly. Challenges are just part of everyday life when working in a complex environment.

Impediments are usually external and require some collaboration with others in the organization to resolve. In these cases, a Professional Scrum Master might be required to resolve them. In any case, impediments must be cleared in order for the team to resume productivity. In Scrum, there are two formal opportunities to identify impediments—at the Daily Scrum and at the Sprint Retrospective. That said, impediments can occur at any time and the Scrum Team should be ready to mitigate them.

> **Tip** Remove impediments—don't manage them! Successful complex product development hinges on the ability to inspect and adapt. If impediments are identified but are not being removed, the Scrum Master must become more actively involved. They may also need some additional authority.

If the Developer experiencing the blockage cannot remove the impediment themselves, another Developer should offer to help. If necessary, the Scrum Master should facilitate its removal and ensure that it happens. As a last resort, the Scrum Master should take ownership of the impediment and remove it.

Impediments that surface during the Daily Scrum should be briefly discussed. Problem-solving discussions detract from the real purpose of the Daily Scrum. Although not all impediments are the same size or complexity, they should be removed as early as possible. A Developer shouldn't wait until the Daily Scrum to raise the issue if they are completely blocked. Remember that it's more than okay to talk to people at any time during the day!

> **Smell** It's a smell when I hear Developers repeatedly telling one another that they have no impediments. Developing a product like software is a complex process fraught with risks and the potential for problems every day. Work like this can regularly yield impediments. In my experience, Developers tend to be optimistic, problem-solving individuals, and in their opinion, nothing ever blocks them. They have lots of (other) work they can be doing. This attitude is more common on teams new to Scrum where Developers might be hesitant to reflect the Scrum Value of openness and to share their problems openly. Professional Scrum Teams know that it's about the team, not the individual. Professional Scrum Developers ask for help, raise impediments early, and are transparent about progress. This attitude represents the Scrum Value of courage.

If the impediment cannot be resolved immediately, an Impediment work item can be created in Azure Boards. You can see an example of this in Figure 6-10. Tracking impediments in this way is not a requirement, especially if the Scrum Team works to resolve them quickly. The team may decide to link the new Impediment work item back to the related Task or PBI work item that is affected by the impediment. In addition, Task work items can be marked as Blocked (or tagged as Blocked) so that they will stand out on the Taskboard. Although these extra steps provide context, they should be used only if the team decides there is value in tracking the additional detail.

As I mentioned in Chapter 3, open impediments can be tracked by creating and running a shared work item query. This query would return all Impediment work items with a State of Open—which are the unresolved impediments. Once an impediment is resolved, its State should be changed to Closed and it will disappear from the query results.

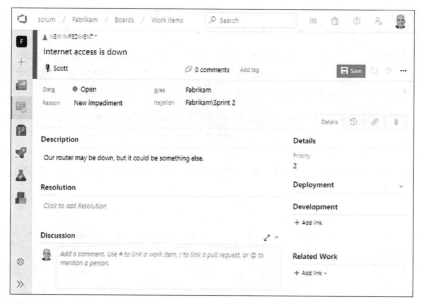

FIGURE 6-10 Here's an example of an Impediment work item.

Tip Managers and other leaders should regularly query the list of open and closed impediments. It's important to make this list transparent to the right people. Doing so provides a view of not only what organizational changes have taken place (closed impediments), but also what improvements are planned (open impediments) and which ones might require their intervention and authority. Seeing indicators that the team is improving can be as powerful as seeing the product improving.

Fabrikam Fiber Case Study

The Scrum Team has very few open impediments. The Developers have become quite good at removing them quickly. Because of this, they don't have many Impediment work items and none that are currently blocking specific work items. Regular team communication keeps everyone aware of the current issues and what might be blocked.

Decomposing Tasks

Work planned by the Developers should be decomposed into units of one day or less. If the Developers are using tasks, then a task should be small enough to be completed in a single day—whether by a single Developer, a pair, or a mob. This constraint forces Developers to create more specific, granular, atomic tasks that elevate efficiency by reducing bottlenecks and risks. By pursuing smaller units of work, Developers can take complex problems and decompose them to a level of detail that they can more easily understand and achieve. Smaller tasks also lead to improved transparency, improved opportunities to swarm, and a more accurate assessment of work remaining. I'll talk more about swarming on tasks in Chapter 8, "Effective Collaboration."

Assuming the one-day task size limit, you might find stubbed-out "epic" tasks in the Sprint Backlog, especially if placeholder tasks were created in Sprint Planning. Similar in concept to epic PBIs, these are tasks that are too large to be completed in a single day. Such tasks should probably be decomposed.

Developers new to Scrum may try to plan the tasks for their entire (eight-hour) day's capacity. Doing so is not realistic, and it usually results in waste. People take breaks, check email, have impromptu meetings, and perform other activities that distract from their Sprint Backlog work. With these factors in mind, Professional Scrum Developers may even want to decompose tasks that are larger than six hours, or even four hours.

Smell It's a smell when someone asks me how to track the original "epic" task. I understand that they want to establish a visual breakdown of the work, which can be helpful for teams new to Scrum. With complex work, the plan is fluid. This is why Scrum considers planning an (ongoing) activity and not a document. In other words, planning is more important than plans. If you need to split a large task into two or more smaller tasks, just rename the original large task, or delete it when you are done. This keeps the Sprint Backlog lean. Also, burndown analytics will be more accurate without a bunch of inert tasks in your Sprint Backlog. Remember, the Sprint Backlog is by the Developers, for the Developers, and nobody else should be seeing it, updating it, or caring about how work breaks down during the Sprint.

As the Sprint progresses and development work unfolds, the Sprint Backlog will become more refined as those "epic" tasks are shed. In other words, the Developers should break down large tasks as they work. For example, there shouldn't be any 16-hour tasks titled "Develop it" in the Sprint Backlog toward the end of the Sprint.

I'm often asked about sequencing tasks when there are interdependencies. For example, "Create acceptance tests" must be done before "Run acceptance tests." The Task work item does have a hidden *Backlog Priority* field, much like the PBI work item. The Taskboard displays Task work items ordered by this field, much in the same way the Product Backlog does. A sparsification background process will assign the Backlog Priority as tasks are dragged and dropped into a different sequence.

Smell It's a smell when I see Developers spending time sequencing their tasks. The smell has a hint of waterfall and also command-and-control. It could be that the work is so new and so complex that the Developers really do need to organize the tasks in a particular order. Hopefully this is not the case, or at least not the norm. In my experience, Professional Scrum Teams are able to figure out which task to perform next through conversation and collaboration, without having to have someone else sequence the work.

The Taskboard

The Taskboard in Azure Boards is a collaborative tool used by the Developers to communicate and execute their Sprint plan. It's pretty awesome. It provides visibility into what work is in progress, what work has yet to be started, and what work is done. Developers can quickly assess progress by viewing the Taskboard. You can think of the Taskboard as an *information radiator*. It's always on and (hopefully) updated regularly.

> **Note** The Taskboard is not a reporting tool. It's not meant to be used by management to hold the Developers accountable for their progress. In other words, it should not be used as a "blame board." If these behaviors start to surface in an organization, the Developers will be less inclined to be honest and transparent about the tasks they are working on. The fear is that they will revert to their old ways and focus on making the burndown look good, even if that means stretching the truth. This is why the Sprint Backlog, how the Developers work, and the tools that they use are for their eyes only.

Unlike the Product Backlog and Sprint Backlog, the Taskboard is two-dimensional, much like the Kanban board. The forecasted PBI work items are listed down the left side, with the associated Task work items listed to the right. Across the board, their associated Task work items can be in one of three states: To Do, In Progress, or Done. The Taskboard also summarizes the total amount of Remaining Work for all tasks within each column.

The Taskboard displays PBI work items that are assigned to the current Sprint. You can view Taskboards from previous Sprints as well. These PBIs can be in any State (other than Removed), although they will primarily be in the Forecasted or Done state—if you are following guidance from both me and Microsoft. In other words, you shouldn't see any New or Ready PBIs in the Taskboard. The PBI work items don't have to have any associated tasks to be displayed, either. Actually, the Taskboard is a great place to start adding associated tasks.

If a PBI work item is assigned to another Sprint but has linked tasks in the current Sprint, you will also see them listed here. Unassociated tasks are displayed on the board in an Unparented section. Hopefully you won't have any of these, because Professional Scrum Developers should only be doing work on the forecasted PBIs in order to meet the Sprint Goal. An exception to this guidance would be the inclusion of a process improvement identified in the previous Sprint Retrospective that, though important, doesn't necessarily have to be visualized as Task work items.

> **Smell** It's a smell if there are any PBI work items listed on the Taskboard that are *not* set to the current Sprint. It's also a smell if any of the associated Task work items are from a different Sprint. It could be that this was a one-time missed forecast and carryover of work, but it might be systemic. Professional Scrum Developers don't carry over unfinished PBIs to the next Sprint as a rule but instead improve their practices to meet their forecasts.

By default, the Task work item cards show the work item *ID*, *Title*, *Assigned To*, *State*, *Remaining Work*, and any *Tags*. These default fields can be removed and new fields added in the Taskboard settings. For example, you may want to remove the *State* field from the cards, because you'll be able to see its *State* based on which column it is in.

Here are a few points of guidance for creating and managing Task work item fields:

- **Title**   Provide a short, meaningful title that quickly summarizes what this task represents. Two- or three-word titles work best ("Create tests," "Create pipeline," "Create artifact feed," etc.).

- **Assigned To**   Leave tasks "unassigned" until a Developer drags it to the *In Progress* column. If an unfinished task is dragged back to the *To Do* column, remove the Developer's name.

- **Remaining Work**   Optional, but enter hours in this field. Actionable tasks have a value of 8 hours or less. Remaining work should be reestimated at least once a day. It's okay if this value increases.

- **Tags**   Optional, but tags help visualize metadata about tasks (for example, blocked). You can also style the cards based on tags (such as showing blocked tasks in red).

The Developers can update the Taskboard to reflect the status of the Task work items visually by dragging and dropping. Anybody viewing the board can then see the progress that the Developers are making against each PBI. Developers can focus and collaborate on the remaining pieces of work. The board leads to increased transparency, honesty, and accountability.

A number of activities can be performed in the Taskboard:

- View the plan (tasks) for developing the forecasted work.

- Group tasks by PBI or by Developer.

- Filter the Taskboard by Developer, type of task, State, Tag, or Area.

- Style the cards based on rules (such as showing aging tasks in orange and blocked tasks in red).

- Assess progress in a number of ways.

- Add associated tasks to a PBI item.

- Make edits to a specific PBI or Task work item.

- Drag a Task work item to reparent to a different PBI work item.

- Change a Task work item's State by dragging between columns.

- Change the *Remaining Work* value.

- Select an older Sprint to view its plan.

- Customize the Taskboard by adding a new column.

In the Scrum process, a Task work item can be in the To Do, In Progress, Done, or Removed state. I did not change this in our Professional Scrum process. The natural progression is To Do ⇒ In Progress ⇒ Done. These three states map to the three columns on the Taskboard. A Task work item can be dragged to one of these state columns. When dropped, the task's *State* field will change to that value and be saved automatically. When you drop a task in the *Done* column, the *Remaining Work* field is set to 0 (blank) and the field becomes read only.

No column is mapped to the Removed state. If you want to remove a task from the Sprint Backlog, you will have to open it and manually change its state to Removed. You can also simply delete it.

> **Note** The Taskboard, like the Kanban board, supports Live Updates. This means that the Taskboard now automatically refreshes when changes occur. As other Developers add, change, or move cards on their Taskboard, your Taskboard will automatically reflect these changes. In other words, you don't need to continually press F5 (refresh) to see the latest changes. Live Updates uses SignalR, an open source library developed by Microsoft that allows server code to send asynchronous notifications to client-side web applications.

> **Smell** It's a smell whenever I see someone outside the team—or even the Product Owner or Scrum Master—changing the Sprint Backlog. Only the Developers are allowed to add, delete, or change the contents of the Sprint Backlog. The Taskboard is fun to use, and sometimes Scrum Masters, Product Owners, or stakeholders want to "play" with it. This is fine—as long as they are making changes according to the Developers' wishes and not causing them to lose focus.

> ## Fabrikam Fiber Case Study
>
> The Developers use the Taskboard to some degree in each Sprint. They refer to it throughout the day to see what work remains to be done, to see what their colleagues are working on, and to take on new work themselves. Scott (the Scrum Master) uses the Taskboard to keep an eye on the big picture. He watches the amount of work in progress while looking for bottlenecks. Scott's goal is to coach the Developers to deliver the forecasted PBIs—ideally in the order that Paula wants them.

Viewing Tasks by Developer

The default view in the Taskboard is Task work items grouped by PBI work item. This arrangement provides a simple visualization of the work in progress for each of the forecasted items in the Sprint Backlog. By default, all PBI work items on the left appear with a blue border and Task work items appear with a yellow border. These colors align with the colors of the respective work item icons.

As an alternative, the Taskboard can be configured to see the tasks owned by a particular Developer, or yourself (@me), or those tasks that are unassigned. From the Person drop-down, select the Developer and their card(s) will highlight. You can see an example of this in Figure 6-11.

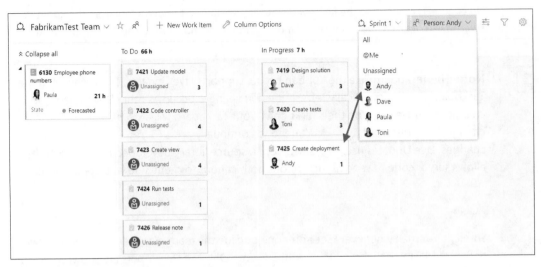

FIGURE 6-11 You can highlight tasks by individual Developer.

A Professional Scrum Team knows the power of self-management. This means, among other things, that they do not assign work ahead of time. The Developers also know that it's important to finish work before starting new work and that multitasking is a myth. Therefore, no more than one *In Progress* Task work item should be owned by any given Developer at any given time. In other words, viewing tasks by developer should not be very helpful for a team practicing Professional Scrum. Even using the *Unassigned* highlighter should be redundant because those tasks should, by definition, be the only ones that are in the *To Do* column. It's probably best to just leave this feature set to *All*. In other words, you shouldn't need to use it. That said, the Developers may find some value in showing all of the Done tasks highlighted by Developer—especially for plans with numerous tasks.

There is another way to view tasks on the board by Developer: the Group by *People* view. You can use this view to pivot the Taskboard to display tasks by Developer, rather than by PBI. This view shows the Developers down the left side of the screen and their associated tasks horizontally to the right, as in Figure 6-12. The first row lists those tasks that are *Unassigned*.

Ordering cards is disabled when your Taskboard is grouped by people. This means that you cannot drag tasks vertically within the same Developer (or unassigned) in this view. You *can* drag a task to another Developer or drag a task horizontally between the states. A nice side effect of dragging a task to a Developer is that the Taskboard will automatically set that Developer as the owner of the Task work item.

My aforementioned guidance applies to this view as well. There may not be a lot of utility here for a Professional Scrum Team since only one Task work item should be owned/in-progress at a time per Developer. That said, the Developers may find some value in showing all of the Done tasks grouped by Developer—especially for plans with numerous tasks.

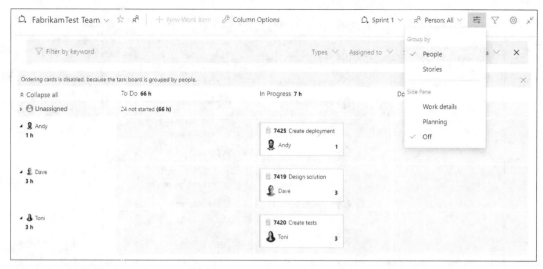

FIGURE 6-12 You can group tasks by individual Developer.

Adding a New Task

As the Developers identify new work or want to decompose larger pieces of work, new Task work items can be created. Ideally, this is done at the beginning of the Sprint, primarily as a result of Sprint Planning. The plan will emerge throughout the Sprint. New tasks will be added. Tasks will be changed. Tasks will be split. Tasks will be deleted. If this is *not* happening throughout the Sprint, then perhaps your work is not that complex. The Developers always know more today than they did yesterday.

Over time, the Developers will improve in their ability to identify and create the plan—in the form of Task work items—earlier in the Sprint. That said, it will probably never be the case that all tasks can be identified in Sprint Planning, nor should it. The important thing to remember about the Taskboard is that it should honestly and accurately reflect the remaining work in the Sprint, to the best of everyone's knowledge, at that moment in time.

There are several places in Azure Boards to create a new Task work item, but if the Developers are using the Taskboard regularly throughout the day then it makes sense to create one from that page. Next to each PBI work item, in the *To Do* column, is a large icon you can click to add a new item, as you can see in Figure 6-13. When you click this icon, a new task card appears, allowing you to add a title. This can be done in rapid succession as the Developers brainstorm a plan in real time. The Task work item's *Area, Iteration,* and *Related Work* link are all defaulted to the associated PBI work item's values.

After the Task work item has been saved, you can quickly update the other visible fields (*Assigned To, Work Remaining,* etc.) on the card. Remember, it's best to leave the Task work item unassigned, unless you or another Developer will begin working on it right away. Clicking the title opens the Task work item form so that you can set other fields, such as *Description*.

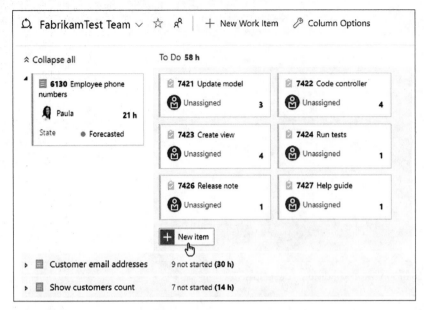

FIGURE 6-13 You can add a new Task work item in the Taskboard.

> **Tip** It's best to add tasks as the result of a conversation with all Developers, as in Sprint Planning. If, during development, this is not possible, it's valuable to have another Developer (or two) review new tasks and changes to the plan. This is especially important if the plan has substantial complexity. Developers should also quickly assist in estimating the effort required to complete the task. These estimates can be adjusted later when more information is known. Changes to the plan should be discussed at the next Daily Scrum.

Taking Ownership of a Task

During the Sprint, the Developers work the plan in order to achieve the Sprint Goal. What this looks like is very context specific and will vary from team to team and product to product. If the Developers are using tasks to represent their plan, then they will complete those tasks in order to achieve the Sprint Goal by developing and delivering the forecasted PBIs.

Professional Scrum Developers may detail out the tasks for the first few PBIs but may also broadly identify tasks—at least at a high level—to be broken down later for the remaining PBIs in the forecast. Teams just getting started with Scrum might find themselves still working out the plan (that is, identifying tasks) well into the Sprint. All teams will do this to some degree. Being able to envision and capture the plan before starting the actual work is a skill that comes with the experience of working together as a team, on the same domain, using the same tools and the same practices.

Task ownership is not a required outcome of Sprint Planning. In fact, it's important to leave To Do tasks unassigned so that Developers who have *availability* can pick a relevant task to work on next.

In Scrum, work should not be directed or assigned. In other words, when creating or updating a task, don't assign it to anyone who doesn't request—or at least wish to own—that piece of work.

You should also resist the urge to assign tasks to the *ideal* Developer for the task ahead of time. Doing so will decrease collaboration and the opportunity for other Developers to learn. When the time is right, the Developers decide who will take on that task. This decision takes many factors into account, including the background, experience, and availability of the candidate Developer. More than likely it *will* be the ideal Developer because that's a recipe for a great product. Professional Scrum Developers know that selecting and taking ownership of a task should not be based on what task they want to do next, or what task they feel they are best suited to do, but what task needs to be done. If possible, they should take ownership of the next most important task, as determined by the team, required to complete the current PBI before moving on to the next one. This is a hard but important discipline.

> **Note** Fellow Professional Scrum Trainer Benjamin Day says preassigning a task is like "licking a cookie and putting it back on the plate"—nobody else will want to touch it after that. If you exit Sprint Planning with your name on a bunch of tasks (or even PBIs), you have essentially licked those cookies. A fresh plan of tasks is like having a plate of fresh cookies. They look delicious, so you grab one and start eating. You finish that cookie and decide that you want another one. But you find out that, in the interim, someone else came through and licked a bunch of them. In that rogue cookie-licker's defense, they're licking just the cookies that they fully and honestly intend to eat. Nevertheless, what are the chances of you eating one of those licked cookies? Probably zero—same as the chances of you working on a task with someone else's name on it.

Remember that the purpose of the Daily Scrum is to identify the plan for the next 24 hours. This means that Developers, at least once per day, synchronize on what tasks each will be working on or pairing up to work on. The Developers can then go and take ownership of those Task work items in question. Taking ownership, as well as changing ownership, can happen at any time during the day, and even multiple times per day.

> **Tip** Having Azure Boards up during the Daily Scrum can distract from the Developers' ability to collaborate and create a plan. Worse, it can turn the Daily Scrum into a "board walk" status meeting. Someone—possibly the Scrum Master—should keep an eye on this, and even suggest turning devices off as an experiment. If conversation, collaboration, and planning improve, the experiment will have proven itself.

Ironically, in Azure Boards, ownership is tracked using the *Assigned To* field. When the time is right, the Developer who will be performing the work should change the entry in the task's *Assigned To* field to their name. They can ask another Developer to do it as well. In the Taskboard, you can take ownership of a task by clicking in the "assigned to" area of the card (by default, this area shows the label *Unassigned*). From the resulting drop-down list, select a user or reset the task back to Unassigned, as you can see in Figure 6-14. Task ownership can also be set by opening and editing the Task work item directly.

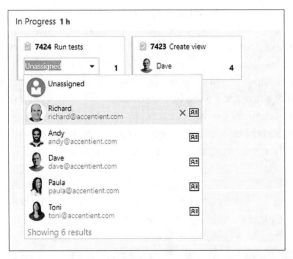

FIGURE 6-14 A Developer can take ownership of a task in the Taskboard.

When selecting a Developer from the drop-down list, please keep these considerations in mind:

- A work item can be owned only by a user who has been added to a project.

- When you're selecting a user from the Taskboard, the drop-down list reflects members of that team only.

- You may have to first search for a specific Developer before being able to select them. Over time, the drop-down list will display those users that you have most recently selected (your Developers).

- A work item can be owned by only one user at a time. If a pair or mob is working on the task, then just pick one of them to be the owner. If it's a large "epic" task, then break it down first. If this one-owner limitation concerns you, I have to say that is a smell of a command-and-control (C&C) mindset or possibly valuing output over outcome. Let's focus on getting done, achieving the Sprint Goal, and delivering value instead of tracking who did what for the sake of management.

- The system shows the display name and adds the username when required to clarify identical display names.

Tip When you take ownership of a task, you should also reestimate the remaining work. The current number might be a team estimate or the estimate from another Developer who previously worked on the task. Now that you know that *you* will be doing the work, you should do another estimate.

Smell It's a smell when I see a Developer regularly having more than one in-progress task at a time. There are situations where this can occur, such as when the original task gets blocked or the work "merges" with the new task. If a task is blocked, it should be put back into the *To Do* column with an appropriate note or tag or put into a custom *Blocked* column. In general, Developers should strive to limit their work in progress—ideally to one—in order to maximize their focus and productivity.

I'm often asked by teams new to Scrum (and Azure Boards) if the Developer should be the one who physically sets ownership of their task, or if someone else—like the Scrum Master—should do it. My answer is to let the team decide. Yes, this might be a helpful practice for a new Scrum Team to adopt in order to shed any of its old command-and-control (C&C)-style behavior. If no such behavior existed, then I say let the Developer closest to the keyboard (or the one with the best keyboarding skills) get the job done. If a question, disagreement, or problem arises, remember that Azure Boards tracks who adds or changes a work item, what exactly was changed, and when.

Fabrikam Fiber Case Study

For the most part, each individual Developer manages their own tasks. Occasionally, during a Daily Scrum or other planning meeting, the Taskboard may be displayed on a large screen, with one of the Developers driving. They will create and manage the tasks at the team's behest. Professional Scrum Teams are good about regularly rotating drivers.

Blocked Tasks

During the Sprint, tasks can become blocked. Dependencies can surface, causing progress on an existing task to be put on hold. This type of issue is normal when dealing with complex work. The question becomes how to signal to others that it has occurred. You shouldn't simply create or pull a new task into the *In Progress* column without doing something with the original, blocked task.

It's also important that the plan be updated in some way to reflect the blockage. After dragging the blocked task back to the *To Do* column and removing the owner's name, you can perform any of these optional activities:

- Reestimate the remaining work.

- Add a note to the *Description* or *Discussion* field.

- Set the *Blocked* field.

- Tag the Task work item as *Blocked*.

If you mark blocked tasks by using the *Blocked* field or with a tag, then those work items can be styled in a way to make them more transparent on the Taskboard. Figure 6-15 shows a Blocked style rule being created to display those cards with a slight red color on the Taskboard.

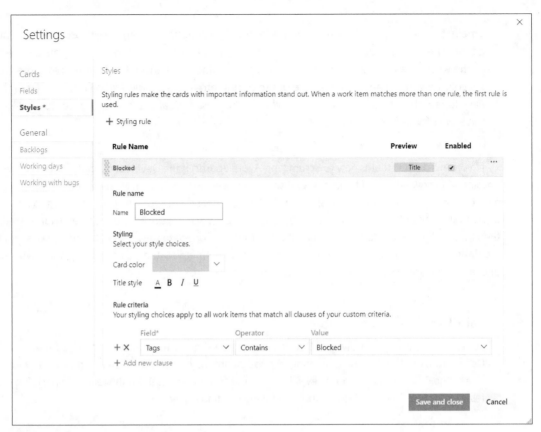

FIGURE 6-15 You can style blocked tasks on the Taskboard.

As you can see, it's very easy to style cards on the Taskboard. There really is no limit to how much bling you can apply to your cards. In addition to styling cards when they are blocked, there are some other styles that a Scrum Team may wish to consider. Styling rules also apply to the PBI work items down the left side. Table 6-1 provides some styling rule ideas, with example criteria.

TABLE 6-1 Styling rules of interest to Scrum Teams.

Rule	Example criteria
High-value PBIs	Business Value > 500
High-effort PBIs	Effort > 5
High-effort tasks	Remaining Work >8
PBI assigned to a specific area	Area Path Under Fabrikam\Mobile
PBI or Task contains specific tag	Tags Contain Blocked
Stale tasks	Changed Date @Today-3
Task assigned to a specific Developer	Assigned To = Richard
Title contains a keyword	Title Contains Foo

Another option for visualizing and managing blocked Task work items is to add a custom column to the Taskboard. You can access this feature through the Column Options link at the top of the Taskboard. I will cover this in the next section.

Customizing the Taskboard

Azure Boards allows Developers to add, rename, and remove columns on the Taskboard. For example, they could add a *Blocked* column to the Taskboard to hold those Task work items that are blocked—as an alternative for (or in addition to) tagging and styling those tasks.

Similar to the Kanban board, each column must map to a category state. For the Task work item type there are three category states: To Do, In Progress, and Done. At least one column must map to the To Do state and one column must map to the Done state. In other words, the *In Progress* column could be removed—which would be weird.

Figure 6-16 shows a *Blocked* column being added in-between the *In Progress* and *Done* columns and mapped to the In Progress category state. This means that blocked Task work items remain in the In Progress state when dragged to this column. I could have put this column in any location and also chosen a different category state, but this location works for most situations. The Developers can decide which mapping and column location makes the most sense for their work.

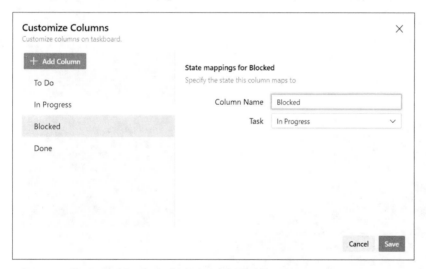

FIGURE 6-16 You can add a *Blocked* column to the Taskboard.

> **Tip** The Developers may want to consider adding more than one *Blocked* column to the Taskboard. This way, they can differentiate between tasks that are blocked from *starting* from tasks that are blocked while *in progress*. Keep in mind that Taskboard column names must be unique.

Inspecting and Updating Progress

At any point in time in a Sprint, the total work remaining in the Sprint Backlog can be summed. By tracking the remaining work throughout the Sprint, the Developers can manage their progress. Professional Scrum Developers know this and regularly inspect their progress toward the Sprint Goal and how progress is trending toward completing the forecasted PBIs overall.

To help with this, Azure Boards provides an integrated, real-time Sprint burndown chart that is embedded in the Taskboard. Unfortunately, the thumbnail is too small to read and be of any use. Once you click the thumbnail, a larger, interactive burndown chart is displayed. You can open it up further in the Analytics page to gain full access to the chart's controls, as you can see in Figure 6-17. The burndown chart visualizes the amount of remaining work within the Sprint. You can use the burndown chart to inspect progress of the work in the Sprint Backlog over the Sprint. This chart helps answer the question: Is our team on track to complete the forecasted PBIs by the end of the Sprint?

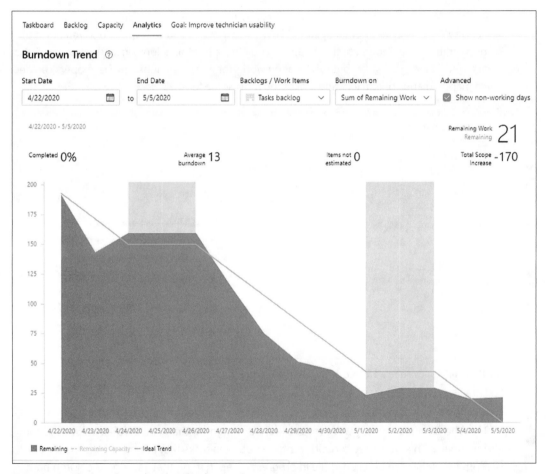

FIGURE 6-17 The burndown chart is a visual way to inspect progress.

The burndown chart provides an easy way to monitor progress during the Sprint by showing work remaining over time. Work remaining is the vertical axis and time is the horizontal axis. Remaining work is calculated as a sum of the *Remaining Work (hours)* field but can also be calculated as a total count of tasks remaining. The burndown can also show a count of PBIs remaining or a sum of Effort/ Size points—as an alternative way to monitor progress. In addition, the burndown chart calculates and displays the average burndown rate and added scope over the course of the Sprint.

Based on historical burndown and scope increase, the burndown chart shows a projected work completion date. Using the burndown chart, the Developers can stay on top of their progress and see the immediate impact of their work against the forecasted work. The burndown chart provides the following useful metrics:

- **Percentage work complete**   The percentage of work completed based on the original scope.

- **Average burndown rate**   Average work completed per interval or iteration.

- **Total scope increase**   How much work was added since the beginning of the Sprint.

- **Number of work items not estimated**   The current number of items that do not have a value burning down on the sum of a field (Remaining Work or Effort/Size).

- **Projected burndown**   Shows how fast the Developers are burning down the work, based on historical burndown rate.

- **Projected scope increase**   Calculated based on historical scope increase rate.

- **Projected completion date**   Calculated based on the Remaining Work and historical burndown and scope increase rates. If the projected completion date is before the Sprint end date, it draws a vertical line on the interval when the work should complete. If the projected completion date is beyond the Sprint end date, it displays the number of additional intervals/iterations needed to complete the work.

Note In Scrum, "projects" are always on time. This is because a project is just another name for a Sprint, and Sprints always end on time—on a specific date. Whether or not all of the expected scope (forecasted PBIs) is delivered at the end of the Sprint is the more important question.

Looking at the burndown chart, the Developers can not only get immediate insight as to their progress, but also learn about their behavior and cadence. Most burndown lines aren't straight because Developers never move at a fixed velocity. Because of this, burndown charts also help Developers visualize the risks to their release. If the Sprint end date maps to a release date, Developers may need to reduce scope or extend the release date. Burndown charts can also indicate that progress is greater than anticipated, providing the wonderful option of collaborating with the Product Owner to add new PBIs to the Sprint.

Although the burndown is an obvious choice for inspecting progress, the Developers don't have to use it. Progress can be assessed by direct observation of the Taskboard. Here are some approaches:

- **Sum work remaining**   The total hours left for all To Do and In Progress tasks can be compared against the Developers' availability for the remainder of the Sprint.

- **Count incomplete tasks**   The total number of To Do and In Progress tasks can be compared against those that are Done as well as the Developers' availability for the remainder of the Sprint. This approach works well when tasks are all generally the same size.

- **Sum PBI effort/size remaining**   The total effort/size points left for all Forecasted PBIs can be compared against those that are Done as well as the Developers' availability for the remainder of the Sprint.

- **Count incomplete PBIs**   The total number of Forecasted PBIs can be compared against those that are Done as well as the Developers' availability for the remainder of the Sprint. This works when PBIs are all generally the same size.

As you may have noticed, each of these approaches is available as a configuration option when viewing the burndown. If it's easier to view the burndown, then go for it. Otherwise, just know that you can inspect progress by looking at the Taskboard itself—which Developers tend to do multiple times per day anyway.

If you want to inspect progress by counting or summing tasks, then it's important for the Developers to regularly update the board—at least once daily. The Developers should make sure the tasks are in the correct column, are owned by the correct Developer, and contain accurate Remaining Work estimates—assuming the Developers are estimating and tracking hours by task.

Developers should inspect progress at least daily—possibly at the Daily Scrum—in order to project the likelihood of achieving the Sprint Goal. Professional Scrum Developers will update their progress several times throughout the day so that progress can be assessed more often than once daily. This means that Developers should update their Remaining Work to regularly—and honestly—reflect the time required to complete the task.

You can update the *Remaining Work* field quickly by clicking the number on the card and selecting a new one from the drop-down list that appears, as shown in Figure 6-18. Any change to the *Remaining Work* field takes place immediately. There is no need to save your changes.

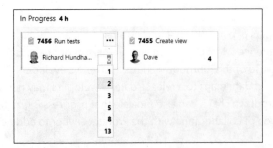

FIGURE 6-18 You can set the *Remaining Work* field directly on the card.

Tip If the number you want is not in the drop-down list, you can just type it into the tiny field. This can be difficult on a new task where there isn't a visible number. You will have to guess where the number would be displayed and click that area on the card. If all else fails, you can open the Task work item and change the *Remaining Work* field on the form.

The drop-down list provides some interesting results. The list population scheme is based on the findings of studies that Microsoft performed internally. The Azure Boards team analyzed the hour deltas as they were reduced over time so that whenever a Developer updated the *Remaining Work* field, they tracked the average change from the original number to the new number. After studying six months of data, they found approximately 80 percent of all deltas were within a small range of the current number. Interestingly, they also found that about 2 percent of the numbers were over 20 hours to begin with. Regardless of these findings, or how the drop-down control behaves, Developers can choose to record whatever number is accurate for them.

Note If the task has no (zero) Remaining Work, then you are presented with a Fibonacci sequence (1, 2, 3, 5, 8, 13). I think Microsoft got confused here and is suggesting standard story points rather than hours. Regardless, there might be a chance one of those numbers will work. Otherwise, you can change it to whatever you want.

Smell It's a smell when I see a Developer tracking their *actual* hours worked. In complex product development, outcome is more important than output. Hopefully the members of the Scrum Team know this. The Scrum Master may have to "remind" management of this truth. Also, Scrum is a *team* effort, and tracking actual hours turns it into a collection of individual efforts. If management is really after the knowledge of whether the team is ahead of schedule or running behind, they could ask the Product Owner and/or inspect the Done working product at the Sprint Review. If the Developers, without any external influence, wish to track actual hours in an effort to improve their learning, that's fine.

Fabrikam Fiber Case Study

Each Developer updates their remaining work estimates at least daily, and sometimes several times throughout the day as new tasks are worked and new assessments are made. Since the Developers tend to swarm on many PBIs, they also keep an eye on the upcoming, unassigned tasks to see if those estimates still feel right. If changes are recommended, they discuss them as needed but at least once a day at the Daily Scrum.

When all associated tasks are done, and the Definition of Done is met, the PBI is considered done. A Developer should set the PBI work item's State to Done at that point. The Product Owner doesn't have to be the one to flip this switch—any Developer can do it. In fact, unless it's explicitly stated in the Definition of Done, the Product Owner has no say over whether a PBI is Done or not. After a PBI is completed, the Developers should begin swarming on the next most important item in the Sprint Backlog.

Smells

You can learn a lot about how a team is working as a team—or not working as a team—by observing the Taskboard. This, of course, assumes that the team is using the Taskboard in the way I've recommended in this chapter and is keeping it regularly updated—which can be challenging. I'd rather a team spend their time making the product awesome than keeping their Taskboard in good shape—but it is a balance. In addition, the *observer effect* can come into play. In physics the observer effect is the theory that the mere observation of a phenomenon (such as team behavior) inevitably changes that phenomenon.

As a culmination of all the guidance in this chapter, take a look at Figure 6-19 and see if you can identify the smells.

FIGURE 6-19 Can you identify the smells on this Taskboard?

Here are the smells that you should have found:

- **Toni has licked two cookies** There are two tasks in the *To Do* column with Toni's name on it. Assuming Toni has the most experience in running tests, it's likely that she will perform those tasks when it comes time; however, she may not be available or there might be more important work for her to help with first. Ownership should be deferred until the last responsible moment.

- **Missing estimates** One To Do task and two In Progress tasks are missing an estimate of remaining work. Assuming the Developers have agreed to estimate/track remaining work, this should be done consistently for all tasks in the *To Do* and *In Progress* columns. This will increase accuracy as the Developers inspect progress.

- **Wade must be a cyborg** Since humans are not able to multitask, how can Wade have two In Progress tasks? Perhaps he had finished one or became blocked and didn't update the Task-board. Perhaps he merged the work. Wade should take a moment and true-up the Taskboard.

- **The mysterious solution designer** The done Design Solution task has no name on it. Although the team can probably figure out who did the work, it's best to leave the Developer's name on it.

- **Why risk it?** The most serious smell—and the one that most new teams miss—is that Toni has started work on the second PBI before the first PBI was done. She would say that she is the "tes-ter," so therefore she does the testing tasks. Remember, in Scrum, there are no titles. Also, it's more important that the team deliver value—by getting PBIs done—than delivering output by just doing a bunch of tasks. Toni needs to be more T-shaped.

> **Tip** If the Developers struggle with the dysfunctional behaviors behind these smells, one or more style rules can be created to highlight those cards. I covered how to do this earlier in this chapter. More importantly, they should discuss this issue in the Sprint Retrospective and come up with some experiments to try.

Closing Out a Sprint

Ideally, when Sprint Review comes, the team has achieved the Sprint Goal, completed all forecasted PBIs according to the Definition of Done, and all tasks are in the *Done* column. This may or may not happen. If it does, then there will be nothing a team has to do to "close out" the Sprint—just make sure all of the States are set to Done. On the other hand, when the Sprint Backlog contains PBIs that are incomplete—whether or not they were started—there will be some extra steps in order to close out the Sprint.

Here's a list of activities that a team should do to close out a Sprint:

- **Move PBIs back to the Product Backlog** Any PBI that was not finished according to the Definition of Done should be moved back to the Product Backlog. This can be done by displaying the Planning pane and dragging those PBIs back to the root iteration (for example, Fabrikam Team Backlog). State values should be changed as well.

- **Reestimate those PBIs** The Developers will have learned more about those items, the domain, the codebase, the tools, and so forth. This new knowledge may result in new estimates. In other words, those incomplete PBIs may decrease (or increase) in size based on what the Developers have learned during the Sprint.

- **Salvage the plan** Some or all of the tasks may still be relevant in a future Sprint. If so, the team can use the bulk-modify *Move to iteration* feature of the Sprint Backlog (not the Task-board) to change the Sprint on those tasks so that they appear in the new Sprint Backlog and Taskboard. You can see an example of this in Figure 6-20. Salvaged tasks should not be moved until Sprint Planning of that new Sprint so that the plan can be created at the last responsible moment.

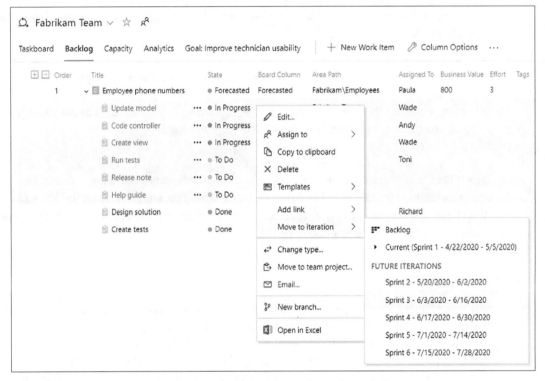

FIGURE 6-20 Use the bulk-modify feature to move incomplete tasks to a future Sprint.

 Smell It's a smell when I see a Scrum Team continually pushing incomplete work to the next Sprint. Although this might reduce waste, it also decreases agility because the Product Owner may want to do something completely different in that next Sprint. Instead, they are "held hostage" by the Developers' need to complete incomplete work from the previous Sprint. If this is a regular occurrence, then it should be a regular topic at the Sprint Retro-spective. In Chapter 8, I will explore some practices and behaviors that a team can adopt to limit the risk of having incomplete work.

Canceling a Sprint

A Sprint can be canceled before its timebox is over. A Sprint would be canceled if the Sprint Goal becomes obsolete. This might occur if the organization changes direction or if business or technology conditions change radically. In general, a Sprint should be canceled if it no longer makes sense given the circumstances. But, due to the short duration of Sprints, cancellation rarely makes sense. Only the Product Owner has the authority to cancel the Sprint, although they may do so under influence from the stakeholders or others on the Scrum Team.

When a Sprint is canceled, any done PBIs should be reviewed. These items may still be released, if the Product Owner desires. As previously mentioned, all incomplete PBIs should be reestimated and returned to the Product Backlog. Depending on the reason for the cancellation, some PBIs may no longer be needed and can be removed from the Product Backlog.

A Sprint cancellation can be traumatic to the Scrum Team because it interrupts their cadence. Cancellations are very uncommon. In fact, in my career I've experienced only two.

Sprint Planning Checklist

Azure Boards has first-class support for tracking the forecasted PBIs and the plan for delivering them. Developers will plan and execute their work in different ways, and Azure Boards provides several methods to assist them. For example, the Sprint Backlog can be viewed and inspected in many ways—all in order for the Developers to manage their work and assess their progress toward the Sprint Goal.

Here is a checklist for Sprint Planning:

Scrum

❒ Prior to Sprint Planning, know your team's Definition of Done, availability/capacity, past performance (velocity, throughput, etc.), and any actionable improvements identified in the previous Sprint Retrospective.

❒ Prior to Sprint Planning, ensure the Product Backlog is refined, in good shape, and contains some ready PBIs at the top.

❒ Prior to Sprint Planning, the Product Owner should have a business objective or goal in mind, which may influence (or simply become) the Sprint Goal.

Azure DevOps

❒ Ensure the top items in the Product Backlog are ready.

❒ Ensure the current Sprint has been created and selected for your team.

❒ Set the Iteration to the current Sprint and the State to Forecasted for those PBIs selected for the Sprint.

❒ Add associated Task work items to each forecasted Product Backlog Item to represent the plan for delivering it (optional).

❏ If the PBI was planned in an earlier Sprint, bring forward any older Task work items that could still be part of the plan. Do this by setting their *Iteration* field to the current Sprint.

❏ If representing the Definition of Done with Task work items, ensure that they are each added to each forecasted PBI as applicable.

❏ Ensure that all Task work items have a team-estimated *Remaining Work* value (optional).

❏ Ensure that all Task work items are unassigned.

❏ Ensure that any "epic" tasks (such as tasks greater than 8 hours) are decomposed into two or more smaller tasks.

❏ Avoid setting the *Activity* field in the Task work item to reinforce the team's cross-functional, self-managing, T-shaped behaviors.

Chapter Retrospective

Here are the key concepts I covered in this chapter:

- **The Sprint** A timeboxed event of one month or less during which a done, usable, and releasable product is developed and delivered. The Sprint is the container for the other Scrum events.

- **Sprint Planning** The first event in the Sprint where the work to be performed in the Sprint is forecasted and planned.

- **Sprint Backlog** An artifact that contains the set of forecasted PBIs plus a plan for delivering them in order to realize the Sprint Goal.

- **Forecast** Those PBIs that the Developers feel that they can deliver during the Sprint given current conditions.

- **Plan** How the Developers will work to deliver the forecasted PBIs. The plan is typically represented using tasks but can also be represented using failing tests or other artifacts.

- **Sprint Goal** An objective for the Sprint that can be met through the delivery of the forecasted PBIs. Since there is no first-class support for Sprint Goals in Azure Boards, a team may want to use an extension.

- **Past performance** A measure of how much work the Developers have been able to historically complete per Sprint. Velocity is a common measure and is the average number of size/effort points done per Sprint. The flow-based metric of Throughput would be another past performance metric to consider.

- **Availability** The Developers' available time for the Sprint. Availability is one of the inputs to Sprint Planning that influences how many PBIs are forecasted. Availability is a better word to use than capacity.

- **Tasks** Ideally, tasks are created early in the Sprint. It's to be expected that additional tasks will be identified and created later in the Sprint as well.

- **Decomposition**   Generally, tasks should be small enough to be accomplished in a day or less. Larger "epic" tasks can be created during Sprint Planning as the Developers continue to brainstorm the plan, but such tasks should be decomposed later in the Sprint.

- **Impediments**   Issues that slow or block a team's ability to achieve the Sprint Goal. Impediments should be removed rather than managed but may require the Scrum Master's involvement. Create an Impediment work item as a last result.

- **Taskboard**   The Taskboard is a great way to visualize the work in progress, as well as the work yet to be done in the Sprint. Progress can be inspected in many ways on the Taskboard.

- **Taking ownership of a task**   Developers take ownership of Task work items by setting the *Assigned To* field to their username. This should only be done when that Developer begins working on the task. Work is never assigned in Scrum.

- **Changing the state of a task**   A Developer can change the state of a task by dragging it to the respective column in the Taskboard. The board should accurately reflect the work the Developers are doing.

- **Updating remaining work**   Each Developer should reestimate the remaining work for their tasks at least once per day. Use the Taskboard to update these values quickly.

- **Burndown**   A chart that shows how quickly the Developers are completing tasks or PBIs. A burndown chart is a great tool for inspecting progress in order to update the plan and expectations.

CHAPTER 7

Planning with Tests

Traditional software development directed us to hand off a code-complete application to testers at the end of a lengthy development cycle. As our craft improved, we sought shorter cycles, but we still handed off code to testers at the end of the development cycle. As we've embraced Scrum, we've removed the handoff by forming a cross-functional team able to perform all the required activities. Handoffs are still done, but at least they are to a person on the team with a shared commitment and focus. We're getting better, but there is still room for improvement.

The physics of software development tells us that testers must have a stable, finished piece of software to click, poke, and test. On the surface, this sounds like an honest appraisal. I mean, why bother wasting time testing software that hasn't been written yet, right? Isn't Scrum about identifying and removing waste? Yes, it is, and there is waste to be removed when testing efforts are *delayed*. This is the underlying motivation behind the "shift left" principle in DevOps.

As mentioned in the previous chapter, one of the outputs of Sprint Planning is the plan. While most Scrum Teams use tasks to formulate the plan, the *Scrum Guide* is actually quiet on the subject. This means that the Developers on those Scrum Teams can experiment with different approaches. Some Developers use the columns of a Kanban board. Others use diagrams on whiteboards. Others just use impromptu conversations as the "plan." Personally, that last one sounds difficult for Developers to inspect progress and estimate remaining work.

Another option is for the Developers to formulate their plan using acceptance tests. These tests can be manual or—preferably—automated. They are written by the Developers and run by the Developers. At any point during the Sprint, the sum of the count of failing acceptance tests could be compared against the sum of the count of passing acceptance tests, either per PBI or for the entire forecast, as a way to inspect progress. The reality is that Developers are going to have tests anyway, so why not *start* the Sprint by creating those acceptance tests and let them drive the development?

Starting work by creating failing tests can feel very counterintuitive and weird, but it works. Test-driven development (TDD) proves this every day. In fact, this approach to planning and driving work by creating failing tests is a part of *acceptance test-driven development* (ATDD). Whereas TDD is from a Developer's perspective, ATDD is from a stakeholder's or a user's perspective.

ATDD is still a relatively unknown practice that shifts testing activities to an earlier time in the Sprint—all the way to the beginning. In fact, ATDD encourages the Developers to discuss collaboratively the acceptance criteria with the right people. These conversations yield practical examples that give way to understanding the features and scenarios that become the basis for the acceptance tests and even for coding design. All of this can be accomplished *prior* to any application coding. A benefit of

ATDD is that it provides the Developers with a shared understanding of what they are developing and what *Done* looks like at each step of the process.

This chapter introduces ATDD and shows how to implement it using Azure Boards and Azure Test Plans.

 Note In this chapter, I am using the customized Professional Scrum process, not the out-of-the-box Scrum process. Please refer to Chapter 3, "Azure Boards," for information on this custom process and how to create it for yourself.

Azure Test Plans

Before I jump into creating and executing a plan based on acceptance tests, I'll spend some time acquainting you with Azure Test Plans and its related artifacts and functionality.

Azure Test Plans is the Azure DevOps service that helps teams and organizations plan, manage, and execute their testing efforts. It supports teams of all sizes and types regardless of whether they are practicing manual, automated, or exploratory testing to drive collaboration and quality. The browser-based experience enables the team to define, execute, and chart the results of their tests within the same integrated experience as the other Azure DevOps services, like Azure Boards.

Many teams think that Azure Test Plans is simply about creating and running tests. This is how Microsoft promotes the product and how it tends to be demoed. Although it can do those things, it's also ideally suited for Scrum Teams to use as a way to represent their Sprint plan.

Each Sprint, the Developers create a new test plan for that Sprint. The name of the plan is simply the name of the Sprint (such as "Sprint 2"). This plan contains all the acceptance tests that prove the forecasted PBIs are done, according to their acceptance criteria. For example, if a Sprint's forecast includes eight PBIs, each having five acceptance criteria, then the test plan will contain 40 acceptance tests, give or take. These acceptance tests can be created during Sprint Planning—even if only test names are provided. The design of those tests may even have started long before Sprint Planning, perhaps during refinement.

Organizing Tests

Azure Test Plans offers many ways to organize your tests. By creating and configuring a Test Plan work item, Developers can then organize their tests for the entire product, for a specific area, and/or for a specific Sprint. I recommend keeping it simple and having one test plan per Sprint. In Figure 7-1, I'm using the Test Plans page to create a new test plan named "Sprint 1" that maps to the Sprint 1 iteration.

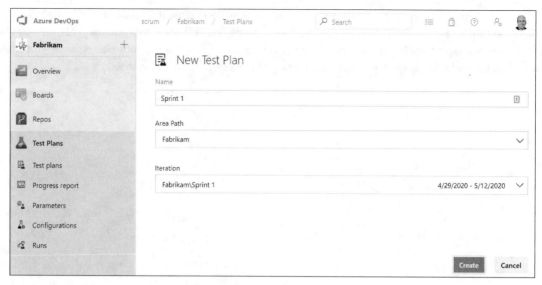

FIGURE 7-1 We're creating a new test plan for Sprint 1.

Developers who create and manage test plans, as well as test suites, require the *Basic + Test Plans* license. This license is included with Visual Studio Enterprise, Visual Studio Test Professional, and MSDN Platforms subscriptions. It can also be purchased separately. If you don't have the ability to create a new test plan, then you probably don't have the right license. Check with your Azure DevOps administrator.

Fabrikam Fiber Case Study

As part of creating the Sprint plan, during Sprint Planning a Developer also creates a respective test plan with the same name as the Sprint.

Once the test plan is created, Developers can create test suites. Test suites are essentially folders within the test plan that hold the acceptance tests. Test suites are optional but recommended. There's nothing stopping a Developer from putting all of the acceptance tests in the "root" of the test plan and using a naming convention to identify the tests in question. With a couple of dozen tests, this gets messy, which is why test suites are a good idea.

You can create three types of test suites:

- **Static**   A simple suite that has only a name.

- **Requirement-based**   A suite that maps to a specific PBI work item and is meant to contain acceptance tests for just that PBI. The work item ID and title form the test suite's name (such as "7476 : Twitter feed"). As test cases are added, they will automatically be linked back to the PBI work item.

- **Query-based**   A dynamic test suite that returns Test Case work items meeting specific criteria.

When defining the test plan, Developers should use Requirement-based suites to contain their acceptance tests. Requirement-based suites can be generated in bulk with a naming convention that relates back to the PBI in the Sprint Backlog. The creation of all Requirement-based suites—for all PBIs in the Sprint Backlog—can be performed easily in one step, as I'm doing in Figure 7-2.

FIGURE 7-2 You can create Requirement-based test suites for each PBI in the Sprint Backlog.

Behind the scenes, test plans and test suites are persisted as work items. Unlike the Test Case work item, the Test Plan and Test Suite work items are special, *hidden* work item types. As I mentioned in Chapter 3, hidden work item types are ones that users can't create manually, nor do they typically want to. For example, it wouldn't make sense to create a standalone test suite without being in the context of a test plan. Instead, Developers will use the dedicated tooling in Azure Test Plans to create a test plan and test suites.

Note For Developers using the Kanban board to track their Sprint work, each PBI card has the ability to add an associated test case. Behind the scenes this will create a default test plan and Requirement-based test suite in which it places the newly created Test Case work item. Creating tests from the Kanban board does *not* require those users to have a Basic + Test Plans license—which means small teams can get started with Azure Test Plans easily. If your team is using the Kanban board to plan and track its Sprint work, then it might make sense to experiment with this feature.

> **Tip** The default sorting order of the test suites is not very interesting. Use drag and drop to reorder the test suites so that they are arranged in a more logical order, such as by Backlog Priority. Unfortunately, you will have to do this manually.

Fabrikam Fiber Case Study

By the end of Sprint Planning, the Developers have created one Requirement-based suite per forecasted PBI. Someone also drags those suites so that they are listed in the same order the PBIs are listed in the Sprint Backlog.

Test Cases

After the test plan and test suites are created, it's time to create the acceptance tests themselves. This can be done in Sprint Planning or at any time during development. In Azure Test Plans, acceptance tests are persisted using Test Case work items. The Test Case work item type allows Developers to further specify acceptance criteria of a PBI in the form of a test, either manual or automated. Manual test cases will have verifiable test steps. Automated test cases are ones that will eventually be associated with an automated test—such as an MSTest, xUnit, or NUnit test—and run in a pipeline in Azure Pipelines.

For manual tests, the Test Case work item may be initially defined at a high level—perhaps containing only a description. Later, more detailed steps may emerge. Once the test cases are executed, the test runs will contain the test outcomes and any test attachments.

Test Case work items, like other work items, can be created on the Boards hub. That said, it makes more sense to create them directly in the Test Plans hub, within their respective test suite. That way, it's easy to organize the test cases logically within the test plan. Figure 7-3 shows a single Test Case work item being created, and as you can see, it's just another work item type.

Rather than just having one test case per PBI (such as "Validate Twitter feed"), the Developers should create Test Case work items that relate to a PBI's acceptance criteria. It could be that a single test case can cover a single criterion. It could also be that multiple test cases might be required to cover a single criterion. In other situations, the opposite may be true and multiple criteria could be covered by a single test case. As a general rule, however, Developers should plan on creating a separate Test Case work item for each acceptance criterion.

Why is this important? Consider my guidance in the previous chapter about creating enough small tasks per PBI so that the Developers can effectively swarm on the work. This applies to test cases as well. For example, if the Developers are able to create three distinct acceptance tests (test cases)—each for a distinct acceptance criterion—they can then develop, test, and deliver that PBI in a more asynchronous manner, thus reducing risk.

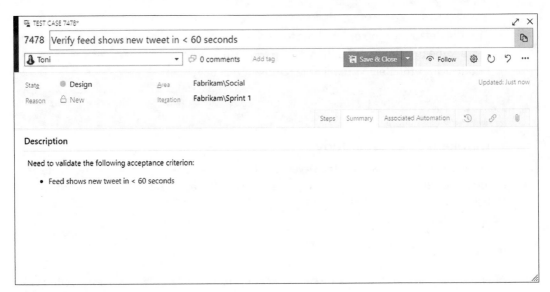

FIGURE 7-3 Test cases in Azure DevOps are just another type of work item.

Smell It's a smell when I see only one Test Case work item per PBI. It's a stench when I don't see any. It could be that the Developers have self-managed around another method to test and verify the acceptability of the forecasted items; if so, that's fine. I might question the transparency of their actions, but at the end of the day Professional Scrum Developers can use whatever tools and practices that they determine bring them value and reduce waste.

It's also a smell when I see a Test Case work item (or Task work item) assigned to the Product Owner. These types of work items exist in the Sprint Backlog and, as such, are owned by the Developers doing the work. If the Product Owner wants to change acceptance criteria or suggest new features, they would need to discuss this with the Developers, and after collaborating on the impact and trade-offs, the Developers change the respective work items. If the Product Owner is also a Developer, then this smell goes away, but another one shows up that I've previously mentioned. Having a Product Owner *also* be a Developer can be problematic in its own right.

Azure Test Plans has a great feature that enables Developers to quickly create one Test Case work item per acceptance criterion. This is the *grid view*, and it allows Developers to quickly add or edit test cases in a two-dimensional grid, similar to Microsoft Excel.

Here are the high-level steps to quickly create test cases from acceptance criteria:

1. Open a PBI work item in the Sprint Backlog.

2. Select and copy (Ctrl+C) the acceptance criteria to the clipboard.

3. Return to the test plan and click the option to add new test cases using the grid.

4. Paste (Ctrl+V) the copied acceptance criteria into the Title column, as I'm doing in Figure 7-4.

5. Clean up the titles and save the test cases.

6. Repeat for the other PBI work items in the Sprint Backlog.

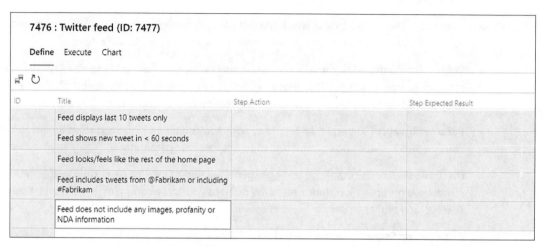

7476 : Twitter feed (ID: 7477)

Define   Execute   Chart

ID	Title	Step Action	Step Expected Result
	Feed displays last 10 tweets only		
	Feed shows new tweet in < 60 seconds		
	Feed looks/feels like the rest of the home page		
	Feed includes tweets from @Fabrikam or including #Fabrikam		
	Feed does not include any images, profanity or NDA information		

FIGURE 7-4 You can quickly create test cases by pasting in a PBI's acceptance criteria.

This copy/paste approach works best when a PBI's acceptance criteria are enumerated as a list, whether it's bulleted, numbered, or simply delimited by line breaks. Unstructured acceptance criteria, such as a paragraph with rambling sentences, won't paste into the grid view correctly and you'll end up with a single test case.

> **Tip** Knowing ahead of time that your team will be creating Test Case work items from acceptance criteria will start to change the way that PBIs and acceptance criteria are specified. This may be an evolution where your team progresses from crafting wordy paragraphs to simple bullets, and finally to given-when-then expressions with sample data. How a Scrum Team captures PBI details, such as acceptance criteria, is perfect for discussion at a Sprint Retrospective. New approaches and experiments can be considered and planned.

As you create or edit Test Case work items, consider the following Professional Scrum guidance while entering data into the pertinent fields:

- **Title (required)**   Enter a short phrase that describes the criteria to test. A naming convention that you could consider using is *Verify [criteria]*. You may want to consider a naming convention where you prefix the PBI's ID and/or short title to further identify it.

- **Assigned To** Select the Developer who is responsible for defining the test and ensuring that it is run. Just as with a task, leave it blank until someone starts working on it.

- **State** Select the state of the test case. States are covered later in this section.

- **Area** Select the best area for this test case. Typically, the area will be the same as the associated PBI.

- **Iteration** This is the Sprint in which the test case will be defined and run. This should be the current Sprint and the same as the Test Plan and the associated PBI work item.

- **Steps** For manual tests, these are the individual test step actions and expected results. Each step can include an attached file that provides more details, such as a screenshot. You can also use a Shared Steps work item to simplify the creation and management of test cases.

- **Parameter values** For manual tests, any parameters defined in the test steps are listed here. You can then provide one or more sets of values for these parameters.

- **Discussion** Add or curate rich text comments relating to the test case. You can mention someone, a group, a work item, or a pull request as you add a comment. Professional Scrum Teams prefer higher-fidelity, in-person communication instead.

- **Automation Status** For automated tests, change this to *Planned*. Later, when you associate an automated test to the Test Case work item, this field will automatically change to *Automated* and the details will appear on the *Associated Automation* tab. This field should remain in the *Not Automated* state for manual tests. I will cover associating automation with test cases later in this chapter.

- **Description (Summary tab)** Provide as much detail as necessary so that another Developer can understand the purpose of the test case.

- **History** Every time a Developer updates the work item, Azure Boards tracks who made the change and the fields that were changed. This tab displays a history of all those changes. The contents are read-only.

- **Links** Add a link to one or more work items or resources (build artifacts, code branches, commits, pull requests, tags, GitHub commits, GitHub issues, GitHub pull requests, test artifacts, wiki pages, hyperlinks, documents, and version-controlled items). You should have one—possibly more—*Tests* links to a PBI work item. If you're using Requirement-based suites, this is done automatically for you.

- **Attachments** Attach one or more files to provide additional details about the test case. Some Developers like to attach notes, whiteboard photos, or even audio/video recordings of the Product Backlog refinement sessions and Sprint Planning.

A Test Case work item can be in one of three states: Design, Ready, or Closed. The typical workflow progression would be Design ⇒ Ready ⇒ Closed. While a Test Case work item is being created, it is in the Design state. After the test case details have emerged—associated automation or manual test steps—the test case is ready to be run, and its state should be changed to Ready. When a test case is no longer required, its state should be changed to Closed. Test Case work items do not have a Removed state, like the other work item types in Azure Boards. Deleting the test case is always an option too.

Note Although test artifacts like test plans, test suites, and test cases are types of work items, the method for deleting them differs from deleting non-test work items. Deleting a test artifact removes it from the test case management (TCM) data store and also deletes the underlying work item. A job runs to delete any child items from both the TCM data store and the work item store. This can include all child items such as child test suites, test points across all configurations, testers, test runs, and other associated history. Azure Test Plans prompts you before deleting child artifacts, as you can see in Figure 7-5.

The final result is the same—all information in both stores is deleted and cannot be restored. Microsoft only supports the permanent deletion of test artifacts. In other words, deleted test artifacts won't appear in the Recycle bin and cannot be restored. You also can't bulk-delete test artifacts. If test artifacts are part of a bulk selection to be deleted, all other work items except the test artifact(s) will be deleted.

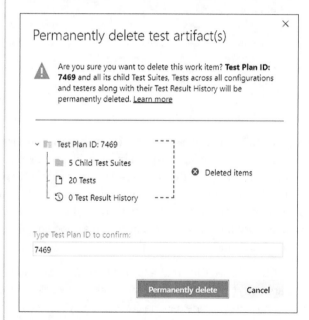

FIGURE 7-5 Confirmation is required before deleting a test plan and related child artifacts.

Inspecting Progress

Observing acceptance tests is a great way to assess progress. Assuming that all forecasted PBIs are expressed as failing acceptance tests, then the more passing tests a team has, the more progress they have made. At the beginning of the Sprint, right after Sprint Planning, there should be zero passing tests. By the end of the Sprint—hopefully—all tests will be passing. At any point along the way, the number of passing tests divided by the total number of tests will roughly equate to progress. This assumes that the work to complete each acceptance criterion, and thus pass each acceptance test, is roughly the same size. It won't be, but it will be close enough to measure progress.

Unfortunately, Azure Test Plans does not provide a first-class way to inspect progress across the whole test plan. In other words, there's no dashboard that shows the outcome of all test cases across all Requirement-based test suites. There is a decent visualization for a single Requirement-based suite, showing test progress for an individual PBI, as you can see in Figure 7-6. Unfortunately, you will have to click through each suite to get an overall assessment of progress.

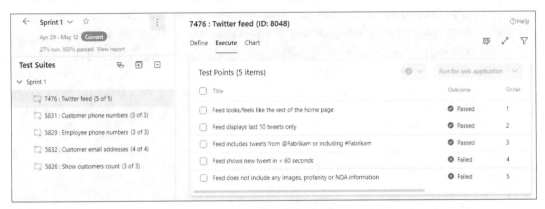

FIGURE 7-6 It's easy to inspect the progress of a single PBI in Test Plans.

There are a few ways to measure progress across all test cases:

- **Show test points from child suites**   This setting allows you to view all the test points for the given suite and its children in one view without having to navigate to individual suites one at a time. This option is only visible when you are on the *Execute* page. You can see an example of this in Figure 7-7. You will need to adopt a test case naming convention to be able to easily identify which test case is related to which Requirement-based suite (PBI). Also, you can't set the order to follow the original Backlog Priority, but Microsoft is considering adding an option that would let you follow the order of the Requirement-based suites.

- **Query-based suite**   By creating a Query-based suite that returns all Test Case work items for the current Sprint, the Developers are also able to see all acceptance tests across all forecasted PBIs. You can see an example in Figure 7-8. A test case naming convention would need to be adopted to identify which test case is related to which PBI. Unfortunately, the order cannot be set to follow the original Backlog Priority. Having a query drive the list of test cases does provide more control, however.

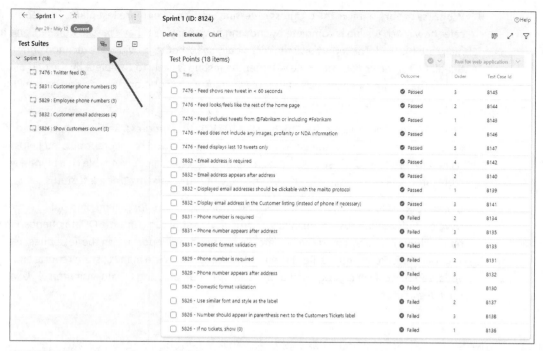

FIGURE 7-7 View all test points across all test suites in the test plan.

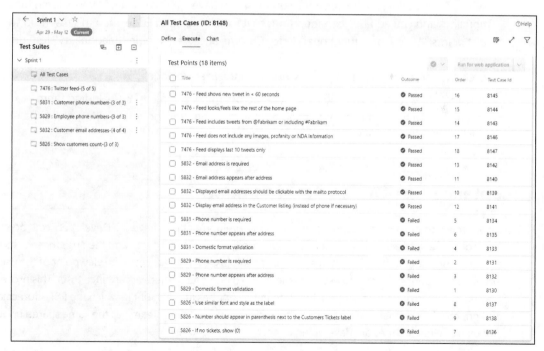

FIGURE 7-8 Use a Query-based suite to list all test cases for the Sprint.

- **Progress report**   Tracks the progress of testing within one or more test plans. The report indicates how much testing is complete by showing how many tests have passed or failed, and how many are blocked. The report also renders a daily snapshot to provide an execution and status trendline. This helps you anticipate whether the testing is likely to complete by the end of the Sprint. The Progress Report is also filterable in many ways. For more information on the Progress report, visit *https://aka.ms/track-test-status*.

- **Charts**   Create and use test results charts to track testing progress. Select the test plan and then view test execution progress. You can choose from a fixed set of prepopulated fields related to results. You can select pie, bar, column, stacked bar, or pivot table chart types. For more information, including specific examples, visit *https://aka.ms/track-test-status*.

- **OData feed**   OData queries are the recommended approach for pulling data out of Azure DevOps and creating custom analytics. Microsoft Power BI can consume OData queries, which return filtered or aggregated sets of data. By querying and reporting on the TestSuites, Tests, TestPoints, TestRuns, and TestResults entities, a team can create many custom reports and visualizations to inspect progress. For more information on using OData, visit *https://aka.ms/extend-analytics*.

> **Note**   Test cases by themselves are not executable. When you add a test case to a test suite, one or more test *points* are generated. A test point is a unique combination of test case, test suite, configuration, and tester. For example, if you have a test case titled "Verify login functionality" and you add two configurations to it for Chrome and Edge, then this results in two test points: "Verify login functionality" for Chrome and "Verify login functionality" for Edge. When tests are executed, test results are generated and visible per test point. The latest execution outcome for a test point is visible on the *Execute* page.
>
> Think of test cases as reusable entities—across suites and plans. By including them in a test plan or suite, test points are generated. By executing test points, you determine the quality and progress of the product being developed.

Acceptance Test-Driven Development

An approach to practicing acceptance test-driven development is to have the Developers collaboratively discuss a PBI's acceptance criteria, compose failing acceptance tests, and then use those tests as a guide to developing the PBI. When examples are discussed, they are from the viewpoint of the user. These conversations and examples are further refined into one or more acceptance tests. This process may have started during refinement, but it definitely finishes in the Sprint in which the PBI is forecasted, and prior to development. ATDD helps ensure that the whole Scrum Team has the same shared understanding of what it is they are developing and what Done looks like.

During the Sprint, the Developers iterate through each PBI, developing and testing and developing and testing until that PBI is done. Depending on the complexity of the PBI and the number of acceptance criteria, the Developers will probably need to create multiple acceptance tests. If the Developers include additional *sad path* and *bad path* tests, then a moderately complex PBI could contain a dozen or more acceptance tests.

> **Smell** It's a smell if the Developers do not have any sad path or bad path acceptance tests. These tests, sometimes collectively called *unhappy path* tests, are ones that pass invalid input in an attempt to root out problems caused by untrained, inattentive, or malicious users. This issue is something the Scrum Team should discuss at a Sprint Retrospective in order to ratchet up their acceptance-testing practices as well as product quality.

Acceptance tests should be created before coding begins. As I've previously mentioned, they can simply be Test Case work items with simple names, used as placeholders for future automated tests. When it comes time to code the automated test, you can achieve this by pairing a Developer who has strong coding skills with another Developer who has a background in testing—in my experience, those two Developers can then work together to craft high-value automated acceptance tests. Remember, all of these will be failing tests until that facet of the PBI is properly coded. For example, if a forecasted PBI has six acceptance criteria and the Definition of Done includes creating both happy and unhappy path tests, then least 12 failing acceptance tests should exist before any coding on the PBI begins.

As development progresses, more and more acceptance tests will start passing. When the last test passes and the Definition of Done is met, the Developers are finished with that PBI. The work can be inspected at the Sprint Review, and more importantly, the release of the Increment will include this new PBI. Remember, it's the Product Owner's decision if and when to release the Increment.

I'm sometimes asked how ATDD is different from behavior-driven development (BDD), specification by example (SBE), test-driven requirements, example-driven development, executable requirements, functional test-driven development, story test-driven development, or flavor-of-the-month-driven development. I tell people that each of these practices, regardless of their nuance differences, have the same goal: to enable better stakeholder collaboration and express abstract business needs in a more understandable and testable format.

ATDD can also provide added value for distributed teams. Those Developers collocated with the Product Owner and stakeholders can have the critical, high-fidelity conversations in order to refine the acceptance criteria into acceptance tests. These tests, written in a natural language, provide the dislocated Developers with clarity so that they can focus on passing the tests. Compared to traditional requirements, these executable specifications provide substantially more value and reduced waste. Furthermore, having a simple yet concrete goal of "make our tests pass" helps Developers who struggle to self-manage find focus in their day.

The approach I have outlined so far is just a partial explanation of an ATDD implementation. If the Developers want a *true* executable specification—one where the framework actually passes the specification data to the test runner to execute—they should implement a proper acceptance-testing framework. With these frameworks, if the specification is changed then it will automatically affect the test.

In the approach using Test Case work items that I have outlined in this chapter, the link is only a *logical* one. If the specification is changed, the underlying test will not be changed automatically.

Tip It can be difficult to find a Product Owner, domain expert, or other stakeholder who is interested in learning and using an acceptance-testing framework. Unfortunately, I've found that it's more common that the Product Owner or stakeholders just tell the Developers what they want and leave the decision of selecting the testing framework up to the Developers. If this were a straight "how" decision, I would agree, but acceptance testing is also about understanding the criteria, features, scenarios, samples, and so forth. These are all "what" items that must involve the Product Owner and possibly the stakeholders. As with any tool or practice, the Developers should experiment and try out a framework for a few Sprints and then embrace, enhance, or abandon it after discussing its value during Sprint Retrospective.

Fabrikam Fiber Case Study

The Developers have just started practicing ATDD. Shifting the testing activities left, prior to coding, was difficult at first, but they quickly realized the benefit of not having to delay testing or refactoring their tests. Some tests are still manual, but the Developers continue to improve their technical excellence and will be writing more and more automated acceptance tests in the future.

The Developers are also evaluating SpecFlow, a popular .NET acceptance testing framework. Compared to the other frameworks, it is far easier to integrate testing with the entire team. Spec-Flow stories are written in plain language, which is appreciated by Paula and the domain experts. The fact that SpecFlow integrates with Visual Studio is also a plus. Visit *https://specflow.org* for more information.

Test-Driven Development

Within ATDD, the Developers can employ any development practices that they choose to. Any practice should strive to minimize waste while allowing the Developers to develop something fit for the desired purpose. Beyond those basic rules, Developers should encourage one another to try new approaches to designing, coding, and testing. The usefulness of these experiments can be discussed at the Sprint Retrospective and then related practices can be embraced, enhanced, or abandoned in future Sprints.

The most popular ATDD "inner loop" practice is test-driven development (TDD). TDD suggests coding in short, repeatable cycles where the Developers (or pair, or mob) first write a failing unit test. The failing test specifies a small unit of desired functionality in the PBI. Next, the test is made to pass by adding the minimum amount of code required. Finally, the code is refactored to patterns or to meet any standards, such as the Definition of Done. The cycle repeats for the next unit of functionality—by starting with another failing unit test.

Tip ATDD can sometimes be confused with TDD. It's akin to confusing ADHD with ADD, but I digress. One way to keep the two sorted in your mind is that unit tests (TDD) ensure that the Developers build the thing right, whereas acceptance tests (ATDD) ensure that the Developers build the right thing.

One of the tenets of TDD is that you do not write a single line of application code until you have written a test that fails in the absence of that code. Advocates of TDD explain that this practice will force requirements to be made clearer and misunderstandings and mistakes to be caught earlier. Developers will also gravitate toward architectures and design patterns that are more testable and easier to refactor. Another nice side effect of adopting TDD is that the Developers will typically end up with more code coverage than before!

The strongest argument in favor of TDD is that it uses tests as technical product requirements. Because the Developer must write a test before writing the code under the test, they are forced to understand the requirements and filter out any ambiguity in order to define the test. This process, in turn, directs Developers to think in small increments and in terms of reuse. As a result, unnecessary code is identified and removed as a clear design emerges.

TDD enables continual refactoring in order to keep the code clean and also to keep technical debt at bay. Having high-quality, fast, repeatable unit tests also provides a safety net—much like having high-performance brakes on a car. Both enable the operator to go fast and take risks. For example, assume Developers have high-quality unit tests that cover a high percentage of their code. When refactoring or experimenting, Developers can immediately see failing test results caused by any side effects. A nice safety net like this provides confidence, as well as the ability to code faster, by reducing the number of side effects and bugs that can be introduced accidentally.

Fabrikam Fiber Case Study

The Developers know TDD, understand its value, and are comfortable practicing it. They have decided as a team that they don't see value in using it for all coding. If a specific scenario involves a lot of design work or involves working on a critical or highly complex area of the application, the Developers will pair up and use TDD to design their way through it.

Automated Acceptance Testing

Professional Scrum Developers agree that automated testing is awesome and is a must-have for software development. Even one of the Agile Manifesto's principles demands "continuous attention to technical excellence..." This applies to automated *acceptance* testing as well. Short of the Product Owner or a stakeholder manually inspecting the work—assuming that is in the Definition of Done—almost every scenario that requires human verification could be covered through an automated acceptance test. It may not be easy, but by adopting an automated acceptance-testing practice, the

Developers will be able to use these tests throughout the Sprint for ATDD, as well as later, for regression testing. More on that in a bit.

> **Note** Some Developers feel that, though possible, there would be a diminishing return on investment for automating all acceptance tests. An example would be a situation where the Developers want to automate the acceptance of user interface (UI) controls being lined up, font types and sizes being consistent, and so on. Manual or exploratory testing would be better suited for this. My guidance is that if you don't have an automated test and have instead opted for a manual acceptance test, it had better be for a good reason—and not just because it's "easier." Let my guidance soak in and then discuss it in an upcoming Sprint Retrospective.

As I've already mentioned, a Test Case is just another work item type. Test cases can be very lightweight, such as having only a title and a description. These work items would merely serve as extra points of documentation. Some Test Case work items might morph to become manual tests, including the actual test steps and expectations. Other Test Case work items can be associated with an automated test, such as a unit test. These are the ones that ATDD practitioners should be using.

Azure Test Plans does not support an end-to-end, automated ATDD solution out of the box. It does, however, provide the foundation for building one yourself. By using Test Case work items, automated tests, and Azure Pipelines, Developers can practice ATDD by using automated acceptance tests.

Next, I'll show you how that's done. You'll follow these high-level steps:

1. Create a Test Case work item in the current Sprint's test plan. Make sure it's associated with the correct test suite (and thus the correct PBI). You can set the *Automation Status* field to *Planned* to help identify these tests.

2. Use Visual Studio to create an automated acceptance test and associate it to the Test Case work item.

3. Check in/push the test project code into Azure Repos.

4. Create a build pipeline in Azure Pipelines to generate a build that contains the test binaries that support the acceptance test.

5. Create a release pipeline in Azure Pipelines to run the automated test.

6. Configure test plan settings and select the respective build and release pipelines.

7. Create a build.

8. Run the test case.

After creating the Test Case work item, you will then associate it to the automated test. This is done in Visual Studio and assumes, of course, that you have an automated acceptance test. Supported testing frameworks include MSTest, xUnit, and NUnit. Other test types that use these frameworks, such as Selenium and SpecFlow, should also work. For more information, read the FAQ at *https://aka.ms/test-case-automation-faq*.

With the Visual Studio test project open and a connection to the Azure DevOps project established, you can associate the automated test to the Test Case work item. This is performed in the Test Explorer window, as you can see in Figure 7-9. You will need to know the identifier (work item ID) of the test case.

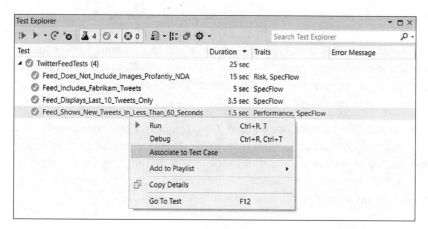

FIGURE 7-9 Use Test Explorer to associate an automated test to a Test Case work item.

After associating the automated test, you will see that the Test Case work item's Automation Status field entry changes to *Automated*. At this point, the automated test name, storage (assembly filename), and test type become visible on the Associated Automation page of the Test Case work item as well. Clicking the Clear button on that page will remove the associated automation, should you need to reset things. Only one automated test may be associated with each test case. For more information, visit *https://aka.ms/test-case-automation*.

> **Tip** When it comes to a naming convention for your automated tests, my recommendation is to have one! There are many ways to name your test projects, assemblies, namespaces, test classes, and test methods. Some Developers like to follow a strict BDD format, whereas others are fine just using clear names that describe the context and expected behaviors. For help getting started, visit *www.stackoverflow.com* to view the latest conversations on the subject.

Next, ensure that the test project exists in Azure Repos and the latest changes, including the passing test code, have been pushed. A pipeline must also be created that builds the test project containing the acceptance test. This pipeline does not need to run the automated accepted tests—it only has to generate the test binaries. More than likely, the build pipeline is not the right environment to run an acceptance test anyway. Also, it's probably better if acceptance tests weren't run as part of a build.

For continuous integration (CI) builds to be effective, they should provide *fast* feedback to the Developers—who tend to be impatient. Having a CI build run lengthy integration and acceptance tests is counterproductive to the goal of fast feedback. Refer back to Chapter 2, "Azure DevOps," for overviews of Azure Repos and Azure Pipelines.

As a corollary, in the context of continuous delivery, continuous integration must mandate running and passing *all* integration and acceptance tests in production-like environments. I distinguish these two types of CI builds by referring to them as CI (unit tests only) and CI+ (all automated testing).

You'll also need to create a release pipeline, which will support the actual running of the acceptance test. This pipeline will use the build pipeline's artifacts as its source and will need at least one stage to represent the testing environment where the acceptance test is run. That computer will need the correct version of Visual Studio installed in order to run the automated test. As an alternative to installing Visual Studio, the pipeline can leverage the *Visual Studio Test Platform Installer* task to install the prerequisite software as part of the workflow.

The release pipeline definition must include a *Visual Studio Test* task, configured to use the *Test run* to select the tests. This setting instructs Azure Test Plans to pass the list of tests selected for execution to Azure Pipelines. The Visual Studio Test task will look up the test run identifier, extract the test execution information such as the container and test method names, run the tests, update the test run results, and set the test points associated with the test results in the test run.

> **Tip** The default pipeline names can be less than clear. Make sure to use simple, meaningful names for the pipeline and pipeline artifacts, such as the stages. This will help other Developers quickly identify the build, release, and deployment environment.

If the acceptance test is a UI test, such as a Selenium test that runs on a physical browser, you must ensure that the agent is set to run as an *interactive* process with auto-logon enabled. Setting up an agent to run interactively must be done before deploying the release to a stage. If you are running UI tests on a headless browser, the interactive process configuration is not required. For more information on configuring interactive agents, visit *https://aka.ms/pipeline-agents*.

With the pipelines created and a build generated, you'll need to return to the Test Plans hub and configure the Sprint's test plan settings. Select the build pipeline that generates the build that contains the test binaries along with a specific build number to test. You can also leave it set to *<latest build>* to let the system automatically use the most recent build when tests are run. Also, select the release pipeline and stage within which to run the test. You can see my selections in Figure 7-10.

To run the automated acceptance test, select the test suite that contains the automated test case and go to the *Execute* page (as opposed to the Design page). Choose the test(s) you want to run and click one of the Run options. Selecting *Run With Options* provides the most control, allowing you to override the defaults and select a different build pipeline, release pipeline, or release stage. You can see the notification that is displayed after running an automated test case in Figure 7-11.

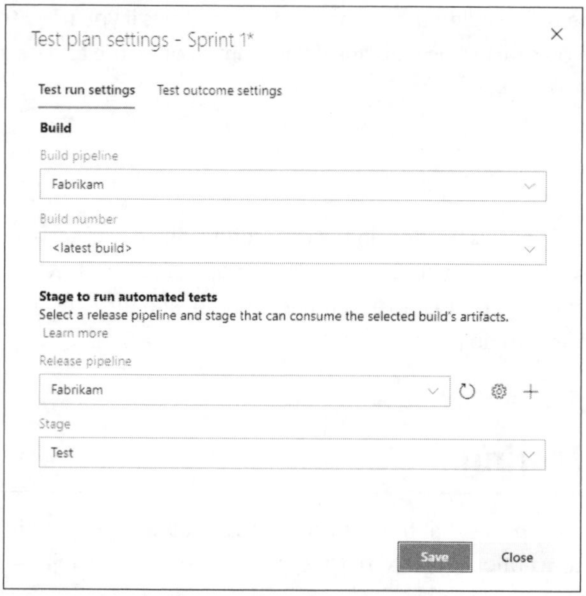

FIGURE 7-10 Configure a test plan to run automated tests in a pipeline.

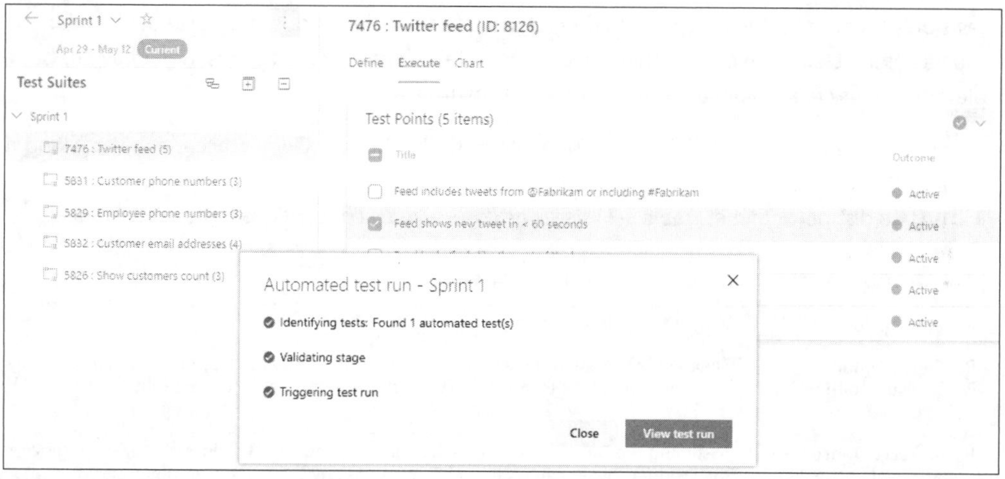

FIGURE 7-11 Run an automated test case in Azure Test Plans.

Assuming the pipelines were configured correctly and the test binaries were built and deployed to that stage, the system will create a release for the selected release pipeline, create a test run, and then trigger the deployment of that release to the selected stage. The Visual Studio Test task will execute and, when it has completed, provide Developers with a pass or fail outcome for that acceptance test.

After triggering the test run, you can go to the *Runs* page to view the test progress and analyze any failed tests. Test results have the relevant information for debugging failed tests such as the error message, stack trace, console logs, and any attachments. You will notice that the test run's title contains the release name (for example, TestRun_Fabrikam_Release-42). The summary includes a link to the release that was created to run the tests, which helps in finding the release that ran the tests if you need to come back later and analyze the results. You can also use this link if you want to open the release and view the release logs.

Fabrikam Fiber Case Study

Paula wants to move to a continuous delivery (CD) model in the near future. She wants each PBI to be released to the production servers as they meet the Definition of Done. The Developers know that this is only possible through automated acceptance testing, and they are investing in tooling and training to be able to do just that.

Acceptance != Acceptance Testing

As I visit with software development teams, I realize that there is confusion about the concept of *acceptance*. For example, a common misconception I hear is that acceptance is performed by the users (known as user acceptance testing). In Scrum, this is never true—only the Developers do the work, and this means *all* work, including testing. Another common misconception is that having passing acceptance tests is equivalent to the PBI being Done. This is not necessarily true. Having passing acceptance tests only proves that the acceptance criteria have been satisfied. It does not necessarily mean that all aspects of the Definition of Done have been completely satisfied. Other items might have to be completed, such as creating documentation or a release note.

If one of the items in a Definition of Done relates to the Product Owner "accepting" or "liking" or "loving" the work created by the Developers, then acceptance testing and Product Owner acceptance will be two distinct activities. Table 7-1 lists some common misconceptions about acceptance.

TABLE 7-1 Common misconceptions about acceptance.

Misconception	Why it's a misconception
Passing acceptance tests is equivalent to the PBI being Done.	The Developers are done when the Definition of Done has been met, which hopefully includes acceptance testing and Product Owner acceptance or delight.
Passing acceptance tests is equivalent to the PBI being accepted.	Assuming the Definition of Done includes some form of Product Owner acceptance, only the Product Owner can accept a PBI, whereas any Developer can run or pass acceptance tests.
Acceptance tests must be run by the stakeholders (users).	In Scrum, only the Developers do the work, which includes testing. If the stakeholders want an opportunity to provide feedback, they should be given that chance, especially at the Sprint Review. They don't, however, get access to the "red button" indicating that a particular PBI is not Done.
Acceptance tests must be manual tests.	Almost all scenarios can be verified using automated tests. For the sake of regression testing, having fast, high-quality, automated acceptance tests is highly recommended. Continuous delivery demands this.
Acceptance can occur only at Sprint Review.	Assuming the Definition of Done includes some form of Product Owner acceptance, this can occur at any time during the Sprint and should take place at the earliest opportunity. In fact, a PBI can even be released to production at any time during the Sprint. Sprint Reviews are about stakeholder *feedback*. In fact, it's a smell if a Product Owner is inspecting PBIs for the first time during a Sprint Review.

I believe that a Professional Scrum Team is one where the Product Owner is regularly involved with the Developers. I like seeing a Product Owner have regular engagement with the Developers and inspecting the PBIs and Increment throughout the Sprint. Unfortunately, I still meet teams where the Product Owner is just another stakeholder at Sprint Review, seeing the functionality for the first time along with everyone else. This pains me.

Even having a strong Definition of Done where the Product Owner must "accept" an item is no guarantee that they will. Putting it off until Sprint Review is even riskier. Professional Scrum Developers know this and will pursue a more collaborative style of working to ensure that the Product Owner provides their feedback earlier in the Sprint. This is especially important if the Definition of Done uses verbiage like "Product Owner likes..." or "Product Owner is delighted by..." This kind of subjective testing is difficult to capture in an executable specification and impossible to automate. This means that Product Owner acceptance, in this form, will always be a carbon-based test, meaning that the Product Owner themselves will need to put eyes and fingers on it.

> **Tip** Product Owners are just people, and people have a hard time specifying their wants and desires. This is especially true with something as abstract (and invisible) as software. This means that you can only count on people telling you what they don't like *after* inspecting it. Scrum embraces this fact, and so should you. For example, if you and your colleagues are working on a PBI and have just finished designing the user interface, have the Product Owner—and even some stakeholders—look at it and give a nod before any additional work is spent wiring it up. What's the harm? Be open and have some courage!

> ### Fabrikam Fiber Case Study
>
> Although Paula is a busy Product Owner, the Developers are fortunate that she works in the same building and makes herself regularly available for questions and feedback. In order to help secure her availability, their Definition of Done includes an item about Paula liking their work. As the Developers make progress on a PBI, they make sure Paula likes what they are doing. The same is true when the Developers are brainstorming complex plumbing or UX designs. Paula wants the product to be the best for her users, and she knows that her regular involvement will produce better results. When it comes time to "accept" the work, the odds are Paula has already accepted it, just not in so many words. It's because of this collaborative ethic and mindset that the Developers believe that a continuous delivery model is within their reach.

Reusing Tests

As I've already discussed, when the Developers set up their testing for a Sprint, they need to create a test plan and then add the appropriate test suites and test cases. Generally speaking, each forecasted PBI should have as many test cases as they do acceptance criteria, multiplied by the types of path testing and configurations that they plan on performing. By design, a single Test Case work item can be associated with multiple PBIs. For example, you might create a generic test case that verifies that a page request returns a response in 5 seconds or less. Since that is such a common acceptance criterion, you might want to reuse this test case for other PBIs in this and later Sprints. You can easily do so by simply adding an existing Test Case work item to a test suite.

When reusing a test case like this, be aware that if you tweak the test case to better support the current Sprint's PBI (such as making it 3 seconds instead of 5), those changes will affect all instances of that test case. This is the nature of having test plans simply *reference* a test case. Microsoft also recognized the need for a true copy and has included the ability to copy test cases and even entire test plans.

When copying a test case, you can specify the destination project, destination test plan (such as the current Sprint), and destination test suite in which to copy the test case(s). An option to include existing links and attachments is also available. You can also copy an entire test plan—test suites, test cases, and all. When doing so, you can choose to reference the existing test cases or make deep copies of them. Figure 7-12 shows the options available for copying a test plan.

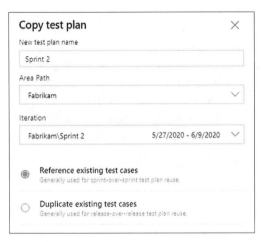

FIGURE 7-12 Azure Test Plans provides the ability to copy an entire test plan.

 Smell It's a smell when I see Developers repeatedly using the Copy Test Plan feature. Perhaps their testing efforts are very similar from Sprint to Sprint and PBI to PBI, but it could also be that the Developers are not completing their work and simply carrying over PBIs to the next Sprint. Ideally there would be no need to copy the last Sprint's test plan, because the new/current Sprint would be all new PBIs with all new acceptance criteria requiring all new test cases. Exceptions will exist, of course.

Regression Tests

Regression testing is the rerunning of acceptance tests for done PBIs to ensure that the Increment still meets the Definition of Done and is releasable. These regression tests could be from the current Sprint as well as previous Sprints. Having a solid foundation of valuable unit tests will help, but because any change to the codebase could potentially cause instability in the Increment, it's important for Developers to continuously perform acceptance regression testing as well. This means that test cases from previous Sprints should be executed in the current Sprint to ensure the integrity of the Increment.

Deciding which of the test cases to run during regression testing is the difficult part. A team may have hundreds of test cases. The team should consider this in Sprint Planning, but also throughout the Sprint as more is learned. As far as *how* to organize the test cases into a regression suite, that's actually quite straightforward, as you will see.

> **Tip** Some Professional Scrum teams will actually add the item "Select applicable regression tests" to their Definition of Done. This will ensure that a done PBI will already have regression tests selected. It could be that some PBIs don't require any regression tests, whereas others may require all of their acceptance tests to be used for regression. Test cases that cover brittle areas, high technical debt areas, and areas of core/critical functionality are good candidates for regression testing. For Sprints with a high number of regression tests, making sure they are fast and reliable automated acceptance tests is important.

Once a team has discussed and identified the test cases that should be used for regression, they should edit each Test Case work item and add a "Regression" tag. If there are multiple sets or types of regression tests, different tags could be used (such as "Regression-A," "Regression-UI," "Regression-Financial"). Figure 7-13 shows that three of the five test cases for the Twitter feed PBI have been tagged for regression.

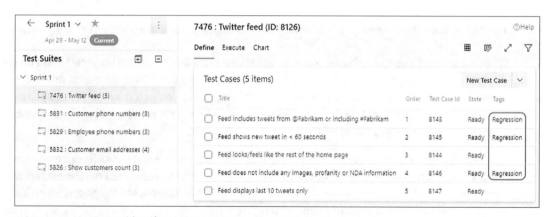

FIGURE 7-13 Use tags to identify existing test cases as regression tests.

The next step is to make those regression tests visible in the new Sprint's test plan. You can easily do so by creating a Query-based suite configured to show all Test Case work items that contain a "Regression" tag. You could further restrict the test cases returned by adding additional criteria (area, iteration, state, automation status, etc.) to the query. Figure 7-14 shows the creation of a Query-based suite for regression tests.

FIGURE 7-14 Use a Query-based suite to list those test cases tagged for regression.

Going forward, this Query-based test suite will always show an up-to-date list of test cases tagged as regression tests. To add or remove a test from the regression suite, a Developer adds or removes the "Regression" tag from the respective Test Case work item. The Developers may want to adopt a test case naming convention that prefixes the test case name with the PBI name or abbreviation to make them easier to identify in a flat, alphabetical list.

One downside to this approach is that the regression test suites are not static. There won't be a history of which specific regression tests applied to which Sprint. For example, if the Developers are currently on Sprint 7 and its regression test suite contains 20 test cases, when they open Sprint 3's regression test suite, it will show the same 20 test cases. If the Developers want to keep a static, historical record of each Sprint's regression tests, they should create a Static test suite and then manually add all of those regression Test Case work items before closing each Sprint. This way, the membership will remain static, even though tags continue to change in future Sprints.

Fabrikam Fiber Case Study

Until continuous delivery becomes a reality, one or more Sprints' worth of finished, tested work can get stacked up, waiting for release. Given the amount of technical debt in the codebase—and the fear it generates—it's important for the Developers to ensure that they don't compromise the integrity of their Done functionality. For this reason, they are adopting a rigorous regression testing approach. They will tag Test Case work items so that they can have a dynamic list of regression tests each Sprint. They will also create a static test suite of regression tests each Sprint so that they won't miss any important regression tests prior to releasing the Increment.

Acceptance Testing Checklist

Every Scrum Team tests their work in some way to ensure that the results are acceptable. These tests can be manual or automated. They can be created before or during development. They can be created by one Developer and run by another. In other words, there are many ways Developers can approach testing. One thing is common across all Scrum Teams—in order to ensure that the Increment is constantly releasable, the Definition of Done should include some form of validation, verification, testing, or other quality checks.

Starting with this assumption, this chapter has proposed shifting the authoring of those acceptance tests earlier and then using them as the Sprint *plan*, which also provides a way to inspect *progress* during the Sprint. This is a completely optional proposition. Although Professional Scrum Developers don't need to plan a Sprint using acceptance tests, I hope by now you can see the benefits. In doing so, I also hope that you see the benefits in using Azure Test Plans with its tight integration with Azure Boards and Azure Pipelines as the framework for supporting acceptance test-driven development.

Here is a checklist for using acceptance tests to represent the Sprint plan, to provide focus, and to inspect progress:

Scrum

- ❏ Capture acceptance criteria in PBIs.

- ❏ Identify which PBIs are forecasted for the Sprint. This is an output of Sprint Planning.

- ❏ Identify acceptance criteria that are common across the forecasted PBIs.

Azure DevOps

- ❏ Create a new test plan for the current Sprint. Ensure the start and finish dates match the Sprint.

- ❏ Create a Requirement-based suite for each forecasted PBI, reordering them according to the Backlog Priority.

- ❏ Create a Query-based suite that includes all Test Case work items containing a "Regression" tag. This assumes your team has previously tagged some Test Case work items as regression tests. Ensure that the list of regression test cases is accurate.

- ❏ Create/add/import any test cases to support acceptance criteria testing for each Requirement-based suite (PBI). The grid view is your friend for quickly pasting in test case titles and even steps. Consider a naming convention to be able to quickly identify the PBI when looking at the test case title.

- ❏ Ensure all Test Case work item Iteration fields are set to the current Sprint.

- ❏ For those Test Case work items that will be automated, set their *Automation Status* field to Planned. This will help identify them in lists and queries.

- ❏ For those Test Case work items that will not be automated, specify test steps and expected results. Consider a minimal amount of detail here—just what is sufficient to assure the tests are repeatable and deliver consistent results. Pairing two Developers is preferred over one Developer writing detailed testing instructions for the other.

- ❏ Use Visual Studio to connect automated acceptance tests to their respective Test Case work items.

- ❏ Use Azure Pipelines to create a build that compiles the test binaries.

- ❏ Use Azure Pipelines to create a release pipeline to run the test in a specific stage (environment).

- ❏ As a team, work down the list of Requirement-based suites (PBIs)—in Backlog Priority order—collaborating/swarming, test by test, to turn those lights from red to green.

- ❏ Use one of the visualization methods to gain insight into the quality of the Increment and the progress of the team.

- ❏ Tag the appropriate Test Case work items as "Regression" tests as late as responsible—preferably during the Sprint, while the context is fresh.

Chapter Retrospective

Here are the key concepts I covered in this chapter:

- **Acceptance criteria** The Product Owner's or stakeholders' definition of success for a given PBI.

- **Acceptance test** A manual or automated test, created and run by the Developers to verify that an aspect of the PBI is done. Acceptance tests typically map to individual acceptance criterion.

- **Test plan** A work item type that Developers can use to organize their tests. For Scrum Teams, a test plan typically maps to a Sprint.

- **Static suite** A simple test suite that only has a name.

- **Requirement-based suite** A test suite that maps to a specific PBI work item and is meant to contain acceptance tests for just that PBI.

- **Query-based suite**   A dynamic test suite that returns Test Case work items that meet specific criteria, such as Tag = "Regression."

- **Test case**   A work item type that Developers can use to further specify a PBI's acceptance criteria in the form of an acceptance test. Test cases can be either manual or automated. Professional Scrum Developers prefer automated acceptance tests.

- **Grid view**   A two-dimensional data-entry view, similar to Microsoft Excel, that a Developer can use to quickly create or edit Test Case work items.

- **Test point**   A unique combination of a test case, a test suite, a test configuration, and a tester. Test points are associated with a test run.

- **OData**   Open Data Protocol is a standard for building and consuming RESTful APIs, such as the data from Azure DevOps Analytics. Tools like Power BI can consume OData via an OData feed.

- **Acceptance test-driven development (ATDD)**   The practice of defining executable specifications in the form of failing automated tests prior to writing any application code. ATDD helps Developers *build the right thing*.

- **Test-driven development (TDD)**   The practice of writing unit tests prior to writing a functional piece of code. When the code is designed to pass the test, refactoring occurs and then another TDD cycle. TDD helps Developers *build the thing right*.

- **Associated automation**   Test Case work items can be associated with an automated acceptance test supported by MSTest, xUnit, or NUnit. The association is performed within Visual Studio and the test is executed within a release pipeline.

- **SpecFlow**   The most popular third-party acceptance-testing framework for .NET.

- **Acceptance**   Acceptance testing and Product Owner acceptance are two separate activities. A Definition of Done should include some form of Product Owner acceptance or delight. The Product Owner should never accept work that hasn't passed acceptance tests.

- **Regression testing**   The preferred practice of running test cases from prior Sprints to ensure the integrity of the Increment remains even as the codebase changes.

CHAPTER 8

Effective Collaboration

There's a buzz—a kind of energy—that you can feel when a high-performance Professional Scrum Team works in harmony to solve a problem. Each Developer gets totally absorbed in their task. They each do their part integrating the design, the coding, and the testing. Slices of working product are assembled and verified. Definitions of Done are met. PBIs are moved to the Done column. Everyone loses track of time. Everyone feels happy and satisfied. They are experiencing *flow*.

Bruce Tuckman wrote about the stages of group development. He identified four stages in the development model: forming, storming, norming, and performing. In the initial, *forming* stage, the individuals come together to form the team. They may not know one another. They may not know everyone's strengths and weaknesses. This leads to the *storming* stage, where each team member competes for their idea's consideration while working together to resolve their differences. This stage can sometimes be completed quickly. Unfortunately, some teams never leave this stage. Once the team members are able to resolve their differences and participate with one another more comfortably, they enter the *norming* stage. Here, the entity of the team begins to emerge. The members can converge on a single goal and come up with a mutual plan. Compromise and consensus decision making occur at this stage. Professional Scrum Teams strive to reach the fourth and final stage, known as *performing*. Not only do these teams function as a unit, but they also find ways to get the job done smoothly and efficiently. They are able to self-manage effectively. In my opinion, very few teams reach this stage, but every one that does has mastered the art of collaboration, can enter and exit conflict gracefully, and can constructively build a better product.

I propose a fifth stage to Tuckman's stages of group development: *swarming*. As you will learn later in this chapter, swarming is a collaborative practice where all Developers work simultaneously on the same PBI. They continue to work on all the related tasks or tests until the PBI is done. Swarming is a proven practice while working in a complex space, such as software development. Positive results can be demonstrated almost immediately.

In this chapter, I will highlight some practices and tools that enable more effective team collaboration. By learning and adopting these practices, a team will increase its ability to reach the *performing* stage of Bruce Tuckman's model, and hopefully the *swarming* stage after that.

Individuals and Interactions

The Agile Manifesto clearly states that though there is value in process and tools, there is *more* value in interacting with individuals. In other words, agile software development recognizes the importance of people and the value that they bring when working together. After all, it's people who build product, not the process and not the tools. If you put bright, empowered, motivated people in a room with no process and inadequate tools, they will still be able to get something accomplished. Their productivity may not be what it could, but they will produce value. They will know to inspect and adapt their processes, always looking for improvement. The group of people becomes a *team* working toward a shared goal. Conversely, if the people don't work well together, no process or tool will fix that. A bad process can screw up a good tool, but bad people can screw up everything.

Software development is a team sport. To succeed in this sport, game after game, the team must share the vision, divide the work, execute, inspect, and adapt as a team. In other words, they must collaborate and learn. Even a team of rock stars (also known as "10x developers") is doomed to fail if they don't collaborate with one another. If the striker on a soccer team has their best game ever—scoring four goals—but the opposing team scores five, it's still a loss. The other team, with more "mediocre" players, probably collaborated better.

> **Note** A *10x developer* is an individual who is thought to be as productive as 10 other developers. In theory, the 10x developer would produce 10 times the outcome of any normal developer. Some claim that the 10x developer concept is a myth. Although there are certainly practitioners out there who are more productive than others, 10 times seems to be a stretch to me. One thing is certain: team members of this echelon are very rare and it would not be prudent to try to assemble a team of them. Instead, do what you can to foster and develop the team you have into a Professional Scrum Team—striving for a 10x *team*!

A few years ago, Ken Schwaber did a series of podcasts where he answered frequently asked questions about Scrum. My favorite question that he answered was, "Do I need very good developers for Scrum?" His answer was insightful: "You need very good developers for software development." You can do Scrum with terrible software developers, and you'll get terrible increments of functionality every Sprint.

When I hear about teams that have tried Scrum and given up because it was "too difficult," I know that they are not talking about the complexity of Scrum itself. These are software developers. They are some of the smartest, most creative problem solvers you'll ever meet. Besides, Scrum is easy to understand. Chapter 1 pretty much covered it, soup to nuts. No, what these people are talking about is the *discipline* of practicing Scrum correctly by using the Scrum Values within an organization that allows them to do so, the right way, every day. Not being able to do that is why they gave up.

I agree with the Agile Manifesto. This has been evident throughout this book as I've pointed out the value of interacting and collaborating with individuals. I have discussed process and tools as well, but I've been vigilant in pointing out that not all application lifecycle management (ALM) tools and DevOps automation frameworks are healthy for a team. Most are. Some, however, can lead to one or more dysfunctional behaviors. For example, social media, tablets, video games, and other devices are appealing and fun, but sometimes the kids (or Developers in this case) need to get outside and interact with others.

Years ago, I was asked to build a web-based work item approval system on top of Team Foundation Server. The client designed it so that email alerts would be sent when a work item changed to a certain state. This was before Microsoft added that functionality to the core product. These emails contained embedded hyperlinks that would redirect the user to a webpage that allowed managers or leads to authorize the state change. It was a sophisticated system—it even knew which users could cover for others if someone was out of the office. My company built it. The client installed it. It did exactly what they wanted, but they ended up not using it. The reason they mothballed it was that it was too mechanical and removed the opportunity for individuals to meet face to face and have a conversation. This was a learning opportunity for me and something I keep in mind whenever I see a shiny new feature in Azure DevOps or Visual Studio. I ask myself, "Does this feature encourage collaboration or discourage it?"

When it comes time to meet and collaborate with members of your Scrum Team or stakeholders, here are some tips to consider:

- Establish the scope and the goal of the meeting, stay focused on those topics, and achieve the desired outcomes.

- Meet face to face, especially if you anticipate a substantive conversation.

- Meet at a whiteboard, especially if you're intent on solving a problem.

- Set a timebox and be prepared to explain its purpose.

- Leave the gadgets in the other room, unless they are required.

- Employ active listening techniques.

- Have a clear facilitator.

In this section, I discuss some of the general—but important—collaboration practices that a Scrum Team can adopt.

Collocate

As I write this, teams everywhere are sequestered away, working in home offices, home basements, and home bedrooms—all thanks to the coronavirus (COVID-19) pandemic. Social distancing policies are keeping us purposely "dis-located." Organizations have switched to remote work models. Are these changes temporary or permanent? Time will tell.

It's important, during times of crisis like this, to lean into the Scrum Values. When the values of commitment, courage, focus, openness, and respect are embodied and lived by the Scrum Team, the Scrum pillars of transparency, inspection, and adaptation come to life and help build trust for everyone—regardless of where they are working.

Successful use of Scrum depends on people becoming more proficient in living these five values. People personally commit to achieving the goals of the Scrum Team. The Scrum Team members have courage to do the right thing and work on tough problems. Everyone focuses on the work of the Sprint and the goals of the Scrum Team. The Scrum Team and its stakeholders agree to be open about all the work and, to some degree, the challenges of performing the work. Scrum Team members expect one another to be capable, independent people.

We can all agree that communication and collaboration provide more value when practiced face to face, rather than remotely. At least I would hope that everyone knows this, because we experience it every day of our lives. When two people communicate face to face, they exchange more than just words. There are facial expressions, body language, and other nonverbal gestures. This kind of side-band data can be just as important, if not more important, than the audible conversation that occurs. For example, the look on a Product Owner's face when you suggest a solution to a problem can short-circuit the need for a detailed explanation. Thank you, collocated Product Owner. You just gave me back 20 minutes of my day!

Remember that Scrum has several formal events (meetings) built into the framework where collaboration can occur. In addition, members of the Scrum Team should be continuously "meeting." These are not traditional meetings, where someone speaks and everyone else listens. These are short, collaborative, timeboxed meetings with the specific purpose of solving a problem. In fact, I wouldn't even call them a meeting, but more of a conversation with a clear outcome. It's important that they occur as needed, with no logistical impediments. For example, if two Developers need to discuss something with the Product Owner but all the conference rooms are booked, they should meet anyway, somewhere, anywhere. To some degree, business formalities, and even etiquette, go out the window during the Sprint when the Developers are in the zone, developing business value.

I may be old school here, still clinging to pre-COVID-19 thinking, but I believe collocation should be a requirement when forming a Professional Scrum Team. It shouldn't just be a nice-to-have feature. It shouldn't be discounted because we can "make remote work." When working in the complex space—as with software development—collocation directly influences complexity, which directly influences the quality of the process, which in turn directly influences the quality of the product.

Just to be clear, when I talk about collocation, I don't mean just being in the same time zone, city, or building. I'm not talking about having always-on cameras and large-screen displays on wheels. Although these options are better than some I've seen, I still prefer to see teams *in the same physical room* or in adjacent physical rooms. The Product Owner should be nearby, too. This way, high-fidelity, face-to-face communication can occur on demand.

Tip When collocation is not possible, bringing a dislocated team together periodically is a proven practice. This builds the sense of team and supercharges them. It reconnects and reenergizes the relationships and commitments. This is especially true at the beginning of a new development initiative, so they know who they are working with. Even better is to bring the team together, each Sprint, for the sake of Sprint Planning. Some of my colleagues have worked with large-scale development efforts where those organizations bring their over-seas teams together once a month for the sake of a scaled Sprint Planning event. In their opinion (and mine), a focused, in-person Sprint Planning can help avoid the waste of 7–8 teams building the wrong thing over the next month. In other words, spending $100,000 on travel is cheap compared to wasting $1 million on building the wrong product.

Professional Scrum Teams know the value of collocation, and they strive for it. That said, there may be cultural, political, or financial reasons for not being able to collocate. This is the reality that I see as I visit larger organizations. The most common justification I'm given when I ask why the team is not collocated is that it "saves money" to have one or more of the functions supported or outsourced remotely, usually overseas. When I hear that, I hope that somebody, somewhere is doing the math, and considering the waste of coordination, handoffs, and rework, not to mention the potential for a decrease in quality of the product. Even if this decrease is not detectable or measurable, the decision makers should consider what the *increase* in quality could be if they were to bring the entire team together.

Note Do I think that Developers working remotely as part of a "dis-located" team *can't* be professional? Of course not. They absolutely can be professional and the team absolutely can collaborate, deliver high-quality software, and create business value. For some tips and tricks, I recommend downloading fellow Professional Scrum Trainer Stefan Wolpers's Remote Agile Guide from *https://age-of-product.com/remote-agile-guide.* Stefan starts with some basic techniques and tools for practicing remote Scrum with distributed teams and also how to apply *Liberating Structures* in the remote realm. For more information on Lib-erating Structures, visit *www.liberatingstructures.com.* An attribute of a Professional Scrum Developer is to inspect and adapt constantly, such as looking for ways to improve their pro-cess. Collocating a "dis-located" team is one of the biggest improvements that can be made, usually resulting in a marked increase of productivity and quality.

As you've probably heard, every company is now a software company and these companies con-sider their custom software to be a strategic advantage over their competitors. I will sometimes ask executives where they would be without their line-of-business (LOB) application or public-facing website. They all agree that it would be a complete disaster. Not only has their staff forgotten how to run the business manually using paper and pencil, but they don't even know where to find the paper and pencils. Next, I ask them why they try to save money by limiting their capabilities and productivity of the team developing that custom software. At this point, I'm either asked to tell them more or I'm escorted out of the building.

Note I once had a conversation with an IT director of a very large organization. He explained to me that the Product Owner worked out of the main office, as did the coders. The testers were overseas—nearly 10 time zones away. He shared with me a problem that they'd been having for the past few months. He said the coders would code a feature and then go home for the night. The testers would come in, download the binaries, begin testing, and run into a bug. This blocked them from doing any further testing until the coders could fix it. The coders would come in the next day, see the lack of progress, fix the bug, and have to wait until the end of the day for the testers to do their thing. Sometimes this dance would take 3–4 days before the work was finished. He asked me how Team Foundation Server could help him. I answered by asking why the testers weren't collocated with the rest of the team. He told me it was because they *save money* by sending the testing work offshore. I'm glad we were having this conversation in person because he was able to see the awesome facial expression I made at that point.

Set Up a Team Room

Having all Developers work in a shared, common room can be a powerful practice. Whiteboards containing plans and design notes are visible to everyone. The Sprint Backlog and a burndown chart can be updated easily and seen by everyone—whether on physical boards or flat screens. During critical design points or bug fixes, this room can become an *incident* room of sorts as the Developers engage in tactical planning and operations. Communication becomes more open and happens in real time. Developers tend to focus their productivity on solving problems, while minimizing time spent on wasteful activities. Team rooms allow everyone, including stakeholders, to feel that buzz that I mentioned in the beginning of the chapter.

However, not every Developer wants to work in an incident room every hour of every day. There needs to be the opportunity to have private conversations, take phone calls, or just take a timeout from the rest of the team. Developers are smart and can self-manage to come up with solutions for these situations. I've seen Developers put on headphones; adjourn themselves to private, quiet rooms; or work away from the office for a short time as needed. Ideally, the managers and the organization trust their people to the point where they can accommodate their needs. If they don't, then that is a smell and possibly a big impediment to self-management. Continually generating business value in the form of working product is a way for the Scrum Team to earn (or re-earn) that trust.

Some personalities and cultures see collocation as an impediment. These Developers may actually be counterproductive in such an environment. Remember that Scrum is about people, and people are human. Their idiosyncrasies map directly to their ability to collaborate and work effectively as a team. The rate at which a team is able to create business value is a function of the team's productivity. Perhaps for these people, being in close proximity to, but not in the same shared room with, the rest of the Developers is good enough. Could it be better? A creative Scrum Master, fostering open and honest Sprint Retrospectives and improvement experiments, might get the Developers to change their minds. Remember that culture is hard to change.

My recommendation is to set up an open-space team room and just try it out. See if management will let the Scrum Team take over one of the conference rooms for a Sprint or two. If, during the Sprint Retrospective, the Scrum Team honestly believes that they were more productive, then the Scrum Master can work with management to create a more permanent, open-space team room.

> **Note** An open-space team room is not the same thing as an *open-plan* office. Open-plan offices are typically inhabited by employees working on different tasks for different projects. Open-space team rooms are inhabited by Scrum Team members working on a common product. Both environments can generate noise, but the type of conversations found in an open-plan office will be more contrasting and, thus, more distracting.

Fabrikam Fiber Case Study

The Scrum Team has been collocated since day one, with Paula in a nearby office. When they are working, they regularly meet and collaborate whenever and wherever it is required. Day to day, the Developers sit near one another in a large, open-space room with a half-dozen whiteboards. Because the Developers use laptops, there's a minimum amount of gear and clutter in the room, and individuals can be more nomadic as they pair up and work.

When one of the Developers needs to focus or requires some personal space, they will put on headphones or go to a quieter room down the hall. When a Developer has to travel or otherwise work remotely, the team will set up a dedicated computer with a Microsoft Teams connection, including video. Microsoft Teams is used anyway to share code and screens between Developers. Scott (the Scrum Master) continues to do a good job in educating the organization. Although the stakeholders know where the team room is located, they know to avoid it when the Developers are working—unless, of course, they're invited in. Scott still has to remind people of this from time to time.

Meet Effectively

Professional Scrum Developers know to only attend the *essential* meetings. To be clear, I'm not talking about the built-in Scrum events, such as Sprint Planning, the Daily Scrum, Sprint Review, or Sprint Retrospective. I'm also not referring to the regular Product Backlog refinement sessions, or those impromptu but important meetings requested by the Scrum Team in order to clarify or get feedback from a stakeholder. Gee, come to think of it, Scrum has built in to it, *all* of the meetings that Developers would need. I hope, in fact, that I've made it clear that these "meetings" are important and that they should be attended by all the involved parties, face to face.

No, I am talking about all the *other* meetings that an organization might require its staff to attend. You know the ones I mean. They are mandatory and *read-only* (they don't ask for your feedback), and they provide zero business value to the product being developed or the team's process. Unfortunately, some of these meetings cannot be avoided. They are a fact of organizational life and a requirement to keep your job and get paid.

When you are invited to such a meeting, try to identify its purpose and expected outcome. This may be stated in the invitation, but if not, you may have to ask the organizer. I know some individuals will not accept a meeting invitation if no clear agenda or objective is given. From this information, hopefully you can determine who the intended audience should be. Will the meeting be technical? Will decisions be made? If you don't fit the audience profile, try to skip the meeting, or send the Scrum Master instead. Being a proxy for the rest of the team at meetings like these is one of the duties of a Scrum Master and a great way to protect the team's focus.

If the tables are ever turned, and you find yourself organizing a meeting, you can follow the same advice:

- Only schedule meetings that are absolutely necessary and that can't be satisfied by one of Scrum's built-in events.

- Outline the agenda and expected outcome in an invitation.

- Keep the meeting as short as possible.

- Establish a timebox to enforce it.

- Send invitations only to those people who need to attend.

- Discourage forwarding the invite to other people.

- At the beginning of the meeting, explain the timebox and possibly its concept.

- Have a clear parking lot and a working agreement for handling off-topic/distracting tangents.

When someone who is versed in Professional Scrum sets up and runs a meeting, they will end up sharing good behaviors and practices, such as transparency, active listening, and timeboxing. This is a fantastic way to get others in the organization more educated on Scrum, the Scrum Values, and some complementary practices. If appropriate, email any retrospective notes to the attendees, including action items. These behaviors may begin to infect the organization as other business units and teams will want to get some of that "Scrum goodness."

Tip One way to keep meetings constructive is to say "yes, and..." instead of "yes, but... " If the current topic or solution being discussed is one in which there is partial agreement, saying "yes, and..." comes across as being constructive, open, and respectful. If someone hears "yes, but..." then they might think their idea is being discounted, or they may feel limited in what can be accomplished. If, however, they hear "yes, and..." they will think that their idea was accepted, or at least understood, and they will be more open to ideas. More importantly, the person will be more open to collaborating on a shared solution, which should always be the goal to prevent discussions from becoming polarized and deadlocked.

Fabrikam Fiber Case Study

Paula and Scott are good at running interference for the Developers. For meetings that are not related to the development of the product, Scott or Paula will attend. For those annoying "all hands" meetings that cannot be avoided, Scott will do his best to be a proxy, letting the other team members know what has transpired.

Listen Actively

In my experience, software developers tend to have short attention spans and be impatient with anyone who is not as smart as they are or who doesn't have the answer that they are looking for. Of course, I could just be talking about myself. But as they say, acknowledging that you have a problem is the first step in overcoming it. For me, *active listening* was that cure.

Active listening is a communication technique where the listener is required to give feedback on what they heard to the speaker. This can be as simple as nodding, writing a note on a piece of paper, or restating or paraphrasing what was said. This demonstrates curiosity, sincerity, and respect for what the person is saying. It also helps alleviate assumptions and other things that get taken for granted. Opening a laptop, scrolling through texts, or otherwise getting distracted by anything else is *not* active listening and may even violate the Scrum Value of respect. This means that active listening is even more difficult over video.

Another part of active listening is waiting to speak. This is my particular problem. I tend to complete the other person's sentence in order to move the conversation along. I think I'm being helpful, but I'm probably coming across as rude. This is especially true for people who don't know me and is very apparent to me when I have a conversation with another individual like myself!

Fortunately, we have techniques that can be used to overcome this particular interpersonal dysfunction. My favorite is to take a stack of sticky notes with me and write down the things that come to mind while the other person is talking. Soon it will be my turn to talk, and I can go back through my notes. See what I did? I solved the feedback and interruption problems with a single solution.

Remember the HARD mnemonic. HARD stands for Honest, Appropriate, Respectful, and Direct. It is a reminder of how you should always communicate with people, especially those who don't know you. Active listening plus HARD communication is a recipe for successful collaboration.

Fabrikam Fiber Case Study

During a recent Sprint Retrospective, Scott (the Scrum Master) brought up his observations made during the Sprint. He witnessed a few Developers having difficulty conversing respectfully with one another (as well as with stakeholders) during a few meetings. As a team, they decided to improve their communication abilities, specifically their active listening skills. Scott did some searching online and found several websites dedicated to the subject. During the next few Sprints, Scott coached the team as they experimented with some of the techniques that they learned.

Collaborate Productively

Collaboration means working with people. This typically means dividing the work between two or more individuals and working together. Both the process of dividing the work and the actual working together with others require concentration and focus. Getting into this productive state, otherwise known as *flow* or the *zone*, can take time. Getting out of that state prematurely, as caused by any kind of interruption, can be considered waste. The irony is that collaboration *requires* interruption, and you will need to get used to it, tame it, and control it in order to collaborate effectively.

We are taught at a young age that it is disrespectful to interrupt others. If your team is working in an open-space team room, it's easy to see when fellow Developers are in the zone—deep in thought. Your instinct should be not to interrupt them. When you're working by yourself, however, it may be harder to know when *you* are in the zone. Stopping to take a mental assessment may actually kick you out of the zone. Professional Scrum Developers know how to minimize wasteful interruptions in order to maximize productivity. Numerous books, blog posts, and whitepapers have been written about being more productive.

Here are a few of my favorite tips:

- **Silence your cell phone**   Turn it to vibrate, turn it off, or leave it in your car or backpack.

- **Exit Microsoft Teams/Slack**   Unless the Developers have established a working agreement, close your collaboration app or set its status to busy.

- **Don't check your email**   Email can be a great productivity tool, but it can waste a lot of your time as well. If you can't or don't want to turn it off, then be sure to disable all notifications. Having an icon appear in the system tray, seeing the mouse pointer change, or hearing an audible alert when a new email arrives can have the same conditioning effect as one of Pavlov's dogs hearing a bell ring. Try to check email only three times a day: at the start of your day, at lunch, and before leaving.

- **Limit Internet searches**   Developers can spend their whole day on the internet if they are not careful. Establish a scope and timebox for any research.

- **Avoid formal meetings**   As I mentioned in the previous section, one reason that Scrum is successful is that it defines the important meetings in order to minimize the need for unimportant ones. A Developer's productivity drops when they are away from their tools. Feel free to attend valuable ad hoc meetings over coffee or at another's desk, but send the Scrum Master to the formal meetings in the team's stead.

- **Stop shaving the yak**   *Yak shaving* is when you perform a task that leads you to another related/semi-related task, and so on, and so on. At any point you can explain or justify what you are doing, but you are also far away from your original goal. Yes, Developers can have complex development environments. These can include multiple versions of software, one or more integrated development environments (IDEs), virtual machines, databases, frameworks, cloud accounts, software development kits (SDKs), testing tools, installers, and so forth. Do yourself a favor: get it working, script it, snapshot it, and forget about it. Endless tweaking tends to have a diminished return on value. Solve today's problem today and tomorrow's problem tomorrow.

- **Use a Pomodoro technique** This is a time management practice that uses a timer to break down work into intervals (such as 25 minutes) separated by short breaks. There are many great Pomodoro/focus-keeping apps in your phone's app store.

- **Just get started** Some planning is required before starting a task, but overplanning can crush productivity.

- **Use active listening** As I mentioned earlier, when your colleague is talking, you should listen to what they are saying and expect the same courtesy when you are talking. Practice this at the Daily Scrum.

- **Realize that life happens** We're all human and have a life outside of our work. When issues emerge, be open and honest about them, and take the necessary time to get your head right. Be appropriately transparent with the rest of your team.

Fabrikam Fiber Case Study

The Scrum Team is always looking to do better. This is evident during their Sprint Retrospectives, where collaboration practices are almost always discussed as improvement is sought. Everyone knows that the best way to improve productivity is to improve how they interact with one another and stakeholders.

Be T-shaped

A Scrum Team must be cross-functional. This means that, for any given PBI, the Developers must be able to do *all* of the work without relying on people who are not on the team. If the Sprint plan is represented with tasks, then it means that each task, for each PBI in the Sprint, can be accomplished by at least one Developer on the team. For example, if one of the tasks requires Microsoft Power BI integration, that means a Developer must know Power BI. It doesn't mean that *all* Developers must know Power BI.

That said, I believe that Professional Scrum Developers should be open to learning and even mastering new skills. This is the notion behind being *T-shaped*, a concept used to describe the abilities of an individual Developer. The vertical line on the letter T represents the depth of a particular specialty (C#, TypeScript, Selenium, UX design, etc.). The horizontal line represents the knowledge base and the ability to be a generalist and collaborate across various skill sets (analysis, design, testing, etc.). You can see a sample T-shaped diagram in Figure 8-1.

Most Developers have probably already mastered more than one skill. Whether or not they claim those other skills—and use them—is the real question. For example, a coder may not write or run Selenium tests because "it's not their job"—even though they are quite skilled at it from previous experience. Remember the "job" of a Professional Scrum Developer is to deliver value in the form of working product—not just writing code, writing tests, designing UI/UX, and so forth.

Developers living the Scrum value of *openness* aren't ashamed to say that they don't have a particular skill. They are also not allergic to doing something different and learning a new skill in the process. Pairing new Developers with Developers who understand this Scrum Value will help build this new openness muscle. Over time, T-shaped individuals who master another skill become Pi-shaped (π) and even comb-shaped—adding increasingly more vertical, specialization "teeth" to the comb over time.

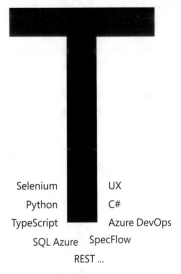

Analysis Content Design Testing UX Writing ...

Selenium   UX

Python   C#

TypeScript   Azure DevOps

SQL Azure   SpecFlow

REST ...

FIGURE 8-1 An example of being T-shaped.

Achieve Continuous Feedback

Professional Scrum Teams love feedback loops, and the shorter and faster, the better. For example, as soon as a coder types a few lines of substantive code, they press F5 to get feedback from the compiler. As soon as they've got a method refactored, they like to get feedback from their unit tests. As soon as they have finished with a task, they like to push their code to Azure DevOps to get feedback from the continuous integration (CI) pipeline. As soon as they have a functional user interface, they like to get feedback from the Product Owner or stakeholders. Quick and continuous feedback like this is healthy for the product, as well as for the team.

Automated feedback provided by unit tests, integration tests, code coverage, code analysis, and acceptance tests is encouraged. Developers can call upon Azure DevOps to provide this feedback at any time, day or night. The results tell the team that they are building the feature correctly. Professional Scrum Developers take advantage of these automated tools to ensure that they are well informed about the progress and quality of their work.

Smell As I've mentioned before, it's a smell if the Developers don't ask for feedback from the Product Owner or stakeholders during the Sprint. Passing unit, integration, and acceptance tests only ensures that the quality of the design and functionality of the PBI have been met. The Developers should still want to make sure that the people requesting the capability are happy with its design, function, and usability. In other words, the Sprint Review should *not* be the first time that the Product Owner sees new functionality being demonstrated. Don't surprise the Product Owner!

Product Owner feedback is just as important as other types of feedback. An engaged Product Owner who knows the product and the wants of its users can quickly give the Developers positive or negative feedback on a PBI being developed. Getting in-person guidance on the usability of a PBI early in its development is very valuable. If the Developers build the wrong thing, it's essentially the same as if they'd introduced a bug into the product. This, of course, would depend on the team's definition of a "bug." Much like the advice for fixing bugs, it's cheaper to build the right feature earlier than to have to rework it later.

> **Note** The Product Owner feedback loop should be as short and fast as possible as well. This is another argument for collocating the Product Owner with the Developers.

I'm often asked if the Developers can reach out directly to a stakeholder, such as a user or customer, who requested the feature in order to collaborate and gather feedback. Yes! I consider it a quality of a self-managing Scrum Team when the Developers reach out to stakeholders directly. At the end of the day, the Scrum Team is trying to create the best possible product for those stakeholders. The Developers need to know the context for features and have conversations about domain knowledge. Since the Product Owner is not omnipotent, it's ideal for Developers to collaborate and co-create solutions directly with stakeholders. This approach alleviates the need for long, rambling requirements documents (which are incorrect and stale anyway). It also enables the Developers to confirm—in real time—that the stakeholders fully understand the PBI in question.

It's probably preferable to involve the Product Owner, or at least make them aware that these conversations are going on. Professional Scrum Product Owners are open to this concept and know, especially at scale, that this will result in less waste and a better product. Keep the Product Owner informed of any potential scope creep or new PBIs. As always, any issues with direct conversations such as these can be discussed at the Sprint Retrospective.

I see Product Owner feedback as falling into four broad categories in Scrum. I have listed them along with some suggested practices and artifacts that can support each, in Table 8-1.

TABLE 8-1 Types of Product Owner feedback.

Type of feedback	When is it given?	How do you do it?	What is updated?
Can you give us more details about this PBI?	Product Backlog refinement, Sprint Planning, any time during development	Collaborate with the Product Owner or stakeholders at a whiteboard.	PBI work item, code, tests, tasks, whiteboard, notes
Do you like this? Is this the behavior you were expecting?	Any time during development	Pair with the Product Owner or stakeholders and inspect the PBI.	Code, tests, tasks, whiteboard, notes
What else do you want this PBI to do or not do?	Any time during development	Pair with the Product Owner or stakeholders and inspect the PBI.	Code, tests, tasks, whiteboard, notes
As a corollary to this PBI, what else do you want the product to do, not do, or do differently?	Sprint Review, Product Backlog refinement, any time during development	Collaborate with the Product Owner or stakeholders to update the Product Backlog.	Product Backlog

The rest of this chapter will discuss effective collaboration practices and tools as they pertain to software development. These are considered complementary practices, which means they are all optional but have been proven useful by other Scrum Teams—especially those using Azure DevOps. Remember that in Scrum, *how* the Developers work, including the practices they follow and the tools they use, is entirely up to the Developers.

Collaborative Development Practices

Even the simplest software product requires a team with many talents. Beyond having the standard capabilities of design, code, test, and deploy, many types and levels of talent can exist within each discipline. Every Developer has a unique background, set of skills, expertise, and personality. Each brings something different to the team. For example, you may have two coders with similar résumés and depth of experience. The way in which they analyze and solve problems can vary radically. If presented with a challenge, each may provide an approach that is fit for purpose according to the acceptance criteria, but they can be very different.

A Professional Scrum Team understands this reality and will even use it to its advantage. These types of teams recognize that everyone has a different way of solving problems, and as long as those solutions fit within the parameters and constraints of the product and the team's practices, they should be embraced. Long, drawn-out discussions and arguments over approaches and coding styles tend to generate little value, and they typically only lower productivity, not to mention morale. If two solutions emerge during discussions, and both are fit for purpose, the Developers should select the least-complex one and move on. The Developers should disagree and commit—and honor the commitment. Over time, every Developer will be able to contribute their design ideas.

In this section, I will explore several contemporary practices that boost the Scrum Team's effectiveness during collaboration.

> **Note** A self-managing Scrum Team should pick and choose from these practices (as well as others not listed in this book) and try them for a Sprint or two or three. Later, during a Sprint Retrospective, the team can decide whether to continue to embrace the practice, to amplify it, or to abandon it.

Collective Code Ownership

Extreme Programming (XP) gave us the notion of *collective code ownership*, which I take to mean collective code *accountability*. With this mindset, individual Developers do not own individual modules, files, classes, or methods. All of those things are owned collectively—by the entire team. In other words, any Developer can make changes anywhere in the code base.

Consider the alternative to collective code ownership, where each Developer owns an assembly, a namespace, or a class. On the surface, that may seem like a good idea. That coder is the expert on that component, as well as its steward and gatekeeper for all changes. Strong code ownership like this can block productivity.

Consider the situation where two Developers (Art and Dave) are working on separate tasks that both need to touch a common component owned by a third Developer (Toni). Dave will have to wait while Art's functionality is coded and tested. A collective code ownership approach would allow Dave to code the feature himself. Rest assured, Azure Repos will track who made what changes to which files and enable a merge (or a rollback) to occur if any problems emerged. Another potential problem with strong code ownership surfaces when refactoring. Modern refactoring tools, like those in Visual Studio, can do this safely, but if the changes to those files cross ownership boundaries, the eventual merge operation will block productivity and may introduce instability.

Adopting a collective code ownership mentality can take time. This is especially true if the Developers are used to having strong code ownership. Pairing, mobbing, and shared learning are ways to break up the turf and the politics. Just as it takes time for the Product Owner and organization to trust the Developers' ability to self-manage, it also takes time for the individual Developers to trust one another.

> **Note** Professional Scrum Developers collectively own more than the code. They also collectively own the Sprint Backlog, the Definition of Done, the Increment, all sizing, all designs, all failing tests, all failing builds, all failures, and all successes!

Tracking Ownership in Azure DevOps

The biggest advantage with collective code ownership is the boost in the social dynamics of the team. Because each Developer has full control over all source code, there are less boundaries and more opportunities to find solutions. Remember that in Scrum, the Developers own all the problems and all the solutions collectively. This includes the artifacts of those solutions, namely the source code.

Should you ever have a need to determine *who* made a specific change to a file, Azure Repos can help. For any folder or file, you can review the complete history of changes, including metadata like who, when, what, and (hopefully) why via a meaningful comment and linked work item(s). You see an example of this in Figure 8-2. Azure Repos uses the information stored in each Git commit to generate a full history.

In addition, you can compare the changes between two versions of a file. The diff view shows lines removed from the older commit and lines added in the new one. You can see an example of comparing two versions (commits) of a code file in Figure 8-3. The UI shows removed text in red and added text in green.

You can also *annotate* (or Git *Blame*) a file to learn who made which changes to which lines of code, when those changes were made, and (hopefully) why. This is done by prefixing each block of code with the most recent Git commit that touched it. Clicking the commit ID will show additional metadata, including the commit message and possibly an associated work item. Figure 8-4 shows the Annotate feature in action.

FIGURE 8-2 Rich history information is generated from Git commits.

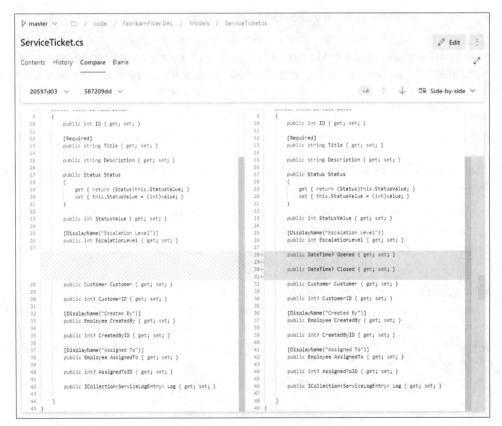

FIGURE 8-3 Azure Repos lets you compare any two versions of a file for differences.

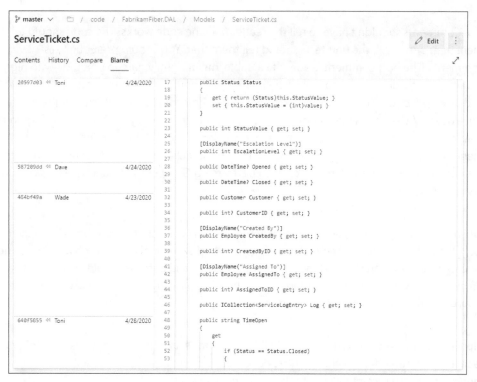

```
master ∨        /  code  /  FabrikamFiber.DAL  /  Models  /  ServiceTicket.cs

ServiceTicket.cs                                                        ✎ Edit    ⋮

Contents   History   Compare   Blame                                          ↗

20597d03 ◁ Toni            4/24/2020    17    public Status Status
                                        18    {
                                        19        get { return (Status)this.StatusValue; }
                                        20        set { this.StatusValue = (int)value; }
                                        21    }
                                        22
                                        23    public int StatusValue { get; set; }
                                        24
                                        25    [DisplayName("Escalation Level")]
                                        26    public int EscalationLevel { get; set; }
                                        27
587209dd ◁ Dave            4/24/2020    28    public DateTime? Opened { get; set; }
                                        29
                                        30    public DateTime? Closed { get; set; }
                                        31
464bf49a    Wade           4/23/2020    32    public Customer Customer { get; set; }
                                        33
                                        34    public int? CustomerID { get; set; }
                                        35
                                        36    [DisplayName("Created By")]
                                        37    public Employee CreatedBy { get; set; }
                                        38
                                        39    public int? CreatedByID { get; set; }
                                        40
                                        41    [DisplayName("Assigned To")]
                                        42    public Employee AssignedTo { get; set; }
                                        43
                                        44    public int? AssignedToID { get; set; }
                                        45
                                        46    public ICollection<ServiceLogEntry> Log { get; set; }
                                        47
640f5855 ◁ Toni            4/28/2020    48    public string TimeOpen
                                        49    {
                                        50        get
                                        51        {
                                        52            if (Status == Status.Closed)
                                        53            {
```

FIGURE 8-4 Annotate (also known as Blame) shows who touched each line of code and when.

Fabrikam Fiber Case Study

Because each Developer is an Azure DevOps Project Administrator, everyone has full control over every aspect of the Azure DevOps project. This includes the ability to view, edit, and even delete files from version control. Should the need arise to see who made a change, the Developers know how to view history, compare, and annotate as needed. Mostly they will use the Annotate feature to *praise* another Developer for good work, rather than blame them.

Commenting in Code

With collective code ownership comes a certain amount of responsibility. Other Developers will need to understand the code and any changes made. If a Developer or a pair of Developers are working on a rather complex part of the code, they might want to add some comments. This can be, for example, a block of comments that gives another Developer enough information to understand the code. The comments can also be regularly sprinkled throughout longer methods. You can think of comments as being messages to the future, and it might be *you* reading those comments a year from now.

Tip Comments shouldn't have to tell the reader *how* the code works. The code should tell them that. Better yet, the unit tests should tell them that. You should prefer unit tests over comments. The best comment is a set of valuable, meaningfully named unit tests with high coverage. Regardless, if the code isn't clear, then you should refactor the code rather than add descriptive comments. You shouldn't add comments for the sake of "training" a future Developer. If/when a new Developer joins the Scrum Team, pairing, mobbing, or referring to unit tests are more preferable ways to learn. Also, don't forget to include meaningful comments and associate with a work item when committing changes in Git. Comments in code are not a replacement for comments when performing Git operations, and vice versa. You need to do both.

When commenting in code, only comment about what the code or unit tests can't say for themselves. If the code is well formed and follows modern patterns and principles, it probably doesn't need comments. When someone looks at the source code, its logic and purpose should be apparent. Keep this in mind while you are coding. Constantly ask yourself how clearly your code and unit tests are telling you, or another Developer, what's going on.

Tip Fellow Professional Scrum Trainer Phil Japikse suggests adding comments to highlight technical debt that the Developers have deliberately generated. You can think of it as pinning a credit card receipt directly to the code that caused the debt.

Remember that comments live inside your source code files, and as such, they become inventory just like the code itself. Comments can even be a form of technical debt if they are wrong or misleading. Be diligent about updating comments or removing them as you refactor and improve your code. Adding more comments isn't necessarily a good thing unless they add value. Perhaps it's time to refactor the code into simpler units rather than adding more comments.

Smell It's a smell when I see a file with the author's name at the top. I understand a Developer wanting to claim credit for their work, but this kind of comment tells everyone else to go away. It could be that the code file is really old and hasn't been touched since the Developers started practicing collective code ownership. If that's the case, someone should remove it. Azure Repos and Git track this metadata through commits, so it is redundant anyway. It could also be an organizational requirement to have predefined headers and require authors to add their names. If that's the case, bring this up at a Sprint Retrospective and consider meeting with the decision makers to ensure that the value delivered by this practice outweighs the waste that it seems to generate.

Fabrikam Fiber Case Study

The Developers use modern frameworks, principles, and practices as they design and code. As a result, there's not a lot of need for comments. Only when they are coding some complex methods do they add comments—mostly so they can quickly make sense of the logic. The Developers also know that when committing changes, they will provide a meaningful comment as well as associate the commit to the relevant Task work item (which links back to a forecasted PBI). Together, these two pieces of metadata provide more than enough context to clarify later *why* the changes were made.

Associating Commits to Work Items

I'm going to make some assumptions here. I'll assume a Developer or a pair of Developers is working on a forecasted PBI in the Sprint Backlog. I'll assume the Sprint plan in the Sprint Backlog is represented as Task work items or Test Case work items. In this situation, the Developers should associate each Git commit with the work item they are working on. Why should they do this? As previously mentioned, by linking work items to other work items or artifacts, the Developers can track related work, dependencies, and changes made over time. By associating work items to Git commits, the Developers bridge work planning with work execution to enable bidirectional, end-to-end traceability. I'll explain.

If a Scrum Team uses the planning tools in Azure Boards in the way I've outlined in earlier chapters, they will automatically achieve the linking of Epic, Feature, PBI, and Task work items. If the Developers choose to use tests for their Sprint plan and follow my advice when using Azure Test Plans as I outlined in Chapter 7, "Planning with Tests," their PBIs will be automatically linked to Test Case work items. On the engineering side, commits are associated with builds and builds with releases. This is done automatically by Azure Repos and Azure Pipelines. In other words, the end-to-end traceability from ideas to plans to work to build to release is performed automatically...almost. There is one critical step that must be performed manually. This step, as depicted by the lightning bolt in Figure 8-5, is the association of each Git commit to the relevant Task or Test Case work item. This manual association will typically be done outside of the browser, in a tool like Visual Studio, Visual Studio Code, or some other IDE.

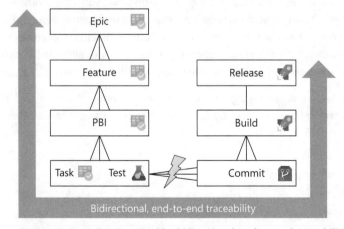

FIGURE 8-5 Azure DevOps provides bidirectional, end-to-end traceability from planning to release.

To associate a Git commit, simply include the work item ID in the commit comment. For example, if you are working on a Task work item with an ID of 42, then your commit comment might read "Added email validation #42" or "#42 Added email validation." When you push the commit to Azure Repos, Azure DevOps creates a Commit type link between the commit and work item #42. This link is visible in the work item form as well as in the commit details. You may need to enable the automatic creation of work item links in project settings since this option may have been disabled.

> **Note** Visual Studio Team Explorer also allows you to associate work items to commits through drag and drop from a query. This is nice because you don't have to memorize work item IDs. You can also associate a commit to a work item—after the fact—in Azure Boards by editing the work item and adding a Commit type link. This is nice if a Developer forgot to do it in Team Explorer.

> **Smell** It's a smell when I don't see commits associated with a work item. It could be that the Developers decided that traceability is not worth the extra steps required. In my experience, more often the Developers didn't know about this capability and that traceability like this was even possible.

Pairing, Swarming, and Mobbing

Of the hundreds of teams I've worked with, most were practicing Scrum—in some form or another. Others were considering it. Regardless, they all wanted to become more agile and enjoy those benefits. When asked if they were "working as a team," they always answered in the affirmative. They would go on to describe that each team member was busy doing *something* and the team usually got *something* done. Oh, and they would stand up every day too and discuss what each person was planning on doing that day. Great, but were they working as a team?

There are many ways to work as a team. Some ways are highly collaborative. Some maximize learning. Some minimize cycle time. Unfortunately, some emphasize individual specializations and output. To the untrained eye, all of these ways of working can look like a harmonious team at work. However, some ways of working actually maximize the risk of not delivering—which is especially true because today's complexity requires combining skills and even cross-learning of skills to succeed.

Over the years, I started identifying and classifying the various styles of teamwork and assessing their impact on the team. I've classified these into four dysfunctional categories and three collaborative categories, as you can see in Figure 8-6.

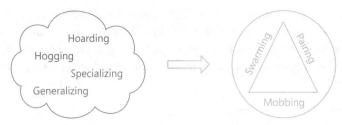

FIGURE 8-6 There are many ways a Scrum Team can work as a team—or not work as a team.

Four ways are dysfunctional in terms of the ability to collectively deliver and to transcend being just a group of individual generalists:

- **Hoarding**   After Sprint Planning, a Developer claims ownership of multiple PBIs and will do all the work on those items in isolation. Intra-team dependencies (such as knowledge) and external impediments get the Developer to cherry-pick among the selected items, ending the Sprint with some items open and few Done.

- **Hogging**   After Sprint Planning, a Developer takes ownership of a single PBI to do all possible work before moving on to do the same with the next PBI, and so on. Dependencies and impediments are similar to hoarding, but the single-piece focus slightly increases transparency and reduces the risk of not getting the item Done.

- **Specializing**   A Developer works those tasks pertaining to their specialty on different PBIs and "maximizes" their specialization across all items. Dependencies are high but remain hidden until toward the end of the Sprint, when late integration greatly increases risk.

- **Generalizing**   A Developer works a few different types of tasks. Not having all the skills required, the person will still start work on a new PBI when blocked. Dependencies may not be as high as specializing, but integration still occurs late, increasing risk.

Teams that start working in any of following three ways enter the "circle of team collaboration goodness," where mutual learning happens as an opportunity for professional development to become more T-shaped, pi-shaped, or comb-shaped. Collaborating in these ways also reduces dependencies and risk. Three ways of working fall into this category:

- **Swarming**   All Developers (alone or in pairs) will work on a single PBI, working multiple tasks in parallel until the PBI is Done.

- **Pairing**   Two Developers collaborate closely on a single task, potentially changing the pair composition in search of the required expertise. Pairs may be part of a larger swarm on a single PBI.

- **Mobbing**   All Developers work on a single task, of a single PBI, until that PBI is Done. As in swarming, there is no jumping or moving on before the PBI is Done. Mobbing is sometimes referred to as *ensemble* programming. To learn more about this practice, visit *https://mobprogramming.org*.

The practices of swarming, pairing, and mobbing are often confused with one another. Let me try to simplify things. Swarming is when all Developers work on a single PBI—alone or in pairs, working separate tasks. Pairing is when two or more Developers use a single keyboard to work on a specific task. Mobbing is when all Developers use a single keyboard to work on a single task. You can see an example of each in Figure 8-7.

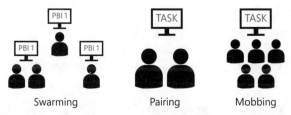

Swarming   Pairing   Mobbing

FIGURE 8-7 Swarming, pairing, and mobbing are all collaborative ways of working.

A benefit of pairing and mobbing is that the driver can focus on the tactical (coding) activities, while the observer(s) are thinking about the broader, strategic solution to the problem, as well as helping with training and coaching. You also get a code review for free. This form of extreme collaboration leads to better and simpler designs and fewer bugs, in shorter periods of time. Pairs and mobs working in close proximity like this are also less prone to get sidetracked from the task at hand. Most important is the extreme learning that takes place!

During pairing or mobbing, knowledge is passed back and forth. Developers can learn new practices and techniques from one another. Pairing a newly hired Developer, or a Developer with a different or weaker skill set, with a Developer who is more experienced will help improve the overall effectiveness and productivity of the team. Figure 8-8 shows the possible outcomes of pairing weaker and stronger Developers together.

Some teams scale this idea using an approach called "promiscuous pairing." In this approach, each Developer cycles through all the other Developers, rather than pairing with only one Developer. Promiscuous pairing ensures that knowledge of the product and its inner workings spreads throughout the whole team and thus reduces risk if a key Developer leaves. Mobbing does this as well.

		Developer B	
		Weak	Strong
Developer A	Strong	Learning	Flow
	Weak	Danger	Learning

FIGURE 8-8 Here are possible outcomes when pairing Developers.

Tip Mobbing maximizes learning across all Developers, and it also reduces the risk of not meeting the Sprint Goal and not completing the forecast. In fact, both swarming and mobbing assume that all Developers are working on the same PBI. As such, they are both fantastic practices to help reduce the risk of having undone work at the end of the Sprint.

Interruptions

From time to time throughout the Sprint (or the day in some environments), a Developer will experience an interruption. In a perfect world, this never happens, but in the real world, it does. Switching context to handle the interruption can be time consuming and wasteful. Professional Scrum Developers work to marginalize this reality by reducing the number of interruptions and by making them less painful.

For example, let's say that you and your colleague are deep in thought, implementing a complex scenario within a PBI, and the Product Owner comes to you with an urgent request. It becomes obvious that your attention is required elsewhere. Since the Product Owner deems it critical to the product and its stakeholders, you *should* drop what you're doing and attend to it. Forget your forecast. It's about saving money, customers, and reputation at this point.

Smell Although the Product Owner has the authority to interrupt the Developers, it's a smell when I see it happening on a regular basis. Perhaps the Product Owner should spend more time with the product and the stakeholders to identify these higher-risk areas. Perhaps the Developers need to improve the Definition of Done. Perhaps the Developers need to build more slack time into their Sprint (such as forecasting fewer PBIs). Perhaps the Developers need to slow down, generate less technical debt, and get more things Done.

Assuming a Developer has their code open and is making changes when the interruption occurs, the reaction might be to finish what they are doing or create a branch. Both of these are valid options. Depending on the urgency and disdain for branching, another valid option might be to *stash* the current code changes. Using the *git stash* command allows you to keep the current changes to your code but not commit them because they are not at a point where you are comfortable doing so.

This command takes the current staged and unstaged changes and saves the work, then returns your workspace to the state of the last commit. This process is similar to shelving in other version control systems. After changes are stashed, you can context switch to your heart's content and attend to any interruption. If you are interrupted again, you can use *git stash* again. Later, you can reapply your stashed changes, as I'm doing from Visual Studio in Figure 8-9. A stash is local to your Git repository. In other words, stashes are not pushed or synced with Azure Repos.

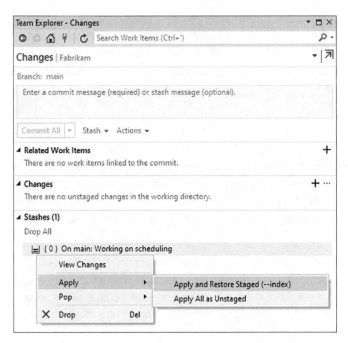

FIGURE 8-9 Manage stashed changes from Visual Studio Team Explorer.

Smell It's a smell if I see more than one stash listed. Maybe you are the kind of Developer who spends more time helping, mentoring, and supporting others and thus are often interrupted. This could explain the various pieces of suspended work. Maybe your environment is so chaotic that even the interruptions get interrupted! Maybe you are just the kind of Developer who gets sidetracked easily, leaving a bunch of half-eaten sandwiches sitting around your house—which is a different kind of smell.

Code Reviews

Code reviews help assure code quality by having other Developers look at the code. This assurance can cover multiple levels of quality. It can assure that the code works, is fit for purpose, is absent of bugs, is absent of technical debt, is readable, meets the Developers' agreed-upon coding standards, and meets the Definition of Done. Additionally, code reviews are learning opportunities for all involved.

Professional Scrum Developers recognize that the candid feedback (colloquially known as criticism) given during a code review is targeted at the code and not the coder. For new Developers, or at least those new to code reviews, there can be a tendency to take this feedback as an insult, even to become defensive. Everyone should be mindful of the Scrum Values of respect and openness during code reviews. Over time, all Developers will have an opportunity to review the work product—and an opportunity to prove that they are human and make mistakes too. Everyone can improve. Code reviews are just another type of shared learning activity, where anyone can learn from anyone.

Tip Code reviews can also catch and enforce coding style and standardization issues. Be careful not to spend too much time with these kinds of topics during a code review, since they can become a rathole. A rathole is any discussion that detours the original purpose of the conversation. Don't get me wrong—discussions about coding styles and standards are very important, but any debate or decisions about changing existing standards or establishing new ones should be deferred until the Sprint Retrospective. Professional Scrum Developers know that matters of style are not absolute. Developers should be allowed to self-manage and use whatever style they determine is fit for purpose. After a team has been working together for a while, their coding standards will begin to emerge—and hopefully merge. These standards may even become part of the Definition of Done.

Also, it may be beneficial to ask people outside the team to review the code and artifacts. These could be people from other teams in the organization or peers who the team respects. It's useful to get the opinion of someone not working directly in the code, especially where new technologies are concerned.

When reviewing someone else's code, you should avoid appearing as a "senior" Developer. Although you may be formerly known as "senior" or "lead," there are no titles in Scrum. All Developers are equal and should focus on sharing and learning. Choose tone and words carefully as you identify issues and improvements in someone else's work. Developers new to Scrum may also be new to the value of openness. Don't aggravate the situation by going on the offensive.

Code reviews don't have to be a formal process. They can happen spontaneously. They also shouldn't be despised or avoided. Professional Scrum Developers actually look forward to code reviews. This is because those Developers know that the code is owned collectively. Problems and criticisms aren't directed at a single Developer; rather, they are learning opportunities for all Developers. Everyone "develops" something, and having a reviewer with a different approach and perspective should be seen as a benefit, demonstrating the Scrum Value of openness.

Smell It's a smell when I see Developers performing code reviews *in addition to* pairing or mobbing. Since you can think of pairing and mobbing as a form of code review—one that happens in real time—I don't like to see added waste in the process by including another stage or gate in the process. It smells like someone outside of the team is imposing code reviews and/or doesn't understand that pairing and mobbing provide the same outcome. That said, if the Developers find value in having an informal code review for the sake of learning, then far be it from me to suggest otherwise.

It's also a smell when I see a collocated team using tools to facilitate code reviews. They should be able to practice these reviews in person. Excuses are usually to the effect of "we're busy." It's obvious that they want to use the asynchronous behavior that the tool provides. I understand that there's a cost to interruptions, and that tools like Microsoft Teams and Slack are good for quick questions. Code reviews are not quick interruptions. They require a full stop, a context shift, focus, and participation in order for the review to provide value. As I've mentioned several times in this chapter, conversations that take place face to face are more efficient, reduce ambiguity and misunderstanding, and provide more value than anything facilitated by a tool.

Professional Scrum Developers build solutions that are fit for purpose while avoiding gold plating. *Gold plating* is any design or work that is above and beyond what is absolutely necessary for the task at hand. For example, if a PBI requires a method that calculates the sales tax for the state of Idaho, and the Developer includes additional logic to handle the surrounding states, that's gold plating. The Developers may try to justify the extra coding as being required down the road for a future Sprint. In order to maximize value and minimize waste, Developers should solve today's problem today and tomorrow's problem tomorrow—in the next Sprint. Code reviews can be a good way to unearth gold plating.

Fabrikam Fiber Case Study

The organization has no policies for code reviews. They leave everything up to the team to decide. Sometimes Developers perform ad hoc code reviews. These are done whenever a Developer hits the wall or needs a better solution for a complex problem. These "reviews" usually become an impromptu pairing session. In addition, the Developers like to invade the conference room and use a big screen. Each Developer in turn shows off their work. With a full room, this approach encourages discussion on design and style.

Branching

Branching is the duplication of a folder or file in a repository so that Developers can make modifications to the branch while keeping the source of that branch stable. The originating branch is typically referred to as the *parent branch,* and *child branches* are those that have a parent. A branch without a parent is referred to as the *trunk* or *main.*

Note As of this writing, the top-level, default branch in Git is named *master.* Git's maintainers will be adding a flag that will let its name be changed. In the next major version, Git will probably start using the term *main* instead of master. GitHub and Azure Repos will follow suit.

Working in branches enables Developers to develop parts of the product in parallel. Branches support the ability to have multiple releases—for different configurations (such as CPU architecture, operating system, or other customizations). Branches also enable Developers to isolate changes without destabilizing the code base, like when they fix bugs, add new features, or integrate versions. These divergent changes are typically merged (integrated) back into the parent branch at a later point in time.

With legacy version control systems, like Team Foundation Version Control (TFVC), working within a branch was optional. The Developers would be able to work in folders, directly on files until they wanted to explicitly create a branch. With Git, you are always working within a branch. In fact, it's impossible to *not* be in a branch.

Developers should limit the number of branches in use. Unfortunately, the software development community has become "branch crazy" with tools, practices, and patterns *encouraging* branching. Creating and working in *development* branches has become mainstream. These would be the branches used by the Developers as they develop individual PBIs. GitFlow and GitHub Flow are both popular branching strategies that define strict branching models. Although there are nuanced differences, both are feature-branch focused and recommend creating a new branch whenever the Developers (or a single Developer) begin working on a new PBI. This strategy sounds good on the surface, but it can lead to dysfunctional team behavior such as hoarding and hogging, which I mentioned earlier. This can lead to delays in getting to Done!

Even with Developers regularly pushing to branches, the integration between the different branches or between the branch and main is delayed. Until the code is in a single shared location such as main, the Developers won't know if something will break during the merge, how difficult the merge will be, if the work will be duplicated, or if the work is incompatible in some way. Automated build and testing will take Developers only so far when working in development branches. Multiple development branches create distance between the Developers, which can cause a decrease in collaboration, a delay in integration, an increase in waste to diagnose and fix a broken build, and an increase in the risk of not achieving the Sprint Goal.

Trunk-Based Development

Enter trunk-based development (*https://trunkbaseddevelopment.com*), an approach to branching that is increasingly gaining more popularity in the community—especially since it has been used at Google by tens of thousands of developers doing concurrent development. Trunk-based development is an anti-branching model that brings Developers together by having them clone, check out, and work directly in main or trunk. The Developers will update, pull, and sync from main or trunk many times a day. They are now integrating their commits with their teammates' commits on an hour-by-hour basis, if not more often.

Similarly, when a Developer (or pair) completes a small piece of work (for example, a task or refactoring) they will commit and push it back to main or trunk. This will cause the team to gravitate to more collaborative ways of working, such as swarming or mobbing. In order to not break the build, any code changes will be pulled, merged, built, and tested in their local repository. Any problems are resolved locally before pushing to main or trunk. Depending on how long it takes to pull, build, test, and resolve issues, this cycle may have to be repeated.

Note Many branching strategies, including trunk-based development, have the rule that anything in the main/trunk branch should always be deployable. Although I understand the desire for this, Azure Pipelines and Azure Artifacts offer cleaner alternatives for ensuring that the latest, stable binaries and artifacts are deployable. In other words, if Developers want to deploy (or redeploy) the latest binaries, rather than pull from main or trunk, rebuild, and deploy, they could just pull the latest stable artifacts from Azure Artifacts or push the latest stable build or release in Azure Pipelines. With Azure DevOps, the need to keep the main/trunk branch stable is no longer a requirement to release on demand.

Fabrikam Fiber Case Study

Individual Developers have become comfortable working in feature branches during development. Creating a separate branch for each PBI has made it easier for them to work on two or more PBIs at a time. Knowing that this way of working is risky—for many reasons—they plan to move to trunk-based development. As a corollary, they will also increase their swarming on a single PBI at a time.

Continuous Integration

Professional Scrum Developers—especially those practicing trunk-based development—have learned how to work smarter, not harder. One way that they do this is by continuously integrating their code changes with other Developers and by running automated tests to verify that those integrations don't break anything. Although these same automated tests can (and should) be run locally—for example, in Visual Studio—Azure Pipelines can probably run them faster and asynchronously, enabling the Developers to work on something else while the build executes. Another benefit is that the tests can be run in a controlled environment that can quickly make transparent any configuration management problems.

A better way to avoid painful, manual merge operations is to do smaller, less-painful merges throughout the day. This is the basis for continuous integration (CI). CI is more than just ensuring that the integrated code builds without error; it also mandates tests being run. In other words, upon a push to Azure Repos a build pipeline gets triggered. Applicable code is compiled, binaries are generated, automated tests are run, and feedback is returned to the Developers *quickly*.

Tip Another way to minimize the pain of manually merging code is to *actively* listen to the other Developers during the Daily Scrum. Remember that the purpose of the Daily Scrum is to synchronize and create a plan for the next 24 hours. This means that each Developer verbally shares their planned work with others. If one Developer hears another Developer mention a task that will be in the same part of the codebase that they were planning on working on, they should consider pairing up and working on their overlapping tasks together. Doing so should alleviate the need to manually merge code changes, as well as increase knowledge and productivity in general.

CI is about reducing risk. When a Developer defers integration until late in the day, the week, or the Sprint, the risk of failure (features not working, side effects, bugs, etc.) increases. By integrating code changes with others regularly throughout the day, the Developers will identify these problems early and be able to fix them sooner because the offending code is fresh in everyone's mind. The practice of CI is a must for a high-performance Professional Scrum Team.

As previously mentioned, any branching strategy—especially trunk-based development—should strive to keep the code in main or trunk pristine—and releasable. One way of doing that is to implement a CI pipeline. The CI build should run as quickly as possible, returning the feedback quickly and clearly, especially if main or trunk is broken. I like using the *CatLight* extension, which puts build notifications right in my tray. You can see how CatLight informs me of a failed build in Figure 8-10. Learn more about this extension at *https://catlight.io*. Remember, Professional Scrum Developers collectively own all things, including working in a way that keeps the build from breaking, but also in repairing a broken build.

FIGURE 8-10 CatLight notifies you when a build fails.

Refactoring, restructuring classes and methods, and changing internal interfaces can be messy work. There may be times that you want to push your not-yet-finished code so that another Developer can begin working with a part of it. You may also want to see how many errors, warnings, and failed tests occur when your changes are integrated with others. Generally speaking, this is fine. This is feedback. This is learning. But you should keep the integrity of the main/trunk code in mind as you do so. If possible, try to adopt a "merge down, copy up" (called "pull, push" in Git) behavior where you pull changes, resolve any conflicts locally, build and test locally, and then push the merged and verified code back to Azure Repos. This strategy minimizes the chances that you will break the build.

Fabrikam Fiber Case Study

The Developers have long been fans of continuous integration and have created several CI pipelines to watch over various parts of their repository. As the Developers move toward trunk-based development, they will explore how to make their CI pipelines faster. A dedicated, self-hosted pipeline with fast hardware will be a good start. Another experiment they plan to try is to enable Test Impact Analysis in the pipeline to speed up testing by only running those tests that were impacted since the last build. For more information, refer to *https://aka.ms/test-impact-analysis*.

Pull Requests

To review, branches are used to isolate work until such time when those changes are to be merged into another branch such as main. A pull request is a mechanism that tells other Developers about changes that have been pushed to Azure Repos with the intent to merge them into another branch. The intention is that the pull request *becomes* the collaborative process that lets the Developers discuss these potential changes and then agree to merge them once everyone approves. This is fine for dislocated teams, such as those collaborating on open source projects, but it's wasted on Professional Scrum Teams, where they collaborate, discuss, and agree in real time.

> **Note** I'm not explicitly anti-pull request, but I am anti-branch. This is because branching, by definition, promotes isolation and not collaboration. I believe that trunk-based development is the approach to take. So, since pull requests require branching, then I guess I'm anti-pull request too. Besides, pairing and mobbing are better alternatives to code reviews (via pull requests) anyway.

If Developers want to self-manage and experiment with pull requests as a mechanism for code reviews, then who am I to suggest otherwise? I would just caution those Developers on preferring a tool over interacting directly with individuals. For example, a dislocated Developer may want someone else to look at their code before integration, and if the Developers have deemed other review practices as less effective, then creating a short-lived branch for the sake of a pull request might be useful. If, on the other hand, it becomes an established policy that every change must go through a pull request, then it becomes a delay that dissuades collective code ownership behaviors.

Chapter Retrospective

Here are the key concepts I covered in this chapter:

- **Collaboration is key**   Software development is a team sport. The Scrum Team members need to communicate with one another, as well as stakeholders, effectively.

- **Collocated teams are more productive**   Scrum Team members working in close proximity are more productive and generate more business value than dislocated teams—those who are geographically, temporally, or culturally distributed. Large, open-space team rooms can be particularly effective.

- **Meet effectively**   Scrum has all the built-in events (meetings) that a team needs. Limit attendance to other meetings, or send the Scrum Master instead—who is supposed to be protecting the team's focus.

- **Listen actively**   Use active listening communication techniques that enable better, more effective dialogue. These techniques support the Scrum Values of respect and openness.

- **Limit interruptions**   Turn off or otherwise neutralize cell phones, email clients, and other communication tools. Limit research, internet searches, nonessential meetings, and yak shaving.

- **Strive to be T-shaped**   Team members who are T-shaped have a deep expertise as well as a wide knowledge base. T-shaped Professional Scrum Developers are open to learning and doing new things, even if they are out of their comfort zone.

- **Developers collectively own the code**   The Developers own every aspect of the code. Everyone can clone, pull, and push code for any aspect of the product. Fear not, Azure Repos will track all changes being made.

- **Comment code only if it adds value**   Only comment when you need to and in accordance with the Developers' working or social agreement. When commenting, be sure to adequately explain your actions to others. When possible, let the unit tests do the commenting.

- **Associate commits to work items**   When performing a Git commit, be sure to associate it with the relevant Task or Test Case work item. This manual step will ensure bidirectional, end-to-end traceability from ideas, to plans, to work, to builds, to release.

- **Consider practicing swarming**   All Developers (alone or in pairs) will work on a single PBI, doing whatever is required, until the PBI is Done.

- **Consider practicing pairing**   Two Developers collaborate closely on the same task.

- **Consider practicing mobbing**   All Developers work on a single task, of a single PBI, until that PBI is Done. Like swarming, there is no jumping or moving on before the PBI is Done.

- **Practice code reviews only if they add value**   Practice these in person, or consider pairing or mobbing as an alternative. Developers should be open to giving and receiving criticism. Avoid using tools to facilitate a code review.

- **Consider practicing trunk-based development**   Simply put, don't work in branches; instead, work directly on main or trunk.

- **Integrate continuously**   Merging is painful, so do it more often and it will hurt less. Stay in touch with your builds so that you will be notified of failed builds as early as possible. Get them healthy again as soon as possible. A failed build is everyone's concern.

Improving

In the first two parts of this book, I introduced you to practicing Professional Scrum using Azure DevOps. The first eight chapters should be sufficient to enable a Professional Scrum Team to start using Azure DevOps to plan, manage, and execute their work in an agile way. I've done my best to nudge you toward the Professional Scrum practices you should consider, as well as the shiny distractions you should avoid.

Going forward, *how* you use Azure DevOps and practice Scrum will be up to you. It will be up to you individually and as a member of a Scrum Team to experiment, inspect, adapt, and improve. That's what the final part of this book is all about—improvement—as a team and at scale. Over the next several chapters, you will be exposed to the concept of flow, including how a Scrum Team could measure and improve it. You will also be made aware of some common challenges facing self-managing Scrum Teams. This is done not to scare you away from Scrum, but to inform you of what impediments and dysfunctional thinking lurk in every organization.

The final chapter in this section will cover scaling and how to control the chaos when multiple Scrum Teams collaborate to build a single product. This will also be an introduction to the *Nexus* scaled Scrum framework as well as how Azure DevOps supports planning and managing work at scale.

CHAPTER 9

Improving Flow

*F*low is such an overloaded term. If we exclude all of the mathematical, scientific, media, software product, software language, and music references and just focus on complex product development, there are still multiple usages of the term. Some use the term "flow" as shorthand for workflow. I mentioned in the previous chapter that there can be flow at the individual (or pair or mob) level, which manifests as people being creative, productive and "in the zone"—even losing track of time while they work. The term "flow," as I will focus on in this chapter, describes how work *flows* through the team's process.

I really like fellow Professional Scrum Trainer Daniel Vacanti's definition of flow. He defines flow as the *movement of customer value throughout the product development system*. This flow can be visualized, can foster collaboration, and allows for the team's optimal process to emerge. This flow is also something that can be measured and—more importantly—improved.

In this chapter I will describe how a Scrum Team can visualize, manage, and improve their flow. I will also show how these teams can use flow metrics in their Sprint events—essentially showing how Kanban can be an effective, complementary practice of a Scrum Team. In fact, if you were to overlay Scrum and Kanban in a Venn diagram and, more specifically where the subset of *Professional Scrum* and subset of *Professional Kanban* overlay—that's where Professional Scrum with Kanban exists. You can see this in Figure 9-1.

> **Note** Professional Kanban would be Kanban practiced according to the *Kanban Guide* with a focus on flow of value by using the Kanban practices and lean metrics to visualize the work, limit work in progress (WIP), manage the flow, and continuously improve through inspection and adaptation of the definition of how work flows. In other words, it's when the team uses the inspections from Kanban to make adaptations in order to maximize the flow of value through the system. You can learn more by visiting *https://prokanban.org*.

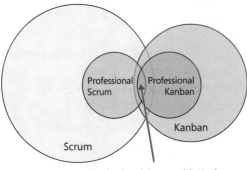

Professional Scrum with Kanban

FIGURE 9-1 Professional Scrum with Kanban sits at the intersection of Professional Scrum and Professional Kanban.

Note In this chapter, I am using the customized Professional Scrum process, not the out-of-the-box Scrum process. Please refer to Chapter 3, "Azure Boards," for information on this custom process and how to create it for yourself.

Visualizing Flow

Professional Scrum Developers *want* to make their work visible. They know that humans process visuals faster than text. They also realize that a visualization that is made transparent to the entire team creates a shared understanding of what work has been done and, more importantly, what work is left to do. In Chapter 6, "The Sprint," I introduced the Taskboard. Initially, the tasks on the Taskboard are in the To Do column (state). As the Developers work on a PBI, the related tasks will transition to In Progress and, eventually, to Done. Keep this in mind as I continue talking about visualizing flow.

In the previous chapter, I discussed some collaborative practices that can improve a team's flow—specifically swarming and mobbing. Both of these practices improve flow as they enable the Developers to focus on a single PBI, without being distracted by other PBIs. This results in a PBI being Done earlier.

It is difficult to visualize flow on the Taskboard. If you omit the outermost *To Do* and *Done* columns, you are left with only one column—*In Progress*—to represent all the different states of work. Only if the Developers were diligent about naming the tasks appropriately would the team be able to determine where they were in the workflow.

For example, let's assume that a typical PBI's workflow looks something like this: designing, coding, testing, releasing. Yes, it's a simplistic, sequential, almost waterfallian workflow, but work with me here. For the Developers to visualize this workflow, they would need to create tasks that essentially map to each of those workflow steps. Generally, this is what a Developers do, although they might have many tasks per workflow step. You can see in Figure 9-2 that there are two design tasks, five coding tasks, one testing task, and two release tasks in our example. I've added tags to help make this apparent.

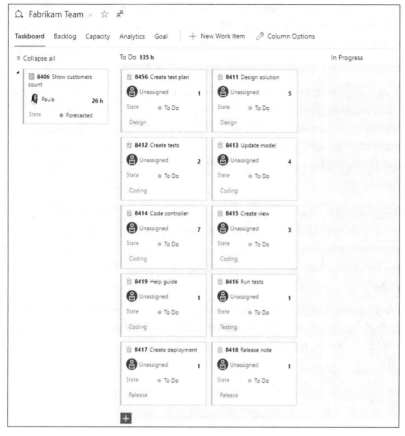

FIGURE 9-2 A PBI may have multiple tasks per workflow step.

If a Developer were to hoard or hog this PBI, they might execute these tasks sequentially, and it would look something like this: design, design, code, code, code, code, code, test, release, and release. As I mentioned several times in the last chapter, Professional Scrum Developers don't execute their work sequentially if at all possible. They will instead find ways to work some tasks in parallel.

As you can see, the Taskboard itself isn't very helpful in modeling a team's workflow. It's intended that tasks will do that. This makes the Taskboard very versatile for representing a variety of disparate workflows—from coding a new feature, to fixing a bug, to upgrading infrastructure, or even to writing documentation. For Developers who would rather have the *board* represent their workflow, the Kanban board may be a better choice for visualizing flow.

The Kanban Board

I covered the Kanban board in Chapter 5, "The Product Backlog," although I did so in a context of a Product Owner using it as a way to visualize a PBI's progression to becoming "ready"—not as Developers using it for development. The latter is the more popular use of a Kanban board—to visualize and manage development workflow. A Kanban board turns a linear backlog into an interactive, two-dimensional board, providing a visual flow of work. As work progresses from idea to completion, Developers update the items on the board. The items in our case are PBIs. This visualization makes transparent the current item's progress—or lack of progress.

 Note Don't confuse using the Kanban board with practicing Kanban. Scrum Teams can use the Kanban board to visualize their work without practicing the other aspects of Kanban. That said, the Professional Scrum community views Kanban as a valid complementary practice. For more information on how to practice Kanban within Scrum, download and read *www.scrum.org/resources/kanban-guide-scrum-teams*.

In Azure Boards, Task work items are not visualized on the Kanban board. Only PBI work items (and Bug work items if you have them enabled) are visible on the Kanban board. The board's columns initially map to a PBI's workflow states. This means that, for our custom Professional Scrum process, the board will initially have these columns: *New*, *Ready*, *Forecasted*, and *Done*. These default states do not make for an interesting or useful Kanban board.

To be useful, Developers would need to expand the default Forecasted column into additional columns, such as *Design*, *Coding*, *Testing*, and *Release*—all of which would map back to the underlying Forecasted state of the PBI work item type. Figure 9-3 shows an example.

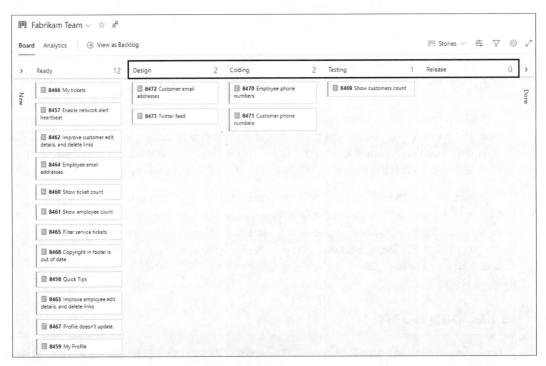

FIGURE 9-3 You can add custom columns to the Kanban board.

Tip When creating a Kanban board, be mindful of having too many columns. Each column is often an opportunity for more work in progress (WIP). Also, make sure that columns are not tied to people. In other words, seeing a column named *Skyler* is a smell. Instead, label that column based on what Skyler does on the team—and then make sure there are other "Skylers" to help out. There also should be a clear "work has started" column and a clear "work has finished" column. For Scrum teams, the finished column is typically titled *Done*. Also, try to avoid columns for blocked items.

When the Developers begin work on a PBI, someone will pull it into the *Design* column. When Design is completed, another Developer will pull it into the *Coding* column, and so on. When the PBI is done—according to the Definition of Done—it is moved into the *Done* column on the far right. Where the Taskboard requires you to see what task is in progress to learn the status/progress of a PBI, you can simply look at the column the PBI is in on a Kanban board to get the same information.

Because each column corresponds to a stage or state of work, you can quickly see the number of items in progress at each state. However, a lag will often occur between when work gets moved into a column and when work actually starts. As a way to address that lag and reveal the actual state of WIP, you can split columns into *Doing* and *Done*, as I've done with the *Design* column in Figure 9-4.

FIGURE 9-4 Columns can be split into *Doing* and *Done*.

The Kanban board is nice for modeling the workflow of most of the work (such as coding a new feature), but not all PBIs will make use of all the columns. For example, a bug fix might skip the *Design* column, and who knows how an infrastructure upgrade PBI would traverse the board. Asymmetrical work like this can be challenging, but before you give up on the Kanban board and return to the Taskboard, please continue reading this section on managing flow and flow metrics, which makes additional arguments for using the Kanban board. For more information on configuring and using the Kanban board, visit *https://aka.ms/kanban-quickstart*.

Managing Flow

Professional Scrum Developers are quite interested in managing their flow. They regularly review and improve the way in which they work in order to deliver Done, working product quickly and efficiently. From a flow perspective, this means removing impediments and making improvements so that value moves in a continuous and smooth way. Sounds great. I'm sure that you would like this, too, but how would you get started? I recommend rereading Chapter 8, "Effective Collaboration," because it covers many collaborative practices that will help you achieve this goal.

Managing flow means always making these kinds of inspections:

- Is work flowing smoothly through the system?

- Are there any impediments that block flow?

- Is a PBI currently blocked?

- Do policies (such as WIP limits) need to be established or changed?

- Are policies being followed?

- Are the Developers starting a new PBI before finishing the previous one?

- Can the Developers work in a different way in order to finish a PBI sooner?

- How can the Developers collect and use flow metrics appropriately?

- Are items aging as expected?

- What experiments can the Developers try to improve their process and flow?

Note *Aging* refers to the amount of time that a PBI remains in progress—usually measured in days—and can be scoped to a particular column (such as *Coding*). By monitoring aging, the Developers can analyze the flow of their work through the board. By tracking a PBI's age, Developers can identify impediments, such as dependencies, before they turn into a larger risk of missing the forecast or the Sprint Goal. The aging per column will vary based on the complexity of the PBI. In other words, some PBIs may spend more than a day in design, whereas others may age longer in testing. By inspecting a PBI's age, Developers can determine whether improvement experiments were successful.

Many of the aforementioned inspections are things that Professional Scrum Developers should already be making during their Sprint Retrospectives. There shouldn't be anything new in that list—with the exception of a couple of new concepts: policies/WIP limits and flow metrics. The rest of this section will be spent covering those concepts.

Smell It's a smell when I see a team create a Kanban board that represents their current (dysfunctional) way of working and then leave it—Sprint after Sprint. Sure, for a large, waterfallian organization, the Kanban board on day 1 of Sprint 1 might have 15 columns—most of which are for external handoffs, such as testing or approval—but hopefully that board would start to collapse over time. For a truly cross-functional, self-managing team full of T-shaped individuals, I'd hope that the Kanban board would eventually collapse to a single *Developing* (also known as *In Progress*) column.

The Kanban board visualizes the Developers' current way of doing things whether it is sequential (waterfall) or more collaborative (flow). The fact that a Kanban board has states that seem like waterfall states isn't the point. If the batch size is small, then it isn't waterfall. What's more, if it's possible to move to a more concurrent collaborative workflow, that's great. Kanban's point is to make a team's current workflow visual so that they can work to simplify it over time.

Note For Scrum Teams practicing Kanban, the Scrum Master should add *flow coaching* to their repertoire of activities. This would include making the team reflect and act, such as following the policies it has created, creating new ones when needed, discussing and acting on exceptions (issues and opportunities), and experimenting to find creative solutions. Much like the "flow manager" on some Kanban teams, the Scrum Master acting as a flow coach should inspire and challenge.

Limiting WIP

Work in progress (WIP) limits constrain the amount of work the Developers undertake at each work state and, thus, in the entire system. Much like a highway at rush hour, you simply do not want to have more traffic (WIP) than the system (Developers) can handle. The goal is for the Developers to focus on completing items before starting new ones. WIP limits encourage this.

Note You might know WIP as work in *process*, not work in *progress*. Technically, both are correct. I prefer *progress* because it describes movement toward a goal, which is what a Professional Scrum Team strives to do in their daily work. Progress also implies improvement (through working and learning). Contrast this to *process*, which simply conjures a series of actions. Fellow Professional Scrum Trainer Peter Götz dives into this deeper at *www.scrum.org/resources/blog/wip-work-inwhat*.

WIP limits should be based on the Developers' capacity and ability to do the work—not on the number of "specialists" for that column. In Azure Boards, this is done by setting a WIP limit on each column. For example, if the Developers set a WIP limit of 1 on the *Testing* column, this policy will require that the Developers can be testing no more than one item a time. As a corollary, upstream work should not proceed if it will build up inventory. This means that Developers who are coding should shift to helping with testing, rather than accrue more code to be tested.

As you can see in Figure 9-5, WIP limits are just a soft constraint on the number of items allowed within the column. If WIP goes above the WIP limit, the number will show in red, providing a visual cue to the Developers. Nothing actually prevents a Developer from pulling more items into a column and exceeding the WIP limit. When WIP drops below the defined limit, that is the signal to start new work.

FIGURE 9-5 Column WIP limits provide a visual indication that there is too much WIP.

Although setting WIP limits is easy, adhering to them takes a commitment by all Developers. Teams new to the concept may find WIP limits counterintuitive and uncomfortable. However, this single practice has helped Developers identify bottlenecks, improve their process, and increase the quality of their products. Limiting WIP helps flow and improves the Developers' self-management, focus, commitment, and collaboration.

> **Note** Once the Developers have placed WIP limits on their Kanban board, they have created a *pull system*. In a pull system, a Developer only starts (pulls) work when there is capacity to do so. For example, only when the *Coding* column has capacity will someone pull a PBI from the *Design-Done* column to *Coding*. Contrast this to a push system, where the PBI that is done with *Design* is pushed to *Coding*, possibly creating an inventory backlog. In a pull system, a PBI is pulled, "owned," and worked on in a just-in-time manner as determined by the team's capacity. This aligns with the guidance I've given previously for using the Taskboard and collaboration. The benefit of implementing a pull system is that it is easy to define what it means for work to have "started." Without establishing or using WIP limits, the Kanban board is only a visualization.

WIP limits should be discussed at each Sprint Retrospective, along with any related challenges the Developers might have. Limits should be adjusted accordingly. Once balanced, WIP limits will ensure that the Developers will keep a productive pace of work without exceeding their work capacity. WIP limits need to be set low enough to have an impact. If the WIP limits don't change behavior, they're not limits.

Much has been written about WIP limits and the intricacies of queuing theory and the underlying mathematics. I recommend starting with Daniel Vacanti's work at *https://actionableagile.com*.

Managing WIP

Limiting WIP is necessary to achieve flow, but it alone is not sufficient. To establish flow, the Developers must actively manage their work items in progress. This can take several forms during the Sprint, and there are many complementary practices that they can use to manage WIP.

Here are some examples of how Developers should manage their WIP during the Sprint:

- Making sure that PBIs are only pulled into the workflow at about the same rate that they leave the workflow

- Ensuring PBIs aren't left to age unnecessarily

- Swarming, pairing, and mobbing to move aging work items

- Responding quickly to blocked PBIs

- Responding quickly to PBIs that are exceeding the expected Cycle Time level (*service level expectation,* or SLE)

As I've said in previous chapters, it's important that the Developers do not start new work before they finish existing work. In other words, the Developers should complete all tasks for PBI #1 (according to the Definition of Done) before starting the first task on PBI #2. By starting work at about the same rate as finishing work, the Developers will be effectively managing their WIP.

Regardless of the way the Sprint plan is formulated (tasks, tests, or Kanban board columns), it's important for the Developers to monitor progress accordingly. This is more than just seeing what they can start on next. It's about ensuring that the current WIP is progressing normally and not aging unnecessarily. If a PBI gets blocked, the Developers should adopt an "all hands on deck" mindset to get it unblocked. As long as it's blocked—for whatever reason—it will age unnecessarily, Cycle Time will increase, and forecasts may be missed. Some blockages may require help from the Scrum Master.

The same reasoning applies to PBIs that are queued, ready to be pulled into a new column. Referring back to Figure 9-5, let's assume that PBI #8472 has been *Design-Done* for three days whereas PBI #8473 has been *Design-Done* for only one day. All things being equal, the Developers should pull the older PBI into *Coding* so that it does not age unnecessarily. This is known as a *pull policy*. This particular pull policy discourages PBIs from unnaturally aging and encourages productivity. There may be times, however, that Developers may want to pull a newer item for valid reasons, such as to manage dependencies. Changes to pull policies should be discussed and agreed on by the Developers.

A service level expectation (SLE) is a forecast based on empirical data on how long it should take any given PBI to flow from start to finish within the workflow (such as Design to Done). The Scrum Team uses its SLE as a gauge to find active flow issues and to inspect and adapt in cases of PBIs falling below those expectations. The SLE itself has two parts: a period of elapsed time (days) and a probability associated with that period (for example, 85% of work items should be finished in four days or less). The SLE should be based on the Developer's historical Cycle Time. Once calculated, the Scrum Team should make its SLE transparent. This is especially important when the Scrum Team moves to continuous delivery.

The ActionableAgile Analytics extension uses a team's historical data in Azure Boards to calculate its Cycle Time SLE. You can see this plainly displayed on the Dashboard in Figure 9-6. To see additional probabilities (50%, 70%, 95%, etc.), you can view the Cycle Time Scatterplot analytic. For more information on the ActionableAgile Analytics extension, visit *https://aka.ms/actionable-agile-analytics*.

FIGURE 9-6 The ActionableAgile Analytics extension calculates many flow metrics.

Note You may not have heard of service level *expectation* (SLE) before. The more popular term in our industry is service level *agreement* (SLA). I prefer SLE because "expectation" suggests a forecast based on evidence as well as a culture of learning and improving. "Agreement" comes from manufacturing, where tasks are generally repeatable with a high degree of accuracy and thus infers a commitment or a promise. Although this is more popular with customers and organizations, it doesn't reflect the reality of operating in the complex space of knowledge work. This subtle renaming is akin to why Sprint *commitment* was renamed to *forecast* almost a decade ago.

If a Scrum Team doesn't have much—or any—historical Cycle Time data, the Scrum Team should make its best guess at an SLE. Over time, as there is more historical data, they can inspect their Cycle Times—using the appropriate analytics—and adapt their SLE. Surprisingly enough, it doesn't take much history to calculate a fairly accurate Cycle Time.

Inspecting and Adapting Workflow

The Developers use the Scrum events to inspect and adapt their definition of workflow, thereby helping to improve empiricism and optimize the value that they deliver. Although improvement conversations are typically relegated to the Sprint Retrospective, workflow inspections can occur at any time during the Sprint. Workflow inspections—that the Developers might make—generally break down into two categories: visualization policies and working policies.

Visualization policies include the adding, merging, splitting, or removing of states (columns) to represent a refined workflow. This can be done as the Developers learn new ways of working and bring more transparency to areas that they want to inspect and adapt.

Working policies include making changes to address specific impediments, such as adjusting WIP limits, adjusting the SLE, or even implementing pull policies. Changes to the Definition of Done might also be considered a change to a working policy.

When the team is setting up the Kanban board, the visualization of the Developers' workflow should include explicit policies about how work flows through each state. This may include one or more items from the Definition of Done. For example, if the Definition of Done includes 12 items, and 4 of those are testing related, then those 4 items might exist in the *Test* column's Definition of Done. A column may also have additional Done criteria that are not in the overarching Definition of Done. Conversely, the Definition of Done may have items that are not listed in any specific column's done definition. Azure Boards supports a Done definition at each column, as you can see in Figure 9-7.

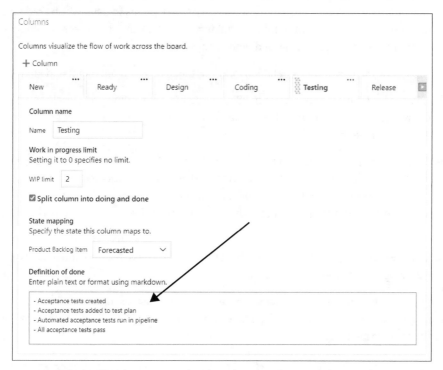

FIGURE 9-7 Each WIP column on a Kanban board can have its own definition of done.

Whether or not the Developers establish definitions of done per column, they should make sure that everyone has a clear understanding of what's required for work to flow through each state. This will help prevent missed communication, additional meetings, and rework. The Kanban board's configuration should prompt the right conversations at the right time and proactively suggest opportunities for improvement.

As with everything else the Developers do, they should use the Kanban board for a period of time, and then inspect and adapt accordingly. During Sprint Retrospective, as well as at other times, they can evaluate the columns, WIP limits, policies, and definitions of done.

Flow Metrics

The metrics of flow are very different from traditional metrics. Instead of focusing on things like story points and velocity, teams can use new metrics that are more transparent and actionable. By transparent, I mean that the metrics provide a high degree of visibility into not just the progress of the Developers' work, but also *how* they work. By actionable, I mean that the metrics themselves will point to some specific improvements needed to improve the Developers' performance. Traditional agile metrics and analytics give no visibility or any suggestion of what to do when things go wrong.

To be truly agile, teams and organizations should adopt the language of their stakeholders, or at least use terminology that is not technical or confusing. Unfortunately, traditional agile metrics are sometimes confusing to stakeholders. Story points, velocity, burndowns, and so forth are all weird terms that might need to be explained. Also, the conversion of story points to days, which I discourage, will still be done (wrongly) in the heads of those stakeholders. Enter flow metrics, where stakeholders will quickly understand basic concepts of elapsed time in days and item counts.

Here are the four flow metrics, within a defined workflow, that a Scrum Team might find interesting:

- **Work in Progress (WIP)** The number of PBIs started but not yet finished. Developers can use this leading indicator metric to provide transparency about their progress in order to reduce their WIP and improve their flow.

- **Cycle Time** The amount of elapsed time between when a PBI starts and when it is Done. In other words, the amount of elapsed time when the PBI enters the first WIP column to when it enters the departure (for example, *Done*) column. Cycle Time is a lagging indicator and best answers the question "How long will it take this PBI to complete?"

- **Work Item Age** The amount of time between when a work item started and the current time. This leading indicator applies only to items that are still in progress. For Developers, PBIs in the Product Backlog are not considered to be aging.

- **Throughput** The number of work items finished per unit of time, such as a Sprint. Throughput is a lagging indicator and best answers the question "How many PBIs will be done by the end of the Sprint or in the next release?"

> **Tip** Although these metrics are less confusing than traditional agile metrics, be careful sharing them with stakeholders. They are primarily technical metrics and may not be helpful when shared beyond the Scrum Team. Worse, stakeholders may misinterpret or infer something that is not true (a commitment, a plan, a budget, etc.). It's better if the Product Owner or Scrum Master reveals the appropriate metrics (such as service level expectation) to stakeholders as necessary.

I've already discussed WIP in this chapter, so the WIP metric should be pretty straightforward. It's simply the number of PBIs that the Developers have started working on but have not yet finished. For Developers practicing strict swarming or mobbing, WIP should be equal to 1. For Developers just getting started with Scrum and not yet practicing any of the collaborative methods I mentioned in Chapter 8, WIP could be as high as the number of Developers or more!

Cycle Time is a telling metric, because it relates to how quickly the Developers are able to finish a PBI once they start working on it. For example, if the Developers pull a PBI into the *Design* column on Monday, then into *Coding* that afternoon, then into *Testing* on Wednesday, and finally into *Release* on Thursday morning, their Cycle Time for that PBI is four days. When calculating Cycle Time, I recommend rounding up. This way, you won't have any zero-day Cycle Times.

> **Note** Some practitioners consider Cycle Time to have started with the first *commit* to the repository and to have ended with deployment to production. On the surface this sounds acceptable, and it could even lend itself to automated calculation via Azure Repos and Azure Pipelines. I see two problems with this approach, however. The first is that work may begin prior to the first commit—and potentially way prior. I'm not just talking about the time it takes to write the code or write the test that is being committed. I'm talking about all the planning conversations, whiteboarding, architecture design, Azure Boards tasking, Azure Test Plans setup, and other non-Git-committable work. Your Cycle Time might miss a whole day (or more) if you begin measuring Cycle Time at the first commit. The second problem is that this approach would require the Developers to work in a new branch per PBI so that any analytics would be sterile—applying only to the PBI in question. Branching per PBI, as you'll recall from Chapter 8, discourages collaboration and increases risk.

By studying the Work Item Age metric, Developers can perform flow analysis. The Developers can visualize how PBIs are progressing toward the *Done* column of the board, including how much time PBIs are spending in the *Doing* versus *Done* sub-columns of each column. This analysis can help the Developers understand where their process is slowing down or stopping altogether—in order to determine why and how to improve it. As improvement experiments are attempted, the comparison of current Work Item Age with historic ones can be used to determine efficacy.

Throughput is a measure of how fast items depart the process (for example, reach the *Done* column). The unit of time for which Throughput is measured is up to the team but is typically per Sprint (such as two weeks) for Scrum Teams. Throughput is different than velocity. Velocity is a measurement of Done story points (or another unit of measure) per Sprint, which is different than Throughput's *count* of Done PBIs.

The Throughput metric answers the important question of "How many PBIs are we expecting to complete this Sprint?" or "How many PBIs might we have done for the October release?" At some point all Product Owners will get questions like this, and they should be ready to provide an answer. Tracking Throughput is one way to be prepared to provide an answer—an answer based on empirical data.

Throughput directly relates to Cycle Time. Longer Cycle Times lead to a decrease in Throughput. As it takes longer to finish an item, the Developers won't be able to finish as many items in a period of time. In other words, a decrease in Throughput means that less work is getting done, and the less work gets done, the less value is being delivered.

Note You may be wondering why I didn't list Lead Time as a flow metric. If you have been exposed to Lean or Kanban concepts, you have probably heard of that term. It is closely related to Cycle Time. Whereas Cycle Time is the amount of elapsed time between when work starts on a PBI to when it's released, lead time is the elapsed time of when the PBI was first requested by a stakeholder (or was added to the backlog) to when it was released. For example, if it took 4 days to develop and deliver it (Cycle Time), but it sat in the Product Backlog for 30 days, lead time would be 34 days. Both terms are dependent on perspective. In other words, lead time to the Developers would be considered Cycle Time to the Product Owner. This is not to suggest that tracking the time it takes for an idea to get into a customer's hands isn't important. It most definitely is. The Product Owner should be concerned if these elapsed times increase. I'm only saying that the Product Owner can use the term Cycle Time and define the context differently from how the Developers do.

Calculating Flow Metrics

The flow metrics we've discussed are fairly inexpensive to gather. In fact, WIP, Cycle Time, and Throughput take very little time to collect and can even be tracked manually using a simple spreadsheet, which is sometimes a team's only option. Once the data is in a spreadsheet—or in a more sophisticated system like Azure DevOps—there are a number of analytics (charts and reports) that can provide these metrics.

Smell It's a smell when the Developers assumes that the Scrum Master will collect all the metrics. If the Scrum Master is knowledgeable, then they may teach the Developers or the Product Owner how to collect, analyze, and report the metrics, but it's not the Scrum Master's job to collect metrics. The good news is that Azure DevOps and ActionableAgile Analytics do the heavy lifting anyway.

The four primary analytics used to calculate these flow metrics are:

- **Cycle Time scatterplot** A representation of how long it takes to complete PBIs. The x-axis represents the timeline and the y-axis represents Cycle Times in days. Dots represent the intersection of a date and number of days (Cycle Time) that it took a specific PBI to complete (move to the *Done* column, for example).

- **Throughput run chart** A representation of how many PBIs are completed over a period of time. The x-axis represents the timeline and the y-axis represents the number of PBIs completed on that date. Throughput histograms can be used to monitor team performance, identify performance trends, and forecast future delivery.

- **Cumulative Flow Diagram (CFD)**   Shows the count of PBIs in each column for a period of time. From this chart you can gain an idea of the amount of WIP, average Cycle Time, and Throughput. The x-axis represents the timeline and the y-axis represents the count of PBIs.

- **Work Item Aging**   Tracks the age of in-progress PBIs. The x-axis lists all columns (states) of the process and the y-axis represents how long each PBI has spent in that column. The dots represent the number of PBIs that spent that many days in that column.

Some flow metrics may be calculated from more than one analytic. For example, Throughput can be calculated from the Cycle Time scatterplot, Throughput run chart, as well as the Cumulative Flow Diagram. Table 9-1 lists which flow metrics can be calculated from which analytic.

TABLE 9-1 Analytics and the flow metrics that they can calculate.

Analytic (Chart/Report)	WIP	Cycle Time	Work Item Aging	Throughput
Cycle Time scatterplot		✓		✓
Throughput run chart				✓
Cumulative Flow Diagram (CFD)	✓	Averages only		✓
Work Item Aging			✓	

All of the core data is already in Azure DevOps, and these metrics are available in various, automated ways. As I've previously mentioned, the Analytics service in Azure DevOps provides the ability to query all kinds of data. Dashboard widgets, in-context reports, OData feeds, and custom extensions are all mechanisms a team can use to query these metrics.

As of this writing, Azure DevOps provides three Analytics widgets that are of interest to our conversation about flow metrics:

- **Cycle Time widget**   Displays the Cycle Time of work items closed in a specified timeframe for a single team and backlog level (such as Product Backlog).

- **Lead Time widget**   Displays the lead time of work items closed in a specified timeframe for a single team and backlog level.

- **Cumulative Flow Diagram**   Displays the cumulative flow of items based on the timeframe, team, and backlog level.

In addition to referencing the easy-to-use widgets and in-context reports, teams can query the Azure DevOps OData feed, pulling pre-aggregated data directly from Azure DevOps. Microsoft Power BI can consume OData queries, which return filtered and aggregated sets of data in the form of a JSON data payload.

For example, running the following OData query in the browser will return the Cycle Time for work item #42:

```
https://analytics.dev.azure.com/scrum/fabrikam/_odata/v3.0-preview/
WorkItems?$filter=WorkItemId%20eq%2042&$select=WorkItemId,Title,WorkItemType,State,CycleTimeDays
```

The returned JSON shows the calculated Cycle Time:

```
{"@odata.context":"https://analytics.dev.azure.com/scrum/fabrikam/_odata/v3.0-
preview/$metadata#WorkItems(WorkItemId,Title,WorkItemType,State,CycleTimeDays)","value":
[{"WorkItemId":42,"CycleTimeDays":5.3333333,"Title":"PBI 42","WorkItemType":
"Product Backlog Item","State":"Done"}]}
```

For information on how to pull Cycle Time using OData, visit *https://aka.ms/sample-boards-leadcycletime*.

Unfortunately, Microsoft's built-in analytics don't provide all the necessary metrics, or at least they don't provide them in an easy-to-consume way. I suspect that a few Azure DevOps practitioners have built their own analytics, consuming OData feeds, to provide the missing metrics, but nothing has been shared with the community that I am aware of. Instead of building your own, let me take a moment (again) to promote the ActionableAgile Analytics extension. Not only is ActionableAgile Analytics the world's leading agile metrics and analytics tool, but it plugs right into Azure DevOps and provides analytics and metrics from within Azure Boards. You can see this in Figure 9-8.

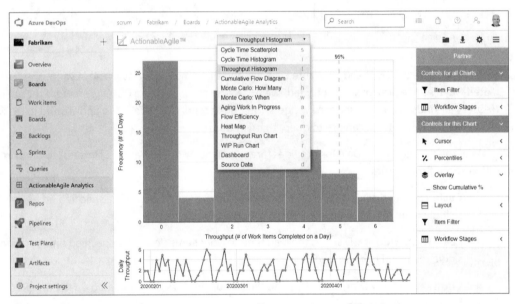

FIGURE 9-8 The ActionableAgile Analytics extension offers many views and analytics.

Flow-Based Scrum Events

The Developers will use the existing Scrum events to inspect and adapt their definition of workflow and thereby help improve empiricism and optimize the value they deliver. What about the inverse? Will the Kanban practices—specifically the various flow metrics—impact Scrum events? Although the events aren't changed by introducing complementary Kanban practices, they will be impacted. Scrum is still Scrum and it doesn't change. The *Scrum Guide* in its entirety still applies. How the team practices Scrum may change.

Practicing Kanban in a Scrum context does not require any additional events. However, using a flow-based perspective and the use of flow metrics in the Scrum events strengthens Scrum's empirical approach. In this section, I will go through each of the Scrum events and discuss briefly how the introduction of Kanban complementary practices might impact them. I refer to each as a "flow-based" event.

The Sprint

Kanban complementary practices don't replace or diminish the need for Scrum's Sprint. Even in environments where continuous flow is desired or achieved, the Sprint still represents a cadence—a regular heartbeat for inspection and adaptation of both product and process. Teams using Scrum with Kanban use the Sprint—and its events—as a feedback improvement loop by collaboratively inspecting flow metrics and adapting their definition of workflow.

Kanban practices can help the Developers improve flow and create an environment where decisions are made just-in-time throughout the Sprint based on inspection and adaptation. In this environment, those Developers rely on the Sprint Goal and close collaboration with the Product Owner and stakeholders to optimize the value delivered in the Sprint.

Some teams might be inclined to toss the Sprint. They see it and its events as a constraint. For complex, plannable work, I see the Sprint as a way to ensure that the Scrum Team sets a Sprint Goal and does some level of planning and retrospecting on a regular cadence. Besides, humans are hardwired for regular cycles. Sprints provide a container for self-management and experimentation. Without Sprints, how would a team plan on communicating with stakeholders and when would a team stop, reflect, and improve? The Kanban answer is "whenever," which I typically see implemented as "hardly ever."

Flow-Based Sprint Planning

Sprint Planning initiates the Sprint by laying out the work to be performed for the Sprint. This resulting plan is created by the collaborative work of the entire Scrum Team. The Product Owner ensures that attendees are prepared to discuss the most important Product Backlog items and how they map to the Product Goal. The Scrum Team may also invite other people to attend Sprint Planning to provide advice.

Flow-based Sprint Planning remains the same although it will have potentially more inputs and historical measures that may make it easier to create a forecast. Flow-based Sprint Planning uses flow metrics as an aid for creating the forecast and developing the Sprint Backlog. For example, the Developers may want to use their historical flow data for predictability.

One of the inputs to Sprint Planning is past performance. Many Scrum Teams use velocity, but in a flow-based context, the Developers can use their Throughput history to help create their forecast. Remember, Throughput is a count of PBIs, not a sum of their story points. A team would need to use a Cycle Time scatter plot, a Throughput run chart, or a Cumulative Flow Diagram to calculate Throughput.

Sprint forecasts can be improved by using Monte Carlo. Monte Carlo simulations can replace traditional practices when forecasting multiple PBIs—such as during Sprint Planning. Figure 9-9 shows the results of 10,000 Monte Carlo simulations in the ActionableAgile Analytics extension. The chart shows the probability of completing a specific number of PBIs in a set period of time (7 days in this case). In this example, there is an 85% chance that the Developers will complete 11 items or more in 7 days. Remember, there is never 100% probability. There is always uncertainty. For a demo, visit *www.actionableagile.com/analytics-demo*. Regardless of what the simulation suggests, the Developers still have the final say on what PBIs get forecasted and pulled into the Sprint Backlog.

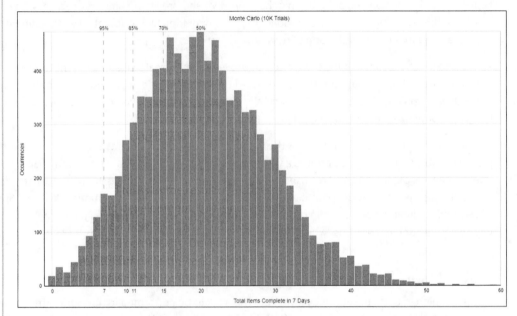

FIGURE 9-9 Monte Carlo simulations help improve multi-PBI forecasting.

Let's assume the Developers have been able to complete between 4 and 20 PBIs per Sprint over the past 8 Sprints. They might calculate their Throughput to be 10 PBIs per Sprint, with a satisfactory level of probability (say, 85%). Assuming the workflow and policies haven't changed for this Sprint, the Developers could simply forecast the top 10 PBIs. This would imply that those PBIs fit the Sprint Goal or that a coherent Sprint Goal could be crafted for those PBIs.

But what about the size of those PBIs? Surely small PBIs will have a lower Cycle Time than large PBIs. While true, remember that Throughput has been calculated on multiple Sprints of *actual* PBIs of varying sizes and complexities. Probability suggests that upcoming work (in the same domain, developed by the same Developers, using the same technologies, with the same tools) should be of similar distribution.

Some Scrum Teams might spend time making sure that each PBI is about the "same size," but that feels like waste to me. I don't think the Product Owner or Developers should be forced to right-size their PBIs. They may want to for other reasons—to keep all Cycle Times to within a few days or less and to improve continuous delivery, for example—but for the sake of forecasting, the Throughput metric is smart enough to handle variation.

As an alternative to traditional right-sizing of a PBI, a Scrum Team may want to perform a quick check to see if each PBI is less than their SLE. This quick "estimate" will tell the Developers whether a PBI might exceed their expected Cycle Time. If so, then they might want to break it down into smaller PBIs. This can be done at Sprint Planning or during Product Backlog refinement, and it should take only a few moments to ask and answer that single question. For more information on Throughput-driven Sprint Planning, refer to fellow Professional Scrum Trainer Louis-Philippe Carignan's article at *www.scrum.org/resources/blog/throughput-driven-sprint-planning*.

Note For Scrum Teams wanting to practice Kanban and enjoy flow-based Sprint Planning, they will probably *not* be creating their Sprint plan in the form of tasks. The flow metrics generated by Azure DevOps assume that the Scrum Team will be using the Kanban board. If you want flow metrics to be calculated by Azure DevOps Analytics or the ActionableAgile Analytics extension, you'll need to use the Kanban Board. The good news is that the Kanban board supports the association of Task work items (as well as Test Case work items) with each PBI, as you can see in Figure 9-10. These tasks don't drive the workflow—they are just help- ful reminders of what activities are required for that PBI.

For Developers who want to continue to use the Taskboard, it would be theoretically pos- sible for an innovative developer to calculate Cycle Time based on the elapsed time between when the first Task work item for a PBI was pulled into the *To Do* column and when the last Task work item was pulled into the *Done* column (or when the PBI was set to the Done state). There would be a lot of nuance to work through here and maybe I'll get around to building something someday—or you can.

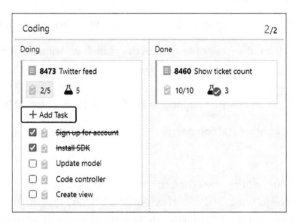

FIGURE 9-10 PBIs on the Kanban board can have associated Task and Test Case work items.

Flow-Based Daily Scrum

A flow-based Daily Scrum focuses on ensuring that the Developers are doing everything they can to maintain a consistent flow. Although the goal of the Daily Scrum remains the same as outlined in the *Scrum Guide*—to focus on progress toward the Sprint Goal and produce an actionable plan for the next 24 hours—the meeting itself takes flow metrics into consideration and focuses on where flow is lacking and on what actions the Developers can take to get it back.

Some Developers prefer to have their Daily Scrum in front of their Kanban board. This way, they can "walk" the board from right to left, focusing on PBIs (not people) while discussing the current state of work and what happened during the last 24 hours, as well as what is likely to happen in the next 24 hours. My only caution is to make sure that Developers are talking to one another (and listening) and not just focusing on the board. Scrum Masters may need to keep an eye on this.

Here are some inspections that can occur at a flow-based Daily Scrum:

- What do we need to get the work closest to the right to Done?

- What work is blocked and what can the Developers do to unblock it?

- What work is flowing more slowly than expected and why?

- What is the Work Item Age of each PBI in progress?

- Have any PBIs violated or are about to violate their SLE, and what can be done to get that work completed?

- Are there any factors that may impact the Developers' ability to complete work today that are not represented on the board?

- Is there any work that has not been made visual?

- Are the Developers pulling work to fulfill WIP limits, or is there extra capacity?

- Have the Developers broken any WIP limits?

- If WIP limits have been exceeded, what can be done to ensure that WIP will be completed?

- If WIP limits have been exceeded, can the Developers swarm to get their WIP under control?

- Are WIP limits set correctly?

- Have the Developers learned anything new that might change the plan for the next 24 hours?

- Are the Developers finishing work before starting new work?

- Are there any impediments to flow?

Although some Developers prefer to use a burndown chart to monitor progress, flow-based teams may find them ineffective. Sprint burndowns typically don't represent flow because they are based on tasks (work invested and work remaining). They don't provide transparency to flow or provide an actionable path toward achieving the Sprint Goal. Instead, the transparency provided by the Kanban board itself is typically enough for most flow-based teams.

The Work Item Age metric provides additional transparency to specific PBIs that are struggling. Developers should use this metric and compare it to their SLE. Otherwise, they can suffer poor flow. Developers using flow-based metrics can use a Cumulative Flow Diagram to provide transparency of their flow. If the Developers really desire to use a burndown chart, they should use one that uses PBI work items as opposed to Task work items. Fortunately, the Burndown analytic in Azure DevOps supports both configurations.

Blocked Work

Blocked work occurs whenever the Developers have to wait for someone or something before work can proceed. This may include waiting for someone to solve a problem or provide information. It may also include waiting for something like software to be installed, hardware to be configured, or a test environment to be provisioned. Blocked work does *not* include PBIs that are in the *Done* sub-column, waiting to be pulled to the next column (for example, *Design-Done* waiting to be pulled to *Coding-Doing*). Blocked work, however, is still counted against the WIP limit.

When a PBI gets blocked—for whatever reason—the Developers should not simply put that blocked item aside and start working on something else. This may be a human response, but it's definitely not flow. This is one reason for not having a "Blocked" column on the Kanban board—to discourage this behavior.

If the Developers want to visually indicate that a PBI is blocked, I recommend adding a "Blocked" tag to the PBI work item. The Kanban board can then be configured to show tags and even style them and the card backgrounds based on rules. You can see the results of these styling configurations in Figure 9-11.

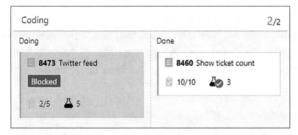

FIGURE 9-11 Blocked work should be made visible on the Kanban board.

While I'm on the topic of styling cards on the Kanban board, another style to consider is one that indicates which cards are aging beyond a certain threshold. Figure 9-12 shows an Aging rule being created to highlight those cards that have not changed within the last two days. Additional styles can be added for different depths of coloration (based on more aging). This won't generate a proper Aging analytic, but it will give the Developers a quick visualization of which PBIs aren't moving. Remember that Work Item Age is a leading indicator of whether or not the PBI will miss the SLE.

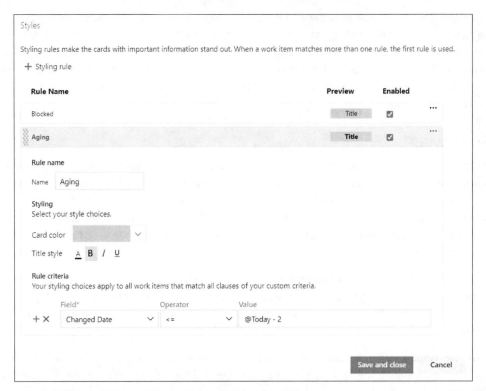

FIGURE 9-12 Styling PBIs that haven't changed in the last two days.

Blocked work impedes flow, accumulates Cycle Time, and should be addressed. The Developers should discuss when aging items are considered to be blocked and for what reasons. They should also discuss if/when to violate WIP limits in these situations. Even though starting new work is almost never the right answer, there may be times when it is. The Developers should also discuss when the item should be kicked out of the system—whether put back on the Product Backlog, renegotiated and re-created, or dropped.

> **Tip** Professional Scrum Developers commit to doing everything they can to unblock work as expeditiously as possible. They closely monitor bottlenecks and workflow policies regarding the order in which they pull items through the system so that PBIs do not age unnecessarily.

Impediments usually break down into two categories: hindrance and obstruction. Both are bad and can limit flow. Although obstruction impediments (or "blockers") need to be mitigated right away, both should be discussed at the Sprint Retrospective, where experiments can be put forth in an attempt to minimize or eliminate them in the future. All repetitive blockers—such as dependencies that go beyond the Scrum Team—should become the Scrum Master's highest priority to help alleviate.

Flow-Based Sprint Review

The purpose of the Sprint Review is to inspect the outcome of the Sprint and determine future adaptations. The Scrum Team presents the results of their work to key stakeholders and progress toward the Product Goal is discussed. During the event, the Scrum Team and stakeholders review what was accomplished in the Sprint and what has changed in their environment. Based on this information, attendees collaborate on what to do next.

Inspecting flow metrics and related visualizations as part of the Sprint Review can create opportunities for new conversations about monitoring progress toward a goal. For example, by reviewing Throughput, the Scrum Team and stakeholders will have additional insight about likely scope and delivery dates.

As previously mentioned, Monte Carlo simulations can be used for forecasting when a set of PBIs might be completed. Stakeholders may appreciate the calendar view visualization provided by the ActionableAgile Analytics extension, as you can see in Figure 9-13. This view is powerful, showing red, orange, and shades of green based on the increasing probability of the remaining items in the release to be done by those dates. Stakeholders may require some explanation of flow metrics, probabilities, and Monte Carlo.

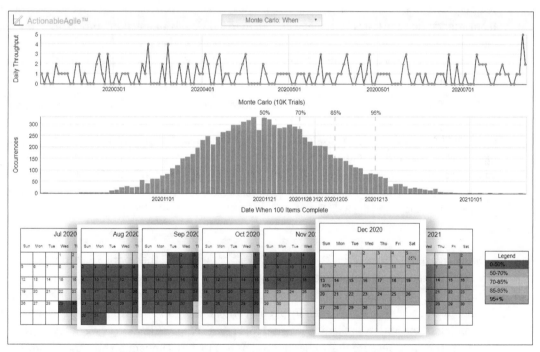

FIGURE 9-13 The calendar view of a Monte Carlo simulation helps set stakeholder expectations.

Another item to discuss at the Sprint Review is the SLE. Stakeholders should know what the number is, what it means, and what the probability is behind it. If the SLE has changed, the Sprint Review is a good opportunity to announce that. Depending on the stakeholders, you may need to differentiate between SLE and SLA—which I contrasted earlier in this chapter. Be sure to remind them that "expectation" does not mean commitment or promise.

Some teams might be inclined to toss the Sprint Review—especially those who are practicing continuous delivery. They might see the meeting as a waste of time and would rather book time with stakeholders as needed. My first word of guidance to these teams would be to remind them that the Sprint Review is *not* an acceptance meeting. Scrum absolutely allows PBIs to be released to production as soon as they are Done—if the Product Owner desires. The Sprint Review is an opportunity for the stakeholders to inspect the Done, working product. It's also an opportunity to hear back from the stakeholders about the state of the market and other factors affecting the product and its goals.

I do like the idea of "booking time" with stakeholders regularly. Having this day-to-day interaction where stakeholders can inspect and adapt is fantastic, and it will result in a higher-quality product. These meetings don't replace the value of the Sprint Review as an opportunity to take a step back and inspect and adapt the whole Increment, the Product Backlog, and the Product Goal with stakeholders based on changes to the business and market.

Flow-Based Sprint Retrospective

The purpose of the Sprint Retrospective is to plan ways to increase quality and effectiveness. The Scrum Team inspects how the last Sprint went with regard to individuals, interactions, processes, tools, and their Definition of Done. The Scrum Team discusses what went well during the Sprint, what problems it encountered, and how those problems were (or were not) solved. The Scrum Team identifies the most helpful changes to improve its effectiveness.

A flow-based Sprint Retrospective adds the inspection of flow metrics and analytics to help determine what improvements the Scrum Team can make to its processes. The Scrum Team can also inspect and adapt its definition of workflow to optimize flow in the next Sprint—just beware of making too many changes at a time. Using a Cumulative Flow Diagram to visualize the Developers' WIP, average approximate Cycle Time, and average Throughput can be valuable during the Sprint Retrospective. This is also an opportunity to discuss what work was blocked, why it was blocked, and how to reduce or remove those blockers in the next Sprint.

Smell It's a smell when I see a team or a Developer wait until the Sprint Retrospective to make inspections or improvements. Although the Sprint Retrospective is the formal opportunity—so that it happens at least once per Sprint—a Professional Scrum Team should consider taking advantage of process inspection and adaptation opportunities as they emerge *throughout* the Sprint. Similarly, changes to a Scrum Team's definition of workflow may happen at any time. Because these changes will have a material impact on how the Scrum Team performs, changes made during the regular cadence provided by the Sprint Retrospective event will reduce complexity and improve focus, commitment, and transparency.

Chapter Retrospective

Here are the key concepts I covered in this chapter:

- **Flow**   The movement of customer value through the product development system.

- **Workflow**   The approach to turning PBIs into Done, working product. The definition of workflow represents the Developers' explicit understanding, which will improve transparency and enable self-management.

- **Kanban board**   An interactive visualization of the team's workflow. Not to be confused with the Taskboard. Azure Boards supports both.

- **Work in Progress (WIP)**   The number of PBIs started but not yet finished. WIP can be assessed for the entire board or for a specific column/state.

- **WIP Limits**   Policies enacted by the Developers to reduce the amount of WIP per column/state. Enabling WIP limits creates a pull system and improves flow.

- **Flow metrics**   Metrics based on how the Developers work, using easy-to-understand units of measure such as days and item counts.

- **Cycle Time**   A flow metric representing the amount of elapsed time between when a PBI starts and when it is Done. Cycle Time is a lagging indicator and best answers the question "How long will it take this PBI to complete?"

- **Work Item Age**   A flow metric representing the amount of time between when a work item started and the current time. This leading indicator applies only to items that are still in progress. For Scrum Teams, PBIs in the Product Backlog are not considered to be aging.

- **Throughput**   A flow metric representing the number of work items finished per unit of time, such as a Sprint. Throughput is a lagging indicator and best answers the question, "How many PBIs will I get by the end of the Sprint or in the next release?"

- **Cycle Time scatterplot**   An analytic that informs how long it takes to complete PBIs. The x-axis represents the timeline and the y-axis represents Cycle Time in days. Dots represent the intersection of a date and the number of days (Cycle Time) that it took that PBI to complete (for example, move to the *Done* column).

- **Throughput run chart**   An analytic that determines how many PBIs have been completed (Done) over a period of time. The x-axis represents the timeline and the y-axis represents the number of PBIs completed on that date.

- **Cumulative Flow Diagram (CFD)**   An analytic that shows the count of PBIs in each column for a period of time. From this chart you can gain an idea of the amount of WIP, average Cycle Time, and Throughput. The x-axis represents the timeline and the y-axis represents the count of PBIs.

- **Work Item Aging**   An analytic that tracks the age of in-progress PBIs. The x-axis represents all columns of the process and the y-axis represents how long each PBI has spent in that column. The dots represent the number of PBIs that spent that many days in that column.

- **Service level expectation (SLE)**   A forecast of how long it should take any given PBI to flow from start to finish within the team's workflow. The SLE itself has two parts: a period of elapsed days and a probability associated with that period. The Developers can use their SLE to find active flow issues and to inspect and adapt in cases of falling below those expectations.

- **ActionableAgile Analytics**   The world's leading agile metrics and analytics tool, with an extension for Azure DevOps.

- **Flow-based Sprint Planning**   Sprint Planning, as defined in the *Scrum Guide*, but using flow metrics (such as Throughput) as an aid for developing the Sprint Backlog.

- **Flow-based Daily Scrum**   Daily Scrum, as defined in the *Scrum Guide*, but with additional inspections to ensure that the Developers are doing everything they can to maintain a consistent flow.

- **Flow-based Sprint Review**   Sprint Review, as defined in the *Scrum Guide*, but including the inspection of flow metrics as a way to monitor progress and predict likely delivery dates.

- **Flow-based Sprint Retrospective**   Sprint Retrospective, as defined in the *Scrum Guide*, but including the inspection of flow metrics and analytics to help determine what improvements the Scrum Team can make to its workflow and processes.

CHAPTER 10

Continuous Improvement

One thing that I hope I have made clear by now is that Professional Scrum Teams know they can always do better. They can build a better product. They can increase quality. They can build faster. They can build with less waste. They can learn new techniques that will help them improve personally and as a team. I use the term *continuous improvement* to categorize all of these goals.

Knowing where to start is a big part of the improvement process. There are tactical improvements, which help the team successfully achieve the Sprint Goal, meet the forecast, and deliver a Done, working Increment while not generating waste or technical debt. Mastering some of the common challenges that Scrum Teams run into will help in this regard. A team can also make strategic improvements, such as learning to become more cross-functional, improving in self-management, and fostering and living the Scrum Values. Individuals can commit to learn more and strive to become more T-shaped. Teams that improve in these areas will see significant increases in learning, capability, and performance.

Note The FBI Sentinel Project is just such an example of a Scrum Team realizing this kind of increase in capability. While working in the basement of the Hoover building, the team finished over 80 percent of the work with just 10 percent of the budget—after a large government contracting agency failed to deliver. You can read more about the FBI case study in *Software in 30 Days* (Wiley, 2012), by Ken Schwaber and Jeff Sutherland. The Healthcare.gov fiasco is another such example where the team that came in to fix the app used Scrum. HealthCare.gov is a U.S. federal government health insurance exchange website operated under the provisions of the Affordable Care Act. The October 1, 2013, launch was marred by serious technological problems, making it difficult for the public to sign up for health insurance. This spectacular failure spawned a team of Scrum developers who rescued the website from disorganized contractors and bureaucratic mismanagement.

In this chapter, I will look at how to handle common challenges, as well as how to identify and overcome various dysfunctions. I'll also explore some healthy behaviors you and your colleagues can adopt to become a high-performance Professional Scrum Team.

Note In this chapter, I am using the customized Professional Scrum process, not the out-of-the-box Scrum process. Please refer to Chapter 3, "Azure Boards," for information on this custom process and how to create it for yourself.

Common Challenges

Many challenges face software development teams, and Scrum Teams in particular. Software development is a complex effort, and anyone who is not in the middle of it will have difficulty understanding that. Even the smartest team members will run into the dilemma of balancing the values of Scrum against getting something out the door.

For example, when an experienced Developer sees the need to refactor a large class, when should they do this? If they put on their propeller hat, their technical side wants to open up the code and start refactoring right now because it shouldn't take more than a few minutes, tops. However, when they put on their Scrum robe, they want to spend time wisely and in ways that provide maximum value. One team member with two urges. Which one wins? The answer is, of course, "It depends."

And in that, you can find the primary goal of this section—to address some of the more common challenges facing Scrum Teams and Developers in order to offer solutions, opinions, and advice so that those teams and team members can learn to self-manage and improve.

Impediments

An *impediment* is anything keeping the team from being productive. Impediments can be environmental, interpersonal, technical, or even aesthetic in nature. Regardless of what the impediment is, or its size, if it's blocking the team from being productive, it should be removed. Just as the sweepers in the game of curling (a popular winter sport in Canada) keep the path of the stone free from bumps and debris, so should members of the team keep the path of productive product development free from impediments.

Scrum has two formal opportunities to identify impediments: each day during the Daily Scrum and at the end of the Sprint during the Sprint Retrospective. However, impediments can be identified at any time during the Sprint. More important, they can be *removed* at any time. The problem is not finding the opportunity to identify impediments, but rather getting team members to be open and honest about their existence—and then find a willingness and authority to remove them.

It's common to hear Developers say that nothing is blocking them. Hearing this repeatedly does not reflect reality and might actually be a smell of an underlying dysfunction. I'm not saying that Developers are patently dishonest. On the contrary, they are often just being optimistic. They have a lot of work to do and can easily find something else to work on. They may not realize that "blocking" can also mean that they are experiencing slow or nonoptimal progress. Also, nobody wants to bother others with something as depressing as an impediment. What they don't realize is that by sharing their problems with the rest of the team, their honesty and openness might invite others to help remove the impediment. The impediment may just disappear sooner than expected.

Identifying an impediment is just the first step. The more important step is to execute a plan to remove it. Some team members will be able to remove certain types of impediments more easily than others. The Product Owner, or management, may have to get involved. Regardless of the level of difficulty or amount of ceremony involved in removing an impediment, it should still be identified. In other words, don't keep an impediment to yourself just because you think it will be difficult to remove. If you see something, say something. Blow your whistle.

> **Smell** It's a smell if I see that a team is not dealing with their impediments. Ideally, any impediment that survives to the next Sprint should be resolved during that Sprint. It's the Scrum Master's job to keep a watchful eye on impediments and appropriately nudge the team to remove them. If the team cannot remove the impediment, then it falls on the Scrum Master to do so. I think of the list of impediments as being the Scrum Master's *backlog*.

Estimation

Estimating the size of PBIs is a team skill that will improve over time. Initially, the team may not have experience working with the domain, the technology, the tools, or one another. They may not have a common baseline to use for relative estimation yet, either. Relative estimation is the sizing of PBIs by comparing or grouping PBIs of equivalent size (effort, difficulty, complexity, etc.). All of this will emerge and improve over time.

Regardless of how experienced your team is, when it comes to agile estimation, always remember the basics:

- **Keep it refined, but not too refined** A big reason for keeping the important items in the Product Backlog refined is to enable more accurate estimation. The Developers should have just enough information in order to estimate a PBI, but no more. Having additional conversations and specifying additional information beyond what the Developers need to estimate is wasteful. Save it until the Sprint where it's forecasted.

- **Estimate as a team** All Developers on the Scrum Team should be involved in the estimation. Each PBI will require different types of activities and skill sets. The entire cross-functional team should be in the room as each item is discussed and estimated. Estimation by proxy— Developers voting on behalf of other Developers not in the conversation— can lead to the wrong estimates and, more importantly, the wrong solution.

- **Be less precise** If you want to be more accurate, be less precise. Initially, consider estimating the size of a PBI using T-shirt sizes. This will help with coarse-grain planning. Later, as it becomes more likely that a PBI will be developed (that is, it floats to the top of the Product Backlog), consider using a more precise unit of measure, such as story points. In Sprint Planning, as well as during the Sprint, the team can be even more precise as it estimates tasks in hours.

- **Be relative** No two PBIs will ever be exactly the same in terms of complexity or amount of effort to develop. This is our world. To mitigate this while estimating, the team should think in terms of how the *size* of one PBI relates to another. This size usually relates to effort, but it can also relate to complexity. It doesn't matter which, as long as the Developers are consistent. By comparing a new PBI with one that has been developed already, the Developers are able to determine if the new one is more work, less work, or about the same. Over time, working consistently as a team, more baseline PBIs will become available that can be used for comparison.

- **Don't translate**   Keep any units of measure abstract. Avoid the temptation by you, or the organization, to translate T-shirt sizes or story points into days, hours, or money. Knowing how many days are in a Sprint allows management to reconcile the work the team delivers or, more importantly, to translate the business value of that work into a monetary value. They already know what the run rate of the Scrum Team is per Sprint. From this information, they can determine the business value per monetary unit, which should be the ultimate metric for any organization. Similarly, the Product Owner can also use the number of days in a Sprint and the Developer's velocity (or Throughput) to assist with release planning and budgeting. Either way, these computations are informative and healthy, and not the dysfunctional type of translation I'm referring to here. Trying to figure out how many hours a "typical" story point equates to or how many dollars a "typical" story point costs is pointless, as well as wasteful.

Note   The *process* of estimation is more important than the outcome. The real value lies in all the information and learning that are gained and uncertainty that is removed as a result of the conversations that estimation demands. In other words, the shared understanding is more important than the estimate number.

When a team new to Scrum estimates features of a new product, in a new domain, using new technology, with new tools, estimates will be way off. As the team normalizes and members become familiar with one another as well as the product and the environment, this should turn around. Eventually, estimation should occur faster and become more accurate. If this is desirable (and it should be), then keep the team together. They will improve.

Smell   It's a smell when I see management break up a high-performance Scrum Team. It's actually more than a smell—it's a downright shame. I know what they're thinking. They're thinking that they can distribute these individuals to other teams within the organization as "seedlings." The seedlings will then grow new high-performance Scrum Teams. Although this may be true, it destroys self-management and generates waste due to the length of time that it requires to become a high-performance team. When a team has progressed through Tuckman's stages of team development (forming, storming, norming, and performing), any change to the team—especially dismantling it altogether—sends everyone back to forming. I contend that the organization will derive more value by leaving that team intact. Bring other team members into the established team as a visitor (akin to a foreign exchange student) or let the Scrum Master nurture new seeds in other teams. This way, you won't dismantle a proven generator of business value.

Tracking Actual Hours

Tracking actual hours spent on PBIs or tasks is not important in Scrum. Although it's not too difficult to track this information in Azure DevOps, I recommend teams resist the urge or request to track or compute actual hours. It can only be used for evil. Once actual hours are tallied by task, or by task activity type, somebody, somewhere will use it as a measuring stick, or a beating stick, to attempt to motivate or "improve" the team's abilities. As I've previously explained, improving because someone else wants you to improve doesn't work. The desire must come from within, not from a spreadsheet.

> **Note** Tracking hours on a timesheet in order to get paid is a different matter. So is the tracking of hours in order to bill a department, a budget, or a client. If management expects this data, provide it. But do so knowing that the totals you are providing have nothing to do with the team's efficacy. If tracking hours detracts from the Developers' ability to develop product efficiently, have the Scrum Master do it in order to protect their focus.

Tracking original task estimates is also not important. The only estimate that Developers might want to track is the amount of remaining work left to be done. In Azure Boards, this relates to the total number of hours for all Task work items in the To Do and In Progress states on the Taskboard. These estimates can change daily and, therefore, should be updated at least daily. Hopefully, the work remaining estimates go down. Some days, however, they may go up. Some days, new tasks are identified that the Developers didn't foresee. This is the nature of a complex effort like software development.

> **Tip** When pressed by management for why original estimates were off, give the honest answer: "What we do is hard." This simplistic answer is 100 percent accurate and can also be used if asked why the build broke, why a bug escaped into production, why the team missed their forecast, why the team did not achieve the Sprint Goal . . . you get the idea.

For example, let's assume the Developers estimated that the development of a new mobile-friendly dashboard would be eight story points and the sum of all the hours of the initial tasks would be 120 hours. Let's also assume that development ended up taking 160 hours to complete the PBI. Should the Developers be concerned about the gap in their estimates? Sure. They should discuss it during the Sprint Retrospective and determine how they might make better estimates going forward. I propose that teams will organically get better at estimating. Should management be concerned about the deviation from the original estimates? Sure, but they should know that their best people are on the job and those people should be given the freedom to self-manage and make improvements themselves. Additional "management" won't make estimates more accurate.

Note Some teams prefer to break PBIs down small enough where they can be completed in a couple of days or less. This removes the need to perform estimation because each PBI would already be quite small. Effort is still required to split a PBI into small pieces, and this effort may be greater than performing agile estimation. Try some experiments and decide for yourself.

Assessing Progress

Assessing progress means knowing how much work has been done and, more importantly, how much is left to do before achieving a specific goal. This is not necessarily the Sprint Goal, but any goal. The goal could be completing a PBI, completing all forecasted PBIs for a Sprint, completing all of the expected functionality in a release, or even achieving the Product Goal. Progress toward each of these types of goals can be measured in a number of ways.

The *Scrum Guide* does not provide guidance on *how* to assess progress other than suggesting that it should be done by the team regularly. Table 10-1 lists the various goals that a Scrum Team will be pursuing, when to assess progress (at a minimum) toward each goal, and some practices that can be used to assess progress.

TABLE 10-1 When and how to assess progress toward various goals.

Goal	When to assess? (at a minimum)	How to assess?
PBI	Daily Scrum	Count related tasks, sum related task hours, count related failing tests, or use Cycle Time and Work Item Age metrics.
Sprint forecast	Daily Scrum	Count PBIs, sum PBI sizes, count tasks, sum task hours, count failing tests, or use Throughput metric of undone PBIs in the Sprint Backlog.
Sprint Goal	Daily Scrum	Similar to assessing progress toward the Sprint forecast, although the Sprint Goal may be independently achievable.
Release	Sprint Review	Count PBIs, sum PBI sizes, or use Throughput metric of undone PBIs planned for the release.
Product Goal	Sprint Review	Similar to assessing progress toward a release, although the Product Goal may be independently achievable.

Assessing Progress Toward a Product Goal

The most popular way to assess the progress toward a Product Goal or a release is to maintain a release burndown chart like the one you see in Figure 10-1. This kind of chart shows how much work remained at the start of each Sprint for a given release goal. Each Sprint appears along the horizontal axis. The vertical axis measures the effort (size of the PBIs) that remained when each Sprint started. The unit of measure is whatever your team has decided to use. Story points are the most common. The data comes from the refined Product Backlog. The more refined the Product Backlog, the more accurate the burndown.

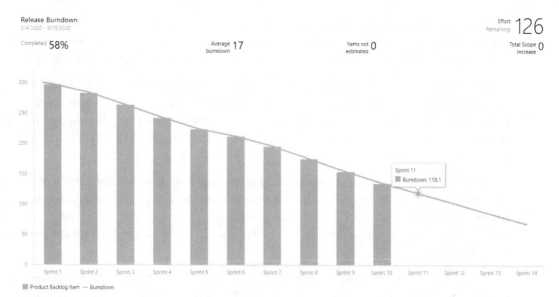

FIGURE 10-1 A release burndown shows progress toward a Product Goal or release.

Note For a long-running development effort, one with multiple releases, the remaining work could actually *increase* over time as the Product Backlog evolves. In cases like this, the release burndown alone won't tell you how quickly the team is completing items in the Product Backlog, especially if the burndown refers to more than one release.

Story maps are another way to visualize progress toward a Product Goal or release. PBIs can be put into rows or tagged to show that they are part of the plan. Whether the Scrum Team uses a physical story map or an electronic tool, such as SpecMap, the Product Owner can ensure the plan is visible for all to see. Refer to Chapter 5, "The Product Backlog," for more details on SpecMap.

Tip Don't discount a conversation with stakeholders as a way to assess progress. As the stakeholders have been inspecting the product Sprint after Sprint, they might be a good gauge as to whether the release is ready or the Product Goal has been met. A stakeholder's feedback can be a powerful assessment.

Assessing Progress Toward a Sprint Goal

The most popular way to assess the progress of a Sprint—toward completing the forecast and/or achieving the Sprint Goal—is to maintain a Sprint burndown chart like the ones shown in Figure 10-2 and Figure 10-3. These kinds of charts show how much work remained at the end of specified intervals during a Sprint. These intervals are typically days, and they appear along the horizontal axis. The vertical axis measures the amount of remaining, identified work (either PBI or Task work items) to complete. The data comes from the Developers' regularly updated Sprint Backlog.

Sprint burndown charts can show the team how much work remains in the Sprint. These charts will often include a trend line. This line represents the ideal rate at which the Developers are able to complete all of the remaining effort, at a constant rate, by the end of the Sprint. It is usually visualized as a plotted line displayed on the chart. By using the trend line, the team can gauge how it's doing and know if it is on track to finish all forecasted work by the end of the Sprint, given the constant rate.

Burndown charts are generated from actual data. Because of this they reflect the reality of the Developers' activity. For example, if the Developers all go away for a three-day training class, the burndown will reflect a flat horizontal line (no movement) for those days. This can also result when new work is added and completed on the same day or if new work is added at the same rate it is completed on the same day.

The Burndown analytic widget can burn down by count of PBI work items, count of Task work items, sum of PBI effort (story points), sum of task remaining hours, or sum of other fields. You can create a burndown for any of the backlog levels: Epics, Features, or Stories (PBIs). In fact, you can burn down by summing any field or by counting any type of work item. The new widget displays average burndown, % complete, and scope increase. You can also choose to burn down on a specific team, which lets a Nexus track Sprint burndowns for multiple teams on the same dashboard. You can learn more about the Burndown analytic and widget by visiting *https://aka.ms/configure-burndown-burnup-widgets*.

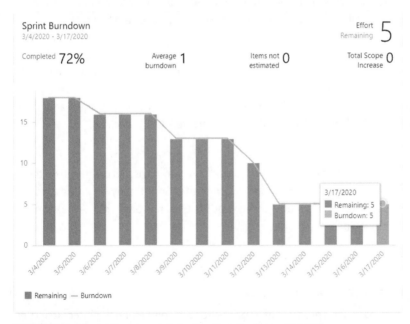

FIGURE 10-2 A Sprint burndown can be based on the forecasted PBIs.

Note There is also a *Burnup* analytic widget, which focuses on completed work. Although progress can be assessed from a burnup, its primary purpose is showing the Product Owner and stakeholders that the Scrum Team is consistently delivering value, Sprint after Sprint.

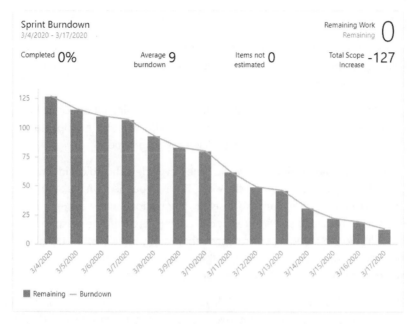

FIGURE 10-3 A Sprint burndown can be based on the remaining hours of tasks.

Burndown charts can also illuminate other facts about how the Developers work:

- **Actual and trend are diverging or are far apart** The Developers will probably miss their forecast. This will happen from time to time, but more often with new teams. As a team norms and performs, their collaboration and estimation practices will improve and velocity will stabilize, resulting in forecasts that become more accurate. On the other hand, sometimes the actual burndown is *below* the ideal one, meaning the Developers will probably complete their forecasted work earlier than expected and will be able to collaborate with the Product Owner about adding more work to the Sprint.

- **The total number of hours is increasing** This occurs when additional work (tasks) are added to the Sprint Backlog during the Sprint. New tasks are to be expected, but a large number of tasks, or new tasks over several days, may indicate poor Sprint planning, a badly refined Product Backlog, or scope creep. This may also be the Developers violating the "You Ain't Gonna Need It" (YAGNI, also known as gold-plating) principle, which discourages the adding of functionality until it becomes necessary and not just because the users *might* need it in the future. Regardless of the origin, unplanned work like this becomes quite evident on a Sprint burndown chart.

> **Note** It can be normal for an experienced team to add additional tasks during the first day or two of the Sprint. This indicates that the Developers did "just enough" planning during Sprint Planning. Additionally, teams that are practicing single-piece flow (swarming or mobbing on a single PBI until it is done before moving to the next one) can witness many small spikes in work while they "task as they go."

- **Actual performance is significantly above the trend line** The Developers have really missed the estimates or the amount of work forecasted. Attention should be given to both during the next Sprint Retrospective.

- **Tasks move to "Done" prematurely** Reactivating a task by moving it from Done back to In Progress will be reflected on the burndown as a bump. Knowing when a Developer (or pair) is truly done with a piece of work is a practice that will improve over time.

- **Tasks are not created for large pieces of work** The Developers should decide when to create a Task work item and when to just do the work. This is dependent on the Sprint length and other team behaviors. The rule of thumb I use is two hours for a two-week Sprint. For example, if a Developer realizes that a Secure Sockets Layer (SSL) certificate is expired while deploying the application to a staging environment, they may want to create a task for the work they are about to do, or they may just want to do the work. Either way, there should be a team working agreement on what triggers the creation of a Task work item.

- **Tasks stay "In Progress" for more than 1 or 2 days** Sprint burndowns don't usually show this level of granularity, but flat spots (or rises) are always a concern. Where this staleness becomes obvious is during the Daily Scrum. Other Developers should be concerned, or at least inquisitive, when one of their team says they are (still) working on the same task for several days. There may be a lack of focus or openness at play here. Keep the burndown, and the team, transparent by updating all of the *Remaining Work* fields of your Task work items at least once per day.

- **Burndown actually burns up** Increasing hours may be due to scope creep, which occurs when a substantial amount of work is added to a Sprint after it is planned. This is to be expected to some degree. Professional Scrum Developers will strive to refine PBIs and effectively estimate work at the beginning of the Sprint.

When an organization first adopts Scrum, the customers and management may still be expecting to see the old management reports and analytics pertaining to development. What they get in Scrum is going to be quite different. There will be a transition period for them to stop using the old reports and begin understanding the new ones. During this period, the customers and management should also learn what it means for the team to be self-managing. This will help explain why the old reports are not necessary anymore. A Professional Scrum Master can clarify to stakeholders that any report or artifact showing progress is primarily for the Developers and that they are being allowed access in order to provide transparency.

Another way to assess progress is to simply *ask* the Developers. At the Daily Scrum, each Developer can indicate how confident they are in being able to meet the Sprint Goal and complete the forecasted work. If the confidence stays the same or goes up, that's good. If it starts to drop, that's an indication that something might be wrong, and one that can be trusted because it comes directly from the front line. This technique for assessing progress can be just as effective as using a burndown.

Renegotiating Scope

In the business world, things can change suddenly. Products being developed to support the business or to sell or support a product or service can quickly become obsolete. Or, thinking about it more optimistically, the product can be modified to take advantage of new opportunities. Regardless of the reason for the change, a Product Owner may determine that one or more forecasted PBIs no longer has value or that other PBIs may have *more* value once the Sprint has begun. This realization can also be initiated by the Developers if they determine that a PBI is not able to be developed to any degree that would be fit for purpose. The Scrum Team should come together whenever forecasted scope is renegotiated.

> **Note** Renegotiating scope can also mean that the Developers have completed all forecasted work and want to add more PBIs to the Sprint Backlog. In this section, however, I'm talking about the more challenging scenarios where the forecasted work was *not* completed, has become irrelevant, or new/different PBIs need to be added to the Sprint Backlog after Sprint Planning.

The best-case scenario when renegotiating scope is that the Developers have not yet started work on the forecasted PBI(s). Hopefully, only an hour or less of the team's time spent planning the work will be wasted. This could include conversations, creating tasks, creating acceptance tests, and so forth. The more complex (and wasteful) scenario is where the scope needs to be renegotiated *after* the Developers have started developing the PBI.

Obviously, the further into development, the more work will be potentially wasted. The Product Owner should take this into consideration before pressing the "red button" to stop development. It may be more efficient (less wasteful) to allow the existing work to finish, creating a new PBI instead. It is the job of the Scrum Master to help the Product Owner understand the trade-offs associated whenever renegotiating scope. The costs are probably irrelevant if the work the Developers are doing is truly without value.

> **Smell** It's a smell if I see a Product Owner *regularly* renegotiating scope. More often than not, it's a dysfunctional Product Owner at work, trying to continuously introduce new, high-priority work. This dysfunction can sometimes be restated as a lack of proper Product Backlog refinement. It could also be that the Product Owner is new to the role and still getting the hang of effectively ordering the Product Backlog. It could also be that the Product Owner is trying to appease too many stakeholders at once. In any case, the Scrum Master should get involved and make sure that changing scope is truly the exception to the process, and not the rule. Nothing frustrates (and burns out) good Developers like an organization that keeps giving priority 1 work, on top of other priority 1 work. If everything is priority 1, then there are no priorities!

Canceling a Sprint

If it is determined that the Sprint Goal becomes obsolete, the Product Owner can cancel the Sprint. In other words, the Sprint could be canceled if the Product Owner determines that there is no chance to realize *any* value in *any* of the forecasted work being developed. For example, if a company suddenly decides to abandon support for a particular platform, the Sprint could be canceled if it only contained PBIs intended for that platform.

> **Note** Only the Product Owner has the authority to cancel a Sprint. The Developers or stakeholders may influence their decision.

Here are some actual examples when a Sprint might be canceled:

- Business conditions change in a way that the PBIs in the Sprint Backlog no longer have value.

- The technology on which you've been building the product proves invalid, and switching to a new technology requires a large amount of new planning.

- The organization undergoes a restructuring where Developers get moved off the team.

- Budgetary issues force a radical change in direction.

- Virus or ransomware attack redirects everyone to work on security.

- The organization buys a startup that just released a product with overlapping functionality with what the Scrum Team was working on that Sprint.

- Critical production support issues arise and take the Developers away to the point where they aren't able to deliver any value in the Increment for the Sprint.

- The organization files for liquidation bankruptcy.

- Patent litigation kills the product.

- New government regulatory requirements negate the Sprint Goal.

When a Sprint is canceled, any Done PBIs should be reviewed to determine if they are releasable. All undone PBIs are moved back to the Product Backlog. If they are still pertinent, they should be reestimated.

When a Sprint is canceled, it means that at least some amount of the work will have to be discarded. Although Scrum minimizes waste through short iterations and just-in-time planning, canceling a Sprint should be the last thing considered. Sprint cancellations consume resources, since everyone has to regroup in another Sprint Planning to start another Sprint. They are often traumatic to the Scrum Team and, as such, are not a good thing.

Tip I know several Professional Scrum Trainers and other practitioners who have gone their whole career without experiencing a canceled Sprint. If your Product Owner is canceling Sprints frequently (even more than once), consider shortening the Sprint instead. I've worked with teams in volatile markets who normalized on three-day Sprints for just such a reason. This is an extremely short Sprint, with a relatively high cost in the overhead of the required events, and should be attempted only by high-performance Professional Scrum Teams.

Undone Work

A common problem that Scrum Teams face is that of undone work. The Sprint is over and some things did not get done. Maybe an entire PBI (or two) didn't even get touched. More likely, however, is that a PBI is in progress with one or more undone tasks, code half-written, tests half-passing, the Definition of Done not fully met, and so on. I refer to that type of undone work as *unfinished* work. Regardless of how much work was accomplished for a PBI, it cannot be released unless it is done—according to the Definition of Done. Not only would the feature not work correctly, but the Developers will have introduced technical debt into the product.

Regardless of the type of undone work, or its level of "undoneness," the guidance is the same:

1. The PBI should not be released.

2. The PBI should not be inspected at Sprint Review. Doing so might set expectations that it is done or close to being done or that it will look/feel like this when it is done.

3. The PBI should be moved back to the Product Backlog.

4. The Developers should reestimate the PBI.

5. The Product Owner will consider developing the PBI in a future Sprint.

There are other nuances to consider when dealing with undone work:

- **Don't give partial credit** Points for partially completed PBIs should not be summed into velocity, even partially. The Developers are either done with a PBI (according to the Definition of Done) or they aren't. The exception to this is when a PBI can be split and a portion completed according to the Definition of Done during the Sprint (see the next bullet).

- **Split and release smaller PBIs** If a PBI can't be delivered in whole, as forecasted, and the Developers determine that it can be split, they should talk with the Product Owner. That discussion might yield a plan to release smaller, logical parts of the PBI—each delivering a slice of value. Splitting on acceptance criteria boundaries is often a good choice.

- **Reestimate the PBIs** Undone PBIs should be refined and reestimated. This increases transparency because the new estimates reflect the Developers' latest thinking. Besides, the new estimates will be more accurate, since the team has firsthand experience. Estimates may be lower than original, because some work was performed. Estimates may also be higher—new complexities may have been uncovered.

- **The Product Owner owns the order** It is always the Product Owner's prerogative to order the Product Backlog at any time, for any reason. This means that an undone PBI from the current Sprint may not necessarily return to the Product Backlog at the top, or even anywhere near the top. In other words, don't assume that undone PBIs will just get forecasted (rolled) into the next Sprint. This takes agility away from the Product Owner.

- **Exclude unfinished work from the Increment** In the event that a Sprint ends with unfinished work, manual effort may be required to exclude the unfinished code and behavior from the Increment. Single-piece flow (limiting work in progress by swarming or mobbing on a PBI) can help alleviate this headache. If this is a regular issue, the Developers can establish a working agreement that they will not advance to the next PBI in the Sprint Backlog until the prior one is Done. For some environments, this is not practical, so other, more engineering-centric solutions are required. The two most common approaches are (1) create a branch per PBI, and (2) use *feature flags*. I've already discussed the risks in using branches in Chapter 8, "Effective Collaboration." I will cover feature flags in the next section.

- **The Product Owner cannot override the Definition of Done** Under no circumstances can the Product Owner, a stakeholder, or the organization override the Definition of Done and say that a PBI is done when it isn't. Releasing unfinished work decreases transparency, increases risk, and can add technical debt to the Increment.

A releasable product is one that has been designed, developed, tested, and otherwise made Done according to the Definition of Done. This may or may not include the actual release of the product. This means that the Developers may have to create build packages, installers, release notes, help files, and other artifacts to assist in the actual release—as well as releasing—before declaring a PBI, as well as the Increment, Done. Whether or not the Product Owner chooses to release is irrelevant. The Developers must complete all work as if the Product Owner was going to release it. This is another advantage of feature flags. They allow for a PBI to be released, but not necessarily enabled. The Product Owner can then decide, at the appropriate time, to enable the capability.

Smell It's a smell when a Scrum Team, including the Product Owner, just assumes that undone PBIs will be carried forward to the next Sprint. That may very well be the case, but assumptions like this diminish agility and can lead to unhealthy behaviors, such as compromising on quality. It's always the Product Owner's decision what work will be considered for the next Sprint. The Product Owner should listen to the Developers, but that is only one source of input to consider. In other words, it's a dysfunction to assume that the Developers will just continue working on unfinished PBIs in the next Sprint.

Velocity is just one input to Sprint Planning. Velocity simply indicates how many PBIs (or the sum of story points) a team usually completes per Sprint on average. It is not a commitment or a target and should only be used as one of several inputs for Sprint Planning. For example, just because the Developers were able to deliver 30 points last Sprint doesn't mean they will again this Sprint. The velocity for future Sprints is always unknown, and Developers can only guess what it will be. These guesses will become more accurate over time because velocity should stabilize as the team improves. It takes courage to stop guessing and trust that the Developers will be able to self-manage and forecast a comfortable amount of work and then deliver the best Increment possible given all the constraints.

Smell It's a smell when I hear the terms *velocity* and *credit* together in the same conversation, or when management tries to compare the velocities of two Scrum Teams. This smells like the Developers are being gamed in some way or are being artificially rewarded somehow. Remember that velocity is just a historical representation of multiple factors from previous Sprints. It is a lagging indicator of the Developers' output. Using velocity in any other way diminishes its value. Don't put too much scrutiny, positive or negative, into those numbers. Developers should focus on achieving the Sprint Goal and delivering business value in the form of working product, and not on increasing or stabilizing its velocity. That will happen organically. Velocity should never be the goal.

Feature Flags

A *feature flag* is a technique where functionality can be selectively excluded, disabled, or enabled from released software. In other words, Developers can modify system behavior without changing code. Feature flags are primarily used as a way to perform a *canary release*—only enabling a new feature for a small percentage of users. This allows the Scrum Team to "test in production"—albeit with a small group of users, thus limiting the impact of any errors or side effects.

Feature flags, also known as feature toggles, feature bits, feature flippers, and feature switches, are not a new concept. Developers have leveraged this kind of practice by wrapping a section of code with an *if/else* statement controlled by logic driven from an external data source. This allows teams to control product feature visibility and utility. There are many commercial SaaS feature flag solutions out there, including Azure App Configuration, LaunchDarkly, Optimizely, and Rollout.io (CloudBees). GitHub has many open source solutions. Honestly, it's not too hard to write your own feature flag implementation, but why would you want to? For more information on open source solutions, visit *https://featureflags.io*.

As I just described them, feature flags sound more interesting to Product Owners—providing a mechanism to perform canary releases, incremental rollouts, blue/green deployments, testing in production, A/B testing, hypothesis-driven development, and even a kill switch. Another, lesser-known use case of feature flags is as a hedge against undone work and in lieu of using branches.

Essentially, Developers can use feature flags to work on and even release the Increment containing an unfinished PBI. As the Developers start work on a PBI, they can be diligent about putting all related code inside respective *if* blocks, driven by a feature flag. In the (unlikely) case of an unfinished PBI, its visibility in the UI and functionality can remain toggled off, allowing the Increment to still be safely released and used. Yes, there will be technical debt, but hopefully it will be short-lived. Improved design will be a positive side effect of working this way—mostly to avoid the multitude of *if* statements.

Ideally, the Developers will finish the partially completed PBI in an upcoming Sprint. Once completed, Developers can toggle the flag on or remove the feature flag infrastructure altogether. This minimizes waste and also ensures the toggled feature doesn't become long-term technical debt in the product.

Be careful releasing a product with feature flags, since they can (by design) cause different behavior in different deployments, which can make the process of triaging a bug more difficult. Remember that the team will naturally get better about completing its forecasted work and may not need such solutions in the future.

> **Note** Don't confuse using feature flags in this way with the practice of *feature flag–driven development* (FFDD), a practice of releasing and iterating features quickly, testing those features, and making lightweight improvements. FFDD is akin to test-driven development (TDD) and Lean UX practices, where features are released to receive market feedback and development iterates on that feedback, making improvements and redeploying. It's a way to test how your features perform in the real world and not just in an artificial test environment. You can learn more by reading this LaunchDarkly blog post: *https://blog.launchdarkly.com/feature-flag-driven-development*.

Regardless of how a team uses feature flags, keep in mind that they are technical debt. They may be cheap and easy to add, but the longer that they are left in the code, the more debt is accrued. It can be hard to keep track of which flags were used for which purpose and which flags are even relevant anymore. They add complexity when debugging, refactoring, and testing—not to mention increasing the difficulty to reproduce a bug. Professional Scrum Teams should limit their use of feature flags, review them often, and be merciless about removing them when they are no longer necessary.

Handling Undone Work in Azure Boards

Azure Boards does not offer any tooling for directly handling undone work. All moving or copying operations will have to be performed manually, and doing so can be time-consuming. Because of this, Sprint Backlogs sometimes end up in a less-than-desired state due to the amount of work required to organize them properly. Teams have moved on to the next Sprint's work, and housekeeping is not a priority. If the Developers are incapable of cleaning up their Sprint Backlog, then it is the responsibility of the Scrum Master to teach or coach them.

When a PBI is not done at the end of a Sprint, the opportunity to work on it as part of the existing body of work is often lost forever. The reason for this is that when the same PBI is planned in a future Sprint, the context for that work may be different. This is why I always recommend that the PBI be reestimated.

When considering how to handle undone work in the context of Azure Boards, there are basically four approaches:

- **Move to the Product Backlog** This is the most common approach and fits with my guidance from earlier in this chapter. The PBI work item is simply moved back to the Product Backlog by changing its Iteration *Path* and *State* fields accordingly. You can also drag and drop to the *Backlog* in the Planning pane. Linked work items remain linked. No record, other than notes in the History, indicates that the PBI was ever in the original Sprint. Linked Task or Test Case work items will still be set to the original Sprint.

- **Copy to the Product Backlog** The PBI work item is shallow-copied and the copy's Iteration *Path* and *State* fields are set for it to appear in the Product Backlog. The original PBI work item and all linked work items remain in the original Sprint in the Forecasted State. The new PBI has no linked work items, so a brand-new plan (Task and/or Test Case work items) must be created during a future Sprint Planning. The old PBI, and any linked work items, are abandoned and left to the historical record of that Sprint.

- **Move to the (next) Sprint Backlog** If the Product Owner wishes, then the Iteration Path of the PBI work item and all linked work items can be changed to the next Sprint. Everything appears just the way it was in the original Sprint. No record, other than notes in the History, shows that the PBI was ever in the original Sprint.

- **Use the Split extension** By installing and using this extension, the team can continue working on unfinished PBIs in the next Sprint by "splitting" the PBI into a new work item in the next Sprint. The team can choose to move any of the incomplete tasks forward to the next Sprint as well, as you see in Figure 10-4. For more information visit *https://marketplace.visualstudio.com/items?itemName=blueprint.vsts-extension-split-work.*

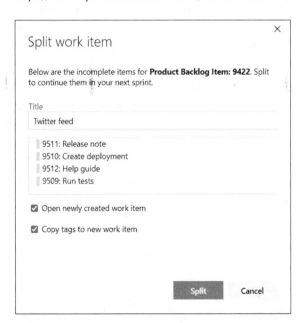

FIGURE 10-4 The Split extension makes it easy to move undone work to the next Sprint.

There will be times that a Scrum Team may want to use each of these approaches. One cannot be prescribed as the "recommended way" without knowing more about how the team works and other factors. That said, I'll repeat myself and say that there is a smell of dysfunction when a Scrum Team constantly rolls undone work over to the next Sprint—whether manually or by using an extension.

You should consider a number of other activities when copying or moving undone work items in Azure Boards. I've listed some of these in Table 10-2.

TABLE 10-2 Considerations when copying or moving undone work items.

Set the PBI's Iteration to the root after moving or copying to the Product Backlog.
Set the PBI's State to Ready after moving or copying to the Product Backlog.
Reestimate and reorder PBIs after moving or copying them to the Product Backlog.
Clear all Remaining Work values for undone Task work items (they will be reestimated).
Clear all *Assigned To* fields for undone Task work items (they should not be owned by anyone).
Set all states and fields appropriately for other linked work items (such as Test Cases).
Copy all (or just the undone) tasks to the next Sprint Backlog if the plan still makes sense.
Copy Test Case and other work items to the (next) Sprint if applicable.
Add appropriate notes to the History tab of the various work items as needed.

Spikes

There will be times that the Developers may be required to develop something that they haven't done before. They may not be able to do so, let alone even estimate the work with any confidence. This could include developing a new capability using a new product, component, framework, system, language, or tool. The Developers will need to learn and practice in order to develop the feature successfully. They also need this experience sooner, in order to be able to *estimate* the size of the PBI and feel comfortable forecasting the item in a future Sprint.

Organizations can't expect their Developers to gain this knowledge on their own, although some Developers will. I know many developers who consider their profession to be their hobby. For these geeks, learning new things is just fun. For the rest of the world, this learning will have to come during company time, on the company dime. But how does this fit with Scrum? I have a couple of ideas. First is for the team to do that investigation and learning informally during their slack time. This assumes the Developers are allowing themselves slack time—by not forecasting to 100 percent of their capacity.

Another idea is to perform a *spike*, which is another word for a technical investigation, proof of concept (POC), or an experiment, the outcome of which is to gain enough knowledge to be able to give the team some confidence in their plan and their estimate. Ultimately, Scrum is about learning from data derived from experiments—so the spike concept fits right in. When you see the look of sheer panic on the faces of your teammates when discussing/estimating a new PBI, it may be time to perform a spike.

Most spikes are small and executed as needed throughout the Sprint. In fact, I wouldn't even call them a spike. They are just part of development. Some Scrum practitioners call them *spike tasks* (as opposed to larger *spike stories*). If a Developer needs to clarify a technical issue, and another team member cannot help, the Developer can create a quick spike instead. Timeboxing should always be used to keep spikes as small as possible and to help maintain focus. The Developers can decide the criteria for if/when to track a spike as a work item in Azure Boards. Transparency should be their guiding light.

> **Tip** A spike is not the same thing as a *tracer bullet*. A tracer bullet is development that cuts vertically through the many layers of architecture. This is sometimes known as the practice of developing thin, vertical slices. *Emergent architecture* is the practice of continually developing in thin slices like this. Tracer bullets can be experimental in nature, like a spike.

When a spike is expected to take a large amount of time, or is required to be accomplished before the Developers are able to estimate a PBI, the Scrum Team should consider treating it like a PBI. Spikes then become part of the Sprint and should therefore be accounted for in Sprint Planning and be present in the Sprint Backlog. This means that the spike should be added to the Product Backlog first and forecast as part of the Sprint. Spikes can even have their own acceptance criteria to help specify the goal or outcome. A plan should be created and tracked, just like any other feature or bug fix. Because spikes don't offer any direct value to the stakeholders (for example, users), there should always be other PBIs forecasted for that Sprint, even if just a few small ones, so that the product, Product Owner, and stakeholders can enjoy some increment of actual value every Sprint.

> **Smell** It's a smell when a spike takes the majority of the team the majority of the Sprint to accomplish. It's a stench when it takes multiple Sprints. I guess it may be possible that the new architecture or technology is so alien that it really does require that much capacity to understand it to the point of being able to use it effectively. In my experience, however, good Developers aren't caught flat-footed like this very often. New tools and technologies tend to be pretty similar to previous ones. Also, don't let the Developers get into the habit of creating a spike for every PBI that they refine. True spikes should be rare.

Fixed-Price Contracts and Scrum

Scrum works well when the stakeholders trust the Scrum Team, the Product Owner trusts the Developers, and all are able to work together collaboratively. If the customer has had enough projects fail in the past, this trust won't be there initially. In their minds, it will need to be replaced with a contractual relationship with the team instead. The customer's hope is that the contract and its clauses and signatures will minimize the customer's risk and provide a legal way of recovering costs if the team fails to deliver. From their perspective, they have only one shot at getting the product they want, so they want to define everything up front and then manage risk by putting a monetary limit and other constraints into the agreement.

The most common of these contract development agreements are known as *fixed-price* or (*fixed-bid*) contracts. They attempt to predict exactly the cost and the time at which the product that's been specified by the customer will be delivered. The common misconception is that it is impossible to use Scrum on a fixed-price contract. In reality, Scrum handles this in the same way that any other process would. Everything the customer wants is detailed (minimally) and sized, generating an idea of the timeframe at which that scope can be delivered.

Here are the common challenges with fixed-priced contracts:

- Price is the most important factor and is often driven by competition, not quality.

- Requirements are vague, wrong, out of date, or missing.

- Team-based estimation is impossible due to lack of empirical data.

- No knowledgeable person (such as a Product Owner) exists.

- The Scrum Team doesn't have any incentive to spend time enlightening the (potential) customer about Scrum, or creating and refining a Product Backlog prior to signing a contract—although I have seen success when a team simply charges time and materials (T&M) to collaborate and create the Product Backlog, establishing trust during the process.

- Quality is not defined, only assumed.

- There is no Definition of Done, or even a basis for one.

- Deadlines are artificial and often impossible.

- Risks are not shared or are outright ignored.

> **Tip** Beware of fixed-price, *fixed-scope* contracts. Scrum + fixed-price + fixed-scope don't mix. This is the whole idea behind having a Product Backlog and an active Product Owner to order the PBIs. If the customer in a fixed-price contract wants to own both the date and the number of features, the only remaining variable is quality, and sacrificing quality never works out. Remember healthcare.gov?

Any fixed-price contract should be *variable-scope*. Not only does this fit better with the nature of complex work, it fits better with Scrum because the team can now apply a consistent Definition of Done and establish an uncompromising baseline of quality for all of the work it does. The team can then start using iterative, incremental development to begin delivering increments of working product every month or sooner. This model provides more value and less risk to both parties—but it is hard to conceptualize and agree to without knowing more about Scrum and the Scrum Team.

Perhaps a better name for a fixed-price contract would be simply a *fixed-budget* contract. The customer knows how much they want to spend, or at least what the ceiling is. This can be easily translated into the number of Sprints they are able to afford. For example, a customer with $300,000 to spend can afford a Scrum Team whose run rate is $30,000 per Sprint for 10 Sprints. By creating an ordered Product Backlog, the customer will get the most important/valuable features (according to them) before those 10 Sprints are up. Therefore, the ideal Scrum contract model should be *fixed-budget, variable-scope*.

Here are two rules to consider when you're using Scrum for a fixed-price project:

- The customer (via the Product Owner) can replace any item in the Product Backlog with another item of similar size, provided the Developers haven't started working on it or completed it yet. If they have, there will be waste, and more importantly, one or more items may not make the cut before the budget runs out.

- At any point in time, the customer (through the Product Owner) can say that they have enough functionality and effectively end the development effort, potentially saving money. Let any waterfallian contract top that!

It's important for both the customer and the Scrum Team to share the risk. This means that the customer must work closely with a knowledgeable Product Owner or *become* the Product Owner—assuming that they know Scrum and the requirements of the Product Owner role. In either case, the customer can be directly involved in ordering the Product Backlog and determine the scope. This removes the risk of the Developers delivering the wrong features or not getting to those "must-have" features before the budget runs out. Some customers, after learning that they will be accountable for this, may decide to walk away and offer the work to a competitor. You should let this happen. In my opinion, this is the right thing to do, rather than running the risk of building the wrong product or a product of questionable quality and value. Remember healthcare.gov?

Common Dysfunctions

Leo Tolstoy told us that "happy families are all alike, but that every unhappy family is unhappy in its own way." This is true of product development teams as well. A certain amount of dysfunction is going to exist, even in high-performance Professional Scrum Teams, and it will always be unique. This is because Scrum is about people, and people don't behave like predictable machines.

Removing a dysfunctional behavior can be difficult. Identifying it in the first place can be very difficult, especially if you are in the middle of it or if you are the cause of it. Part of becoming good at Scrum is the ability to sniff out dysfunctional behavior. At first, this may be the ability to know when your team isn't following the rules of Scrum, according the *Scrum Guide*. But that's not enough. This is why—in this book—I've diligently pointed out many smells that you should be aware of.

It may seem like the *Scrum Guide* has an answer for everything, but it doesn't. In fact, each version of the *Scrum Guide* seems to provide less guidance. The complex world of product development will sometimes put you and your team in a double-bind in the middle of two choices that conflict with each other. Your abilities should transcend from just knowing the rules to knowing (and applying) Scrum's principles and values. Knowing the higher-level reasoning behind agile software development and why Scrum works allows you to identify and resolve such conflicts—and make leaner choices.

Teams new to Scrum may fumble when applying a practice for a given dysfunction. Their heads are down, executing that practice. They are in the *shu* stage. High-performance Professional Scrum Teams have moved beyond rote practices and think in principles. They are in the *ha* and *ri* stages. Their heads are up, looking for dysfunction and ways to eliminate waste and generate more value. It's a state of mind, and it comes with experience. This section serves as a guidebook to the different types of dysfunction that can be found on a Scrum Team, and it offers ideas for removing them.

> **Note** *Shuhari* is a Japanese martial art concept that describes the stages of learning toward mastery. The idea is that practitioners will pass through three stages of gaining knowledge. *Shu* is the beginning stage, where students follow the teachings of the master precisely. The master can be another team member, an external coach or trainer, a talking head on a training video, or the *Scrum Guide*. *Shu*-level students concentrate on how to do the task, without worrying too much about the underlying theory. *Ha*-level students begin to branch out and apply what they've learned. In doing so, they continue to learn the underlying principles and theory behind the practice. *Ri*-level students are now learning from their own practices rather than from other people. Mastery is just around the corner for them.

Not Getting Done

You would think that "done" is when a new feature or bug fix has been deployed and is running happily in production. I would agree. If that's the status of your PBI, then you are definitely done with it. From Scrum's point of view, however, this is not always the case. Done doesn't necessarily mean that the PBI is in production but that it easily could be. This is the concept of being releasable (or *potentially releasable*, as some practitioners still say). In other words, in Scrum, Done can mean released or releasable.

If the Definition of Done for an increment is part of the standards of the organization, all Scrum Teams must follow it at a minimum. If no standards exist, the Scrum Team must create a Definition of Done appropriate for the product. Since the Definition of Done contains primarily development-related practices and standards, the Developers will probably have the most influence over what gets included. The Developers are required to conform to the Definition of Done. If there are multiple Scrum Teams working together on a product, they must mutually define and comply with the same Definition of Done. Individual Scrum Teams can have a more stringent Definition of Done, however. I cover this in Chapter 11, "Scaled Professional Scrum."

What Done *doesn't* mean is that the PBI has been coded but not yet tested. In Scrum, all development activities, including testing, must be finished before a PBI can be considered Done. It's a dysfunction when the Developers are not able to complete their work according to the Definition of Done. Perhaps their definition is too stringent. Perhaps the Sprint length is too short. Perhaps the Sprint length is too long. Nothing focuses the Developers like knowing they have a Sprint Review coming up in a few days!

External dependencies are another huge cause of not getting done. An example would be having to wait for people outside of the Scrum Team to perform review, testing, audit, or sign-off. These might be users or customers performing user acceptance testing or managers who need to approve the work. It's critical that *only* the Developers do the work—*all* of the work—to make a PBI Done. Not only will this lower the risk of not getting Done, but it's also not Scrum until this is the case. Retrospect hard on these impediments until they are removed.

> **Smell** It's a smell when I hear a team using the terms *done done* or *really done*. Historically, these terms have meant that both coding *and* testing had been completed, which implies that there was a done state where just the coding was finished. In Scrum there is only Done—as defined by the Definition of Done—and the team is either there or they aren't. It is a simple Boolean state.

In Chapter 1, "Professional Scrum," I mentioned that the Definition of Done is an auditable checklist that each PBI must go through before it is considered Done. When each item in that definition is "checked off," the PBI is Done and releasable. Some Scrum Teams include an item in the definition verifying the Product Owner's acceptance. Other teams don't and just understand that it's simply a part of Scrum's workflow for being done. Either approach is fine, as long as it is uniformly applied, understood by the entire Scrum Team, and never undermined. Personally, I would strive for the Product Owner's "delight" or "elation" over just their "acceptance."

> **Tip** There are times that the team will not get done. It can be hard to prevent this from happening. It is important not to make it a habit by forecasting too much work. It is also important to craft a reasonable Sprint Goal during Sprint Planning. Having a Sprint Goal is important because, even if some of the forecasted PBIs aren't completed, at least the goal was met and stakeholders will be able to inspect something of value. The Scrum Team should make good use of Sprint Planning by analyzing all of the inputs, checking capacity and past performance (such as velocity or Throughput), and forecasting a comfortable amount of work each Sprint.

Flaccid Scrum

In January 2009, Martin Fowler wrote a blog post titled "Flaccid Scrum." Just as the name implies, his observations of many teams doing Scrum were that they were doing poorly. His typical observation would include a team that wanted to use an agile process, so they picked Scrum. The team adopted the Scrum practices, and maybe even the principles. After a while, progress slowed because the codebase became a mess and the team found itself drowning in technical debt. You can read Martin Fowler's article here: *http://martinfowler.com/bliki/FlaccidScrum.html*.

The fact that these teams were using Scrum was orthogonal to the root cause of the problem. It was just another example of teams and organizations considering Scrum a silver bullet. Scrum is simply a framework for planning and managing complex work. It says nothing about specific development practices exercised within, other than the generic statements relating to improvement. In other words, the Scrum framework allows any number of practices to be used, regardless of how dysfunctional they may be.

I surmise that the products that these teams were developing suffered low quality because the Developers were not inspecting, not adapting, or both. Perhaps the Scrum Value of openness was deficient as well. Remember that Scrum has built-in opportunities to inspect and adapt, at both the product and the process level. The fact that technical debt was building up to critical levels was because the teams either didn't know (weren't inspecting) or didn't care (weren't adapting).

> **Note** I've met with many such teams who love to throw around the Scrum terms *Sprint, Scrum Master, Product Backlog,* and so on. But when it came to being able to deliver business value within a timebox, they couldn't do it. It seems as though they were using the Scrum nouns but not doing the Scrum verbs. This type of behavior is known as "zombie Scrum," which is Scrum without the beating heart of working product. Other names that refer to this kind of Scrum are mechanical Scrum, dark Scrum, Scrum in name only, and ScrumBut. There are nuanced differences between these, but they all relate to Scrum being practiced in a dysfunctional way.

To fight flaccid Scrum, the Developers need to inspect and adapt their technical practices. This is true especially if there is a lot of technical dysfunction and technical debt present. During the Sprint Retrospective, the team should inspect its current practices and, if improvement is required, agree to adopt, continue using, or abandon the practice in question. They can also take this opportunity to ratchet up their Definition of Done and include more stringent criteria for higher quality. Most importantly, in the next Sprint, they can adapt by executing these improvements.

> **Tip** The Professional Scrum Developer (PSD) program was a direct response to the problem of flaccid Scrum. The program consists of a training course, assessment, certification, and a community developed for the most neglected role in Scrum: the Developer. The PSD course was developed in cooperation between Microsoft, Scrum.org, and Accentient. You can find out more by visiting *www.scrum.org/psd*.

Not Inspecting, Not Adapting

Flaccid Scrum came about because of many reasons. Teams were uneducated. Teams didn't have a Definition of Done, or they didn't stick with it, or they didn't try to improve it. Teams weren't able to deliver business value in a single Sprint. Teams weren't inspecting. Teams weren't adapting.

Scrum is based on empiricism, which means that the team members make decisions based on what is. These professionals must frequently inspect Scrum artifacts and their progress toward goals (product, release, Sprint, etc.) to detect any undesirable variances. Good decisions can't be made without data. Conversely, meaningful data is useless unless it is acted upon. Not doing either is definitely a dysfunction.

> **Tip** If I want to know how well a Scrum Team is inspecting and adapting, I will ask about their Sprint Retrospectives. In my experience, the Sprint Retrospective is the first to suffer when times get rough. Sure, the team may meet and discuss stuff, but they may not act on their findings. I contend that "rough times" can be translated as "We were super busy," "We didn't like what we discovered," or "We didn't want to improve." Another inspection—one intended for the Developers—is to review their automated regression test coverage. This is a key indicator of how well they are improving their technical excellence.

For example, a Scrum Team may be very diligent about scheduling and attending their Sprint Retrospective. They may have rich conversations and discuss the high and low points of the Sprint. They may even identify things to do differently in the future. Multiple team members capture this information and then do nothing with it. They have inspected, but not adapted. This is belief in magic. This is flaccid Scrum.

> **Tip** Anything the Scrum Team tries to do differently in the next Sprint should be visible and transparent. There are several strategies for doing this. Some teams use a "retrospective back-log" or "improvement backlog," whereas others add tasks directly to the next Sprint's Backlog to represent the work and time needed to adapt their practices. One option is to further cus-tomize the Professional Scrum process, adding a new *Improvement* work item type, as you can see in Figure 10-5. I spent some time in Chapter 3 going over how to create a custom Profes-sional Scrum process, inherited from the default Scrum process.

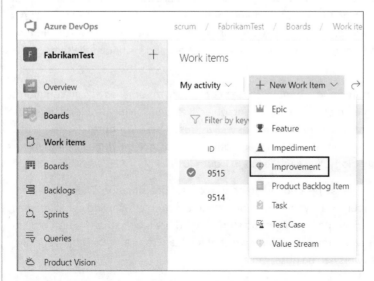

FIGURE 10-5 It's useful and encouraging to track planned improvements and not just impediments.

Smell It's a smell when I see nobody taking notes during the Sprint Retrospective. Does the team not have anything interesting to discuss and record? Maybe they have nothing that can be fixed, and thus no action items. That is sad. It is also untrue. With the organizations I've worked with, every individual and team could improve. This flaccid behavior should be corrected. In the meantime, and at the very least, the Scrum Master should be recording any inspections and then ensuring that the appropriate experiments are conducted and adaptations are made. It's also a smell when I hear the *same* items coming up repeatedly. This is a failure to adapt.

On the other hand, formal inspection should not occur so frequently that it gets in the way of the work. In order to minimize this, each Scrum event is an opportunity for inspection and adaptation:

- **Sprint Planning** The Product Backlog, Product Goal, Increment, Definition of Done, and past performance are inspected, and the Sprint Backlog is adapted.

- **Daily Scrum** The Developers' progress toward the Sprint Goal is inspected, and their plan for the next 24 hours is adapted.

- **Sprint Review** The Increment is inspected, and the Product Backlog is adapted.

- **Sprint Retrospective** The process, practices, Definition of Done, and definition of workflow are inspected and adapted (during the next Sprint).

Developer Challenges

It takes time for the Developers to be able to self-manage, even with the support of the organization. Teams that come from a more formal, waterfallian background are used to the relative safety of the different stages. Hiding behind (the wrong) requirements or in front of (the yet-to-be-run) tests provides a level of safety and cover. Moving to an attitude of transparency—understanding that everybody is on the same team, working toward the same goals, and sharing in the same successes and failures—can take time.

As I've said before, Scrum is about people. These people work together as a team communicating, listening, complementing each other's skills, sharing objectives, and solving problems together. There must be compassion and respect for each other, as well as trust. These attributes will develop and improve over time. Professional Scrum Developers continually balance the three raw ingredients: people, process, and technology. You can see this in Figure 10-6.

People exhibit different behaviors depending on the context of a situation. There is the normal behavior: how team members usually see one another. There is problem-solving behavior: the team members are fully engaged mentally and getting stuff done. There is also stress behavior: quite different from the others and often harder for the rest of the team to deal with. During any given Sprint, each of these behaviors can be observable.

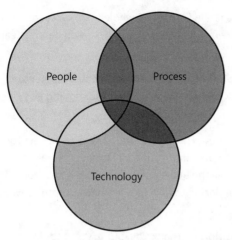

FIGURE 10-6 Achieving high-performance Scrum is a continuous balancing act.

Scrum Teams need to learn to deal effectively with ambiguity. Most of the time, the team won't have all the answers, or even many answers. Effectiveness in the face of ambiguity is a measure of growth. Teams will find themselves in unusual circumstances that cannot be solved by thinking at the practice level; rather, a problem can be solved only by abstracting to the principle level. It's up to the individuals to make judgments and not just copy and paste random bits of practice.

In addition to dealing with ambiguity, a number of other challenges face Developers. Here is a list of dysfunctions—sometimes expressed as "weasel words" —that may be found in any given team of Developers:

- **I don't have all the requirements**   A major theme in Scrum is the ability for the Developers to self-manage and get the job done. If there is missing information, fix that by getting the answers. If you want to add functionality over and above what the Product Owner is asking for, don't. Remember, the Product Owner owns the *what*, and the Developers own the *how*. Collaborating with the Product Owner and stakeholders can help obtain those missing "requirements."

- **I tell other people what to do**   Each Developer needs to be able to self-manage. This means that nobody, not a manager, not the Product Owner, and not even the Scrum Master can order a Developer to work on a particular task or to do a task in a particular way. Work is never assigned in Scrum. If you have a command-and-control personality, perhaps you need to excuse yourself from the team for a while so that they can learn these skills in your stead.

- **I'm quiet and don't like to converse with others**   Effective collaboration requires all parties to communicate. This is more than just actively listening—it's also actively communicating and sharing ideas. Everyone on the team is creative and has ideas to share.

- **I monopolize the conversation**   Stop doing that. Use active listening skills to improve your ability to communicate and collaborate with others. Your teammates will thank you.

- **I'm a coder, not a tester *or* I'm a tester, not a coder**   In Scrum, everyone is a "Developer" regardless of what HR or their business card says or what activity they are working at that moment. Besides, most automated tests are written in code—so you can be both!

- **I'm not to blame; another Developer broke it**   In Scrum, the whole team succeeds or fails. If something broke, it's because the team broke it. The team will fix it. Focus on being a team player.

- **I'll let the Scrum Master remove the impediment**   If you can remove the impediment yourself, do it. If you cannot, or if you can provide more value by doing something else (such as developing product), consider asking the Scrum Master for help.

- **I write great code by myself**   That's nice to know. Having another Developer review your code or pairing with another Developer isn't about you or your code. It can be a learning experience for others, as well as a way to improve the quality of the product. It also helps hedge against the event where you are hit by a bus or decide to quit your job because you've won the lottery.

- **I'll work evenings and weekends to get this done**   Thank you, but that sounds like an unsustainable pace. Typically, this smell is due to time management or over-forecasting dysfunctions. During the next Sprint Retrospective, the team should discuss this and find an alternative approach to reaching its goals, such as forecasting less work in the next Sprint. Slack time is good for experimenting and learning.

- **I can goof off because others will do my work**   Every day, all Developers meet for the Daily Scrum. During this time, each Developer clarifies their plan for the next 24 hours. If, for example, by the third day, a Developer is still working on the same task, the others should take notice. The self-managing team can then find an appropriate solution, such as helping that Developer while explaining the Scrum Value of *openness* to them.

- **Nobody on our team has that skill set**   This may be true, especially as a product delves into new markets requiring new technologies. The reality is that Developers need to acquire the necessary skills. Attend training, watch some videos, perform some experiments, or learn in some other way. Adding a new Developer to the team is an option but can be costly in terms of onboarding and other interruptions. Be sure to account for this disruption at future Sprint Plannings, and add any large spikes to the Product Backlog if necessary.

- **I'll give it to the testers at the end of the Sprint**   First of all, there are no "testers" in Scrum, only Developers. Some Developers will focus on coding tasks and others on testing tasks, but this is not written in stone. For a PBI to be Done, it needs to be tested as well. Waiting until the end of the Sprint to test will increase the risk of not getting Done. If *you* have capacity, *you* can do the testing work yourself. Be T-shaped.

- **Nothing is blocking me**   These are commonly heard words during a Daily Scrum. But they do not always reflect reality. Developers need to learn that transparency and openness start within themselves. If there is an impediment or *even the possibility of an impediment* blocking some work, be sure to let others know, regardless of whether or not you have other things to work on. Identifying actual or possible impediments is not whining, and it is not a sign of weakness. In fact, it is quite the opposite. Identifying impediments is about transparency and creating a *Professional* Scrum Team.

Measuring Performance

The performance of the team of Developers should be measured by what it is able to develop. In other words, rate the team by its ability to turn PBIs into Increments of functionality that are actually released. This can be measured for a specific Sprint or range of Sprints (for example, a release).

Velocity is the measure of how many PBIs (or points) the team is able to deliver each Sprint. This measure can then be divided by the cost of the team to develop that Increment. For example, if it costs $20,000/Sprint to run the Scrum Team and the outcome of the last five Sprints is a release that totaled 100 story points, then the performance of the Scrum Team is $1,000/story point.

> **Tip** From the perspective of the business, velocity is not a good metric on which to measure performance. If the Product Owner tracks business value for each PBI—which they should be doing—this provides a better measure than just the number of PBIs or the sum of their story points. Whether the business value is specified in a numeric range, income/profit, or some specific scale, when this number is expressed over a monetary amount, it will tell the business exactly how much value is being returned on their investment (ROI). Tracking ROI per PBI helps the Product Owner and organization focus on business value for investing and budgeting. To continue with the previous example, the Product Owner could compute how much value was delivered for the $100k. If the PBIs in that release totaled 160 units of business value, then the performance of the Scrum Team is $625/unit of business value.

The performance of individual Developers should not be measured. The team is self-managing and operates as a unit, not as a group of individuals. For a given Sprint or release, some Developers may be a lot more heads-down than others. As such, it may appear to outsiders that these individuals are "working harder" and more worthy of praise or bonuses. Other Developers, those not at a keyboard, may actually be working harder while coaching, mentoring, or designing. Invisible measurements like this are hard to measure and can be missed. A better approach is to rate the performance of the team as a whole and according to the value that they deliver. If it needs to go beyond that, let the team decide who the performers are.

The best way for a manager to see and learn what's really going on within a team is to sit directly with the Developers. Decisions based on indirect information such as dashboards and reports are likely not to be the best decisions. Decisions based on actually going and seeing and understanding the real situation at hand are likely to be much more informed. This practice is called a "gemba walk" or simply "go see."

Working with a Challenging Product Owner

The Product Owner is responsible for maximizing the value of the product and, thus, the work of the Developers. This is a lot of responsibility for a single person, making this the hardest role in Scrum in my opinion. Needless to say, a good Professional Scrum Master can find at least one or two dysfunctions for a Product Owner to improve upon.

One of the biggest dysfunctions a Product Owner can possess is not knowing their role. They must reflect the real value and priorities of the business, customer, or user with respect to the product. This is both the biggest responsibility and the biggest potential risk of the role. The Product Owner is one person, not a committee. The desires of a committee, however, may be represented by the Product Owner.

> **Note** Organizations often struggle to find viable Product Owner candidates among their existing employees. People with technical backgrounds are typically better suited to be a Developer, because they like to get involved with *how* things are developed rather than *what* should be developed. Employees with management backgrounds might be inclined to practice traditional project management techniques, even instilling command-and-control practices. Candidates with strong Scrum knowledge typically gravitate toward being the Scrum Master. These are, of course, generalizations.
>
> Successful Product Owners tend to have product management and even marketing backgrounds. They understand terms like ROI, P&L, market segment, and sales channel. Maybe they were even a user of the product at one time. When no such candidates exist, I've seen some organizations advertise the position of Product Owner publicly. It sounds weird, but by bringing in someone off the street with the knowledge of what a Product Owner should do, but absent the knowledge of organizational politics and the "old way" of doing things in the organization, is often a recipe for success. They just need to learn the product and the desires of the customer and users in order to succeed. They must also be given the appropriate authority.

Being able to negotiate the politics of an organization, its committees, and the users can be an exhaustive, full-time job. For the Product Owner to succeed, the entire organization must respect their decisions. The Product Owner's decisions are visible in the content and ordering of the Product Backlog as well as the functionality of the product. No one is allowed to tell the Developers to work from a different set of requirements, and the Developers aren't allowed to act on what anyone else says.

Here is a list of other challenges you might encounter when working with a Product Owner:

- **Injecting their own version of Scrum** There is only one version of Scrum, and it's documented in the *Scrum Guide*. Implementing anything else risks upsetting the established flow. Any allegiance to old waterfallian habits—even if they are disguised under new Scrum terminology—must go. Understand that mental muscle memory takes time to fade.

- **Insufficient acceptance criteria** A good PBI doesn't just stop at a title and description. The Product Owner, through conversations with stakeholders, should evolve the PBI and define what success looks like in the form of acceptance criteria. These should be testable—maybe even written as executable specifications.

- **Absent or doesn't interact with the team**   In order to maximize the work of the Developers, the Product Owner must interact and collaborate. This is especially true during Sprint Planning, Sprint Review, and Sprint Retrospectives, as well as during Product Backlog refinement. The Product Owner should also be regularly available during the Sprint to answer questions, clarify details, make introductions, review work, and provide feedback. For a Product Owner, a good rule of thumb is to spend 1/3 of their time with the product, 1/3 of their time with the stakeholders, and 1/3 of their time with the Developers. These ratios will change with new/challenging products, stakeholders, and Developers.

- **Disrupts the team**   Whether it's to introduce a new piece of work during the Sprint (scope creep) or just wandering into the team room or Zoom meeting and asking how things are going, these intrusions can interrupt the flow and kill focus. The Scrum Master should get involved and help the Product Owner understand this.

- **Provides the solutions**   The Product Owner must allow the Developers to self-manage and come up with their own solutions. So long as it is fit-for-purpose, meets the acceptance criteria, and abides by the Definition of Done, *any* solution should be acceptable.

- **Blocks Developers from talking with stakeholders**   The Product Owner is not omniscient. There will be many opportunities for the Developers to build a better product by collaborating directly with the stakeholders (such as customers and users) during the Sprint. The Product Owner should not act as a gatekeeper for these conversations.

- **Indecisiveness**   The Product Owner has the authority to make decisions pertaining to the product. This includes everything from determining the value of a PBI, to ordering the Product Backlog, to changing the scope of a Sprint, and even canceling a Sprint. The Scrum Master can help the Product Owner understand the trade-offs of these various decisions, but the decisions still have to be made.

- **Not being prepared**   This common dysfunction is especially risky with regard to Product Owners. The plans of multiple people (the Developers) depend on the decisions made by a single person (the Product Owner). If the Product Owner is not prepared, much waste can be generated and morale can suffer. Regular Product Backlog refinement can help the Product Backlog—and Product Owner—be prepared for Sprint Planning. Sometimes a definition of "ready" can be useful, too.

- **Command-and-control behavior**   In Scrum, the Product Owner is not the "boss" in the traditional sense. It is acceptable for the Developers to say "no"—especially when the Product Owner is asking them to do something that is out of bounds for their role or the rules of Scrum. Examples might include generating technical debt, violating the Definition of Done, or releasing undone work. The Scrum Master can be called in to referee if necessary.

- **Expects a commitment, not a forecast**   An important change was made in the *Scrum Guide* in 2011. The Developers *forecast* the work that they believe can be completed (according to the Definition of Done) during the Sprint, but they don't *commit* to it. If a Product Owner expects a commitment, such as assuming the Developers will work nights and weekends until the Sprint Backlog is complete, that is an unhealthy, dysfunctional behavior. It ignores the reality of complex work and sets the expectation for an unsustainable pace. The Product Owner must learn the difference between a forecast and a commitment—which any Professional Scrum Master can help explain. For more information, read fellow Professional Scrum Trainer Barry Overeem's explanation here: *www.scrum.org/resources/blog/commitment-versus-forecast*.

- **Multiple Product Owners** The Product Owner is one person, not a committee. The Developers, as well as the stakeholders, should have a "single wringable neck" (or "one throat to choke")—metaphorically speaking. Having multiple Product Owners is confusing to everybody. Pick one, and the others become stakeholders who help the Product Owner create PBIs and order the Product Backlog. Alternatively, the Scrum Team could probably use more Developers.

- **Multiple stakeholders, but no true Product Owner** People often confuse the role of Product Owner with that of a business stakeholder. Just because someone has an influence on the product, or the business that uses the product, doesn't make that person a Product Owner. The Product Owner is a specific role—with authority—defined in Scrum. The Product Owner works closely and collaboratively with stakeholders, as well as the rest of the Scrum Team to maximize the value of the product.

- **The Developers maintain the Product Backlog** Developers typically do not have the vision and insight into the needs of the stakeholders to evolve and maintain the Product Backlog adequately. Although there are exceptions, Developers tend to be better at solving technical problems and creating product. This is why the Product Owner needs to be present and accountable for maximizing ROI in the product—which is done through the content and order of the items in the Product Backlog. There may be times that the Developers get involved, such as helping the Product Owner understand how technical dependencies affect delivery order, but they should not be handed the Product Backlog to take over management. The Product Owner is responsible for maintaining the content and order of the Product Backlog, and passing that task along to others smells like an absent Product Owner.

- **Acting as a Developer** Product Owners sometimes come up through an organization's technical ranks. While developing the product, one of the team may have learned everything there is to know about the product and its use and ended up becoming its Product Owner. This increases the chances that the person will become involved with *how* it should be developed, when they should remain focused on *what* to develop. That said, having a Product Owner also be a Developer—which is allowed in Scrum—is sometimes unavoidable, especially for smaller teams such as startups.

The Product Owner is a full member of the Scrum Team and, as such, should be present at all Scrum events, with the possible exception of the Daily Scrum. The purpose of the Daily Scrum is for the Developers to create a plan for the next 24 hours. The Product Owner shouldn't have any input on the plan, nor do they need to know about those plans once they are made. The Product Owner, however, should be available to the Developers as they work. Being in close proximity and ready to collaborate in person as needed is a recipe for a successful product. Keep in mind that the Product Owner also needs to work with stakeholders during these same hours, so availability may be limited. Collocation or "office hours" may be a solution.

Smell It's a smell when I meet a Scrum Team who still goes by their old titles. When I'm introduced to the Scrum Master, and she tells me her name is Audre and she's the director of IT, I get confused. The rest of her teammates might get confused. too. Remember, in Scrum, there are only the Product Owner, Scrum Master, and Developer roles. What you used to be called or your HR designation isn't relevant anymore.

Working with Challenging Stakeholders

Stakeholders are not an official role in Scrum, but they exist and can be challenging to work with. Remember that a stakeholder is any person who has a direct or indirect interest in the work of the Scrum Team. They may be a customer, a user, a domain expert, a manager, an auditor, an executive, or a member of the public. Unless you are the Product Owner, interactions with stakeholders may ebb and flow. The only time Developers are guaranteed to interact with stakeholders is during the Sprint Review.

Stakeholders may or may not know about Scrum. And what they do know, may not be accurate. They may have read something, taken some training, or watched some videos online—none of which guarantees that what they know is Scrum according to the *Scrum Guide*. Some stakeholders may think Scrum is a "silver bullet" and just by using the nouns during conversations and meetings, all risk will be removed and value will flow quickly and perfectly. This is magical thinking.

It's the responsibility of the Scrum Master to squash this illusion and educate the uneducated that Scrum's success depends on empiricism and the commitment of the people practicing it. Stakeholders are welcome, and encouraged, to watch the great experiment take place, interacting where and when appropriate.

Note Scrum was not designed to keep stakeholders from interacting with the Developer. On the contrary, Scrum is intended to bring the two camps—business and technical—closer together, just in a more structured and productive way. For example, the Sprint Review allows stakeholders to inspect working product and provide rich feedback, which can be captured in the Product Backlog. Most stakeholders are ecstatic that there is now a process that actually allows them to see Done, working product every few weeks. Stakeholders typically welcome this transparency.

Here is a list of challenges you might encounter when interacting with a stakeholder:

- **Doesn't understand the Definition of Done** Since stakeholders do not necessarily know Scrum, they may not understand why something "they saw running in a browser yesterday" isn't done and able to be properly inspected during Sprint Review. You, or the Scrum Master, can explain how the Definition of Done ensures an uncompromising level of quality. This explanation should be provided in business terms familiar to the stakeholder. For example, instead of saying, "Load testing has not been completed," you could say, "We are still unsure how the application performs with more than 10 concurrent users."

- **Doesn't provide feedback**   Some stakeholders are just not that interested in the product. They may be paying for it, or managing the employees who will be using it, but they are otherwise indifferent. If it won't hurt the long-term prospects of the product, consider not inviting them to future Sprint Reviews, or at least inviting other, more interested, parties. Whenever possible, invite a few key users. They tend to be passionate about what the team is doing and provide valuable feedback. Remember, the purpose of the Sprint Review is to obtain stakeholder feedback.

- **Injects their own version of Scrum**   There is only one version of Scrum, and it's documented in the *Scrum Guide*. Anything else risks upsetting the established flow.

- **Absent or doesn't interact with the team**   In order to maximize the work of the Developers, stakeholders (especially domain experts) must be available periodically to help answer questions and provide feedback. This is the minimum bar. For an awesome product, stakeholders should consider collaborating directly with the Developers. This means being open to pairing with a Developer for a period of time as they design, code, and validate functionality.

- **Disrupts the team**   Stakeholders, by definition, have an interest in what the Scrum Team is doing. They may wander into the team room or Zoom meeting and ask how things are going. These intrusions can interrupt flow and kill focus. The Scrum Master should get involved. Stakeholders shouldn't attend the Sprint Retrospective.

- **Provides the solutions**   The stakeholders are free to work with the Product Owner to clarify what is to be developed. The Developers, however, are self-managing and come up with their own solutions.

- **Not able to say "no" to a stakeholder**   In Scrum, the stakeholder is not your "boss" in the traditional sense. Unfortunately, they may be the owner of the company, and absolutely your boss outside of the context of Scrum. In this situation, the Scrum Value of courage only goes so far and the Scrum Master may need to join the conversation.

- **Expects a commitment, not a forecast**   Stakeholders must acknowledge the reality of complex product development and allow the Developers to forecast the work they can do in a Sprint—not force them to commit to it. The Scrum Master can explain the difference and also the fact that what the Developers do is hard.

- **Acts as a manager**   The Scrum Team is self-managing. Nobody, including stakeholders, can tell the team how to do their work, or what they should work on next. That said, stakeholders can be very influential as to what should be worked on next. This kind of feedback is very pertinent, should involve the Product Owner, and should be manifested by the order of PBIs in the Product Backlog.

- **Acts as a Developer**   Some stakeholders may be a developer from another team or another decade. Be cautious of them getting too involved in the development your team is doing. They can easily become a distraction. If, on the other hand, they have the skills you need and the capacity to help, have them join your Scrum Team—even if only as a part-time Developer.

- **Acts as an insurgent** Some stakeholders, for whatever reason, are resistant to change and just anti-Scrum. Maybe they tried it at a previous organization and were unsuccessful. Maybe they prefer waterfall or Kanban. Maybe they hate rugby. Sometimes such a person is necessary to support the successful adoption of Scrum. Hopefully the Scrum Master can help educate and open their eyes.

Working with a Challenging Scrum Master

The Scrum Master is responsible for ensuring that Scrum is understood and enacted. Scrum Masters do this by advising, coaching, and mentoring the Scrum Team members so that they adhere to Scrum theory, practices, and rules. The Scrum Master is a servant-leader for the Scrum Team and a facilitator who supports the team in learning self-management, and understanding and adopting the rules of Scrum and the Scrum Values.

A good Scrum Master brings value to the Scrum Team, and the organization, by helping both adopt and progress toward Professional Scrum in a realistic way. By applying what they know, the Scrum Master can help the team build and deliver product that is of higher quality and value. This is done by maximizing the benefits produced by Scrum. Professional Scrum Masters should be putting themselves out of a job by teaching the team to identify and solve their own problems.

> **Tip** The key to finding a good Scrum Master is seeing them in action. Let the candidate attend a Daily Scrum and tell you what they observe. This way, you can see if they have a nose for sniffing out dysfunction. Their knowledge of the rules of Scrum, as well as their perception of the team's behavior and level of collaboration, should speak volumes about experience and capability.

Beyond supporting the Scrum Team, Scrum Masters can also be responsible for educating the organization and leading the effort to adopt Scrum. This means that they may play the role of mentor, coach, consultant, or even trainer. It also means that the Scrum Master is a walking Scrum salesperson, always pointing out the empirical benefits of adopting Scrum to new people and potential teams. Your Scrum Master should be able to articulate why Scrum works and is healthy for the organization, even to the loudest critics and detractors.

Here is a list of challenges you might encounter when working with a Scrum Master:

- **Doesn't know Scrum** This is a deal breaker. If there is one person on the Scrum Team, or in the organization, who must know Scrum, it's the Scrum Master. Inform management that the Scrum Master needs more training than just reading the *Scrum Guide*. If they don't know about the *Scrum Guide*, I think you've found the problem. That may sound silly, but I recently helped an organization fill a Scrum Master role. Of the four candidates I interviewed, only one knew about the *Scrum Guide*. Consider sending them to a Scrum.org class *(www.scrum.org/courses)*.

Experience will come with time, but since it's required on day one, hire an experienced Scrum Master, even if only temporarily.

- **Doesn't enforce the rules**   A Scrum Master is a coach, but also a referee. They should be confident in "throwing a flag" or showing someone a "yellow card" if the situation calls for it. The rules of Scrum have been fine-tuned over years of use. Those rules only work when they are followed. That said, there is room for adaptation once the core principles are embedded in the organization.

> **Tip**   It may be challenging for a Scrum Master to enforce the rules. I tend to think of Scrum Masters as being firm and resolved, but in practice, this type of Scrum Master can sometimes create an adversarial environment. As an alternative, a Scrum Master should coach their team members to follow the rules of Scrum. If the team wants to step outside the rules of Scrum, the Scrum Master should use assertive questioning and dialogue to probe and discuss. If, after the discussion the team still wants to break the rules, the Scrum Master may want to allow it as a learning experience. Then, during the Sprint Retrospective, the Scrum Master should help the team inspect the results from not following the rules.

- **Focuses too much on rules and practices**   A Scrum Master should enforce the rules, but focusing too much on the rules and practices is a form of zombie Scrum and can create a "cargo cult" mentality. In this dysfunction, the team executes the practices, but doesn't reap any benefits. A Scrum Master should always make sure that the team is getting the most out the Scrum practices and rules. For more information on cargo cult, you can read up on it here: *https://en.wikipedia.org/wiki/Cargo_cult*.

- **Doesn't act as a firewall**   The Scrum Master is the protector of the team's focus. As such, they should block any such interruptions—in any way possible. This might include going to meetings in place of the rest of the team, tracking and providing actual hours worked or other wasteful but "required" metrics to the PMO, or educating others in the organization on how to interpret a burndown chart. A Scrum Master should respect the Developers' flow, and do whatever is necessary to protect it.

- **Acts as a manager**   The Developers are self-managing. Nobody, including the Scrum Master, can tell them how to do their work. The Scrum Master should avoid even suggesting how a team member does their work or what to work on next. The exception to this is when the Scrum Master is asked for help, or if one or more Developers exhibit dysfunctional behavior or has otherwise become an impediment. The Scrum Master should have the authority to implement and enact the rules of Scrum, including removing such impediments. I'll leave it at that.

- **Absent Scrum Master**   The Scrum Master is a servant-leader and, as such, should be collocated with the team, ready to help. The only time the Scrum Master should be unavailable is when they are away educating the organization, removing an impediment, or taking a (much-deserved) vacation.

- **Doesn't manage conflicts**   Since Scrum is about people, Scrum Teams will inevitably experience conflicts. Simple conflicts can (and should) be handled by the people involved. More complex conflicts may require the Scrum Master to become involved. If a Scrum Master is hesitant or doesn't have the social skills required to manage such conflicts, this is a dysfunction.

- **Settles for the status quo**   A Scrum Master should be hungry for improvement. Just as a teacher gets excited when students are learning new things and applying what they've learned, so should a Scrum Master thrive on seeing the Scrum Team living the Scrum Values and improving. Professional Scrum Masters are constantly looking for new practices and techniques to increase empiricism and self-management.

- **Poor communication**   This is more than just the Scrum Master not being able to communicate clearly, but allowing communication dysfunctions to fester and grow in the team. This type of behavior can infect other team members—especially if they see the Scrum Master partaking. A Professional Scrum Master knows how to teach and foster good communication techniques in the Scrum Team. This includes teaching topics such as active listening and the Scrum Value of respect.

- **Has a day job**   Any additional role that a Scrum Master occupies is a conflict that can cause difficulty, especially for newly formed teams. Sometimes this is unavoidable. Smaller teams or startups may require dual roles. One dual role—Scrum Master and Project Manager—should be avoided.

- **Doesn't deal with impediments**   A good Scrum Master will give the team the opportunity to remove their own impediments and then learn from the experience. A dysfunctional Scrum Master will allow impediments to linger. If the Scrum Master can't directly remove the impediment, they should at least find someone in the organization who can.

- **Acts like the team's mother**   Not that there's anything wrong with mothers, but the "mom" Scrum Master type can also be a dysfunction. These Scrum Masters deal with the secretarial and nanny tasks. Examples might include meeting scheduler, Azure Boards typist, Sprint Backlog and burndown updater, stopwatch handler, flow metric gatherer, looking disappointed when you're late to the Daily Scrum, and so on. New and uneducated Scrum Teams might think that this is what the Scrum Master does. I only hope somebody in the organization is aware of this dysfunction, since the Scrum Master probably won't be capable of recognizing it themselves.

- **Acts as a Developer**   In a lot of ways, the Scrum Master is like a firefighter. They sit, waiting to be called upon to answer a question or remove an impediment. Having the Scrum Master involved in the actual development, taking on tasks, tends to distract from the job of helping the team and organization understand and follow the rules of Scrum. Sometimes this is unavoidable—especially for smaller teams and startups. If one person is playing both roles, make sure that they give priority to the Scrum Master duties over the Developer duties.

Changing Scrum

Scrum is simply a set of rules put forth in the *Scrum Guide*. This makes it comparable to the game of chess. Chess has rules, too. One rule in chess is that a player is allowed to have only one king on the board. Scrum's rules dictate only having one Product Owner. There are many other comparisons, but you get the idea. When you sit down to play chess, you either play by the rules or you don't. Same with Scrum. If you want a short-term win, you can cheat and have three kings, but you won't learn how to play the game properly or get good at it. Learning how the chess pieces move is fairly easy, just like learning the rules of Scrum, but mastering chess (and Scrum) is difficult and takes a long time and a lot of practice.

Professional Scrum Masters know how to play the game of Scrum properly. Because of that, they can easily spot those who cheat. To avoid the embarrassment of being called out by a Professional Scrum Master, don't cheat. Even if you could get away with it, why would you? You will improve only if you don't cheat.

The rules of Scrum should be considered immutable. An organization or team should not change them. You should inspect and adapt your behaviors within those rules and improve accordingly. Every Scrum role, artifact, event, and rule is designed to provide the desired benefits and address predictable recurring problems. Feel safe. Scrum will not fail you.

Old Waterfallian Habits

Waterfall development is the name given to a more traditional, sequential design approach to development where one phase is completed before moving to the next. Design is done before programming. Programming is done before testing. And so on. Each phase is performed as though you are not coming back to it. Maximum attention is given in getting it right the first time. This approach to developing complex products—such as software—is very risky, more costly, and less efficient than Scrum.

Unfortunately, waterfall has been in existence for over 60 years. Many technical professionals and managers are familiar with it and have it imprinted in their mental muscle memory. When these people are introduced to Scrum, they may feel compelled to change Scrum, molding it into something they are more familiar with.

Here are some waterfallian habits—or at least smells—that should be inspected, adapted, and avoided in Scrum, along with the reasons why:

- **Running longer Sprints (more than one month)**   Sprint lengths of one month or less provide focus and reduce risk. Longer Sprints increase risk exponentially, even if they feel more comfortable.

- **Defining big requirements up front**   Time spent defining detailed requirements, and especially how they should be implemented (specifications), is wasted when development of those items is delayed or skipped altogether because of the Product Owner's decisions.

- **Establishing separate teams to code and test**   Cross-functional teams are more efficient because they are able to work together with fewer handoffs. Only PBIs that meet the Definition of Done (which includes testing) are releasable. This lessens the exponential buildup of work toward the end of the release.

- **Running infrastructure and architecture Sprints**   Every Sprint must generate an Increment containing business value. This keeps the Developers focused on what's best for the customer or user. Emergent architecture is a practice that can help maintain this focus.

- **Delaying testing until later Sprints**   All aspects of development, including testing, must be done during the Sprint. Delaying testing produces technical debt and undone work that accumulates exponentially.

- **Minimizing change (because change is bad)**   Change is a fact of life in complex product development. Scrum embraces this fact through the use of shorter Sprints and an ordered Product Backlog maintained by an engaged Product Owner.

- **Assigning work (command and control)**   The Developers are self-managing and can create and take ownership of their own work. They are also expected to estimate, forecast, and plan the work as a team, not relying on proxies.

- **Following the plan and conforming to the schedule**   In Scrum, the plan is broken up into Sprints of one month or less in length. Beyond that, there is no firm plan, only a Product Backlog with items ordered in a way that represents what the Product Owner would like developed next.

- **Realizing no product value until the very end**   Every Sprint must generate an Increment containing business value. This means that all development activities, including integration with other teams and systems, must be done at least by the end of the Sprint to realize this value.

- **Always reporting a bug**   The Developers are self-managing and can determine whether or not the unexpected behavior is a bug. They are also capable of just correcting that behavior rather than creating a work item to report the bug. Remember: Failed tests are not bugs.

- **Treating Daily Scrums as status meetings**   The Daily Scrum is for the Developers to synchronize and create a plan for the next 24 hours. It's not meant for other purposes or for others to participate.

- **Never reestimating work**   Professional Scrum Developers understand that they know more today than they did yesterday. Applying this new knowledge to existing estimates (either PBIs or tasks) is a healthy practice that increases transparency.

- **Sacrificing quality**   The Definition of Done, when properly adhered to, protects the quality of the work the Developers do and keeps undone work, and any ensuing technical debt, out of the Increment.

- **Gold plating**   The Developers only need to develop what is "fit for purpose" for a given PBI, and nothing more. In Scrum, the Developers shouldn't try to predict what *might* eventually be needed. The next Sprint's work will be revealed just in time.

ScrumButs

Many organizations have modified Scrum against this guidance. In their minds, they are doing the right thing and adapting Scrum to fit their particular flavor of chaos. This is partly because past approaches required tailoring in order to succeed. Scrum is the opposite in that changing Scrum itself can prevent you from succeeding. These changes and tweaks are colloquially known as "ScrumButs." When a representative is asked if their organization or team is doing Scrum, they answer, "Yes, we are doing Scrum, but . . . ."

In fact, a ScrumBut has a particular syntax:

We use Scrum, but (**ScrumBut**) because (**Reason**) so (**Workaround**).

Here is an example of a ScrumBut:

"We use Scrum, but *we don't have Daily Scrums* because *they are too much overhead*, so *we only have them once a week or as needed*."

ScrumButs are excuses why teams and organizations can't take full advantage of Scrum. ScrumButs mean that Scrum has exposed a dysfunction that is contributing to the problem but is too hard to fix. A ScrumBut retains the problem while modifying Scrum to make it invisible so that the dysfunction is no longer a thorn in the side. For more information on ScrumButs, visit *www.scrum.org/scrumbut*.

Organizations may make short-term changes to Scrum in order to give them time to correct deficiencies. For example, a team's Definition of Done may not initially include all the desired testing because of external dependencies or the length of time to develop an automated testing framework. For these Sprints, transparency is compromised. The Scrum Team should work to restore it as quickly as possible.

> **Note** Several Professional Scrum Trainers feel that "ScrumBut" is too negative. Although they acknowledge that they exist, they prefer using a softer, more optimistic metaphor, such as an "adoption compromise" or simply "ScrumAnd." While these are more diplomatic sounding, they still suggest that compromises are being made to the rules of Scrum during adoption. Hopefully a Professional Scrum Master is tracking them and they will be removed as soon as possible.

Becoming a Professional Scrum Team

No matter where you are in the game of Scrum, you can always improve. Whether you're a part of a new team just getting started and still not sure what a timebox is, or your team has released many Increments of product successfully using Scrum, there are always new things to learn and new ways to enhance your practices.

A Scrum Team should inspect and adapt constantly. This includes the behaviors and practices of the team beyond simply identifying and removing dysfunctions. The absence of a dysfunction is an improvement, but the team can do even better. For example, it may take several Sprints for a dysfunctional Scrum Master to stop providing estimates on behalf of the Developers. Yes, I've actually seen this. It may take even more Sprints for the Developers to understand how to estimate on their own. It may take even more Sprints for these estimates to normalize. It may take additional Sprints for the team to learn that estimation is about conversation and learning, and not about the number.

Improvements can occur only if the culture allows it. The organization and management must allow their teams to experiment, fail, inspect, and adapt. Successful organizations yield successful teams because they allow their people the freedom to explore, learn, cross-pollinate, set up practice communities, and implement their retrospective items. Most of all, the culture must understand that improvement takes time.

In this section, I discuss ways in which a Scrum Team can continue to improve beyond just knowing and practicing mechanical Scrum. I call these teams *Professional* Scrum Teams.

Get a Coach

There may be times when the Scrum Team needs help improving their game. Just like any sports team, a Scrum Team can also benefit from the help of a coach. This kind of coach is an expert in Scrum, both in theory and in practice. They have an in-depth understanding of the practices and principles of Scrum, and they have real experience with Scrum. A coach like this can teach and coach all of the Scrum roles, including stakeholders and the organization itself, effectively. They can teach new patterns and behaviors for increased collaboration and high-performance achievement.

Note Don't confuse a Scrum coach with an agile coach. For teams doing Scrum, they will want a Scrum coach who absolutely knows Scrum. This may take the form of a contracted Scrum Master. Agile coaches may or may not know Scrum, and they may even vary on what or how much they know about contemporary agile practices. These days you can't swing a dead cat within an organization without hitting an agile coach. On the other hand, nobody accidentally becomes a Scrum coach.

A good Scrum coach will also have experience in a variety of organizational settings, which is useful when educating the rest of the organization. A coach can help the organization understand how the changes will affect leadership and team member responsibilities. Mentoring and gradually sharing proven practices about Scrum adoption ensures that the shock to the organization won't be so painful.

Tip When searching for a Scrum coach, pay attention to the candidate's background and if they have experience playing the various Scrum roles. It's hard to find a coach who has played the role of the Product Owner, Scrum Master, *and* Developer. At least make sure the candidate has played the role you need the most help with.

There is a myth surrounding what a Scrum coach does. People think that coaching is purely a soft-touch approach—only providing guidance and the ability for people to discover problems and solutions for themselves à la the Socratic method. People also think that coaches do not tell people what to do. Some coaches fit this mold and, in my opinion, fall short of what is typically needed. The truth is, coaches need to have the difficult conversations, and these conversations are sometimes not nice and not polite. This is because coaches help people identify and overcome unpleasant things. One minute the coach will need to be compassionate and understanding and the next minute authoritative and uncompromising. People skills are necessary, but so is being assertive and resolute.

Build a Cross-Functional Team

The team of Developers is a cross-functional body of people possessing all the different skills required to turn PBIs into an Increment of Done, releasable functionality. The Developers need to know all the skills necessary to turn the PBIs into something that the organization defines as Done. Those Developers will need to acquire the skills of analysis, design, coding, testing, content development, deploying, operations, and so on.

It may take several Sprints for the Developers to even *know* what skills they have or are needed. When Scrum was first adopted in an organization, hopefully all of the analysts, coders, and testers were united onto teams. Since each of them played a role in the development of the product, they became known as a Developer. As self-managing and collective ownership attitudes emerged, the backgrounds and titles previously held by those team members became blurred and hopefully forgotten.

> **Note** The opposite of a cross-functional team is a dysfunctional one.

What is a cross-functional team today may not be so tomorrow. Over time, the team, not the management, may determine that additional team members are required. To satisfy this need, new Developers may be added, or the current ones trained, in order to support new tools, technologies, or domains being considered. The opposite may also become a reality—fewer Developers may be needed because they have acquired more cross-functional skills and capabilities.

> **Tip** In Scrum, the Product Owner provides the vision and Product Goal(s) for the product. This should be reflected by the PBIs, and their order, in the Product Backlog. An ordered Product Backlog serves as a roadmap for the planned features. It also serves as a roadmap for the planned technologies and new domains, which can serve as a "heads-up" for what functionality (skills) the Developers will need in the future.

Making unnecessary changes to the team of Developers will cause problems. When the problems that these changes cause are less than the problems caused by not changing, then it is worthwhile. You should be aware and prepared for the difficulty that a new team member will have when being introduced into an existing team. If you refer back to Bruce Tuckman's stages of group development, any changes made to the makeup of a team will cause the team to revert to the *forming* stage of his model. Just think of the problems associated with a child when their family moves to a new town with a new school and the effort that has to occur for that child to fit in and be productive. Developers are not much different.

> **Note** Don't confuse cross-functional teams with cross-functional *individuals*. These are akin to T-shaped individuals. Scrum demands cross-functional teams of Developers. This means there must be at least one Developer who is capable of performing each type of task in the Sprint Backlog. For example, if there are Python tasks that must be accomplished, there must be at least one Developer who can code in Python. Professional Scrum Teams endeavor to have cross-functional individuals as well. This means that if Python tasks are becoming more prevalent in the upcoming Sprints, one or more Developers should acquire that knowledge. Having a cross-functional team of cross-functional Developers is a recipe for meeting goals and delivering an awesome product.

Achieve Self-Management

Scrum relies on self-managing teams to handle the complexity inherent in product development. A self-managing team will approach a challenge, and, based on the goals and requirements, decide how best to develop a solution while taking advantage of each team member's various strengths. It takes a certain mindset and aptitude to be able to self-manage like this. But, compared with traditional practices—such as a chief architect creating the initial design or a project manager assigning work—self-management is a revolutionary improvement.

Every Developer on a self-managing team will work, independently, in pairs, or in a mob toward some shared goal. Everyone collaborates to reach the goal, valuing the team's outcome over individual productivity. Members of the team trust each other and are interested in each other's work, providing constructive feedback where appropriate. Self-managing teams are able to get work done and develop a valuable Increment. They are not blocked by impediments. They communicate any issues appropriately to achieve transparency.

An organization must *allow* its Developers to self-manage. This comes with time and the earning of trust. That trust is a direct result of the Developers being able to deliver increments of business value regularly in the form of working product. Education, provided by the Scrum Master, can help the organization see that this is a reason to trust the team. Once that trust is in place, hopefully the Developers will be given more leniency to make their own decisions and plans, and then be allowed to execute them.

Improve Transparency

Transparency, along with inspection and adaptation, are the three pillars of Scrum, or any empirical process control approach for that matter. The importance of being transparent—as a Developer to the rest of your team, or as a team of Developers to the Product Owner, or as the Scrum Team to the organization—cannot be overstated. Significant aspects of the development process must be made visible to those responsible for the outcome. Transparency requires those aspects be defined by a common standard so that observers share a common understanding of what is being seen.

In Scrum, being transparent means that all observers should understand the basics of the framework, as well as the artifacts they may be looking at: the Product Backlog, Definition of Done, burndown charts, task boards, the Increment, and so on. The output and data reflected in these artifacts are like beacons of light. They shine brightly into all corners of the Scrum Team's activity, leaving nowhere for slackers or other waste to hide.

Some Developers might be reluctant, or at least uncomfortable, about this "nowhere to hide" quality of Scrum. Nobody wants to work in a glass house, even though doing so means that a Developer will be more productive and exhibit healthier behaviors because they never know who might be watching. Being able to admit mistakes and ask for help will assist everyone in becoming more comfortable with this quality of Scrum. Besides, making mistakes and learning from them is a good way to improve.

Take the task board, for example. Whether we are talking about a physical one or an electronic one like in Azure Boards, it is a great information radiator and source of transparency. The board reflects the Developers' current plan. It shows where they are, what they've done, and what they still have to do. The transparency of the board is not created for the sake of reporting, but rather for planning and general awareness. This awareness enables the team to manage themselves.

Professional Scrum Developer Training

Professional Scrum Teams who want to build and deliver great products require the understanding and integration of Professional Scrum, a well-functioning team, proven practices, and modern tools. Scrum.org's Professional Scrum Developer (PSD) course is the only course available that teaches how this can be done.

The PSD course teaches students how to work in a team, using Professional Scrum, contemporary software development practices, and Azure DevOps to turn PBIs into Done, working product. All of this is done as iterative incremental development within the Scrum framework. The course was developed in partnership with Microsoft, Scrum.org, and Accentient. You can learn more about the origins of the Scrum Developer program, as well as Scrum.org, at *www.scrum.org/about*.

The PSD course is suitable for any Developer on a Scrum Team, including architects, programmers, designers, database developers, business analysts, testers, and others who will be contributing to the work. Product Owners, Scrum Masters, and other stakeholders are welcome to attend this class, as long as they keep in mind that all attendees will be expected to participate as a Scrum Developer—delivering Done, working product within mini-Sprint timeboxes.

As with all Scrum.org courses, the curriculum and materials are standardized and regularly enhanced by courseware stewards through contributions from the community of Professional Scrum Trainers and students. Only the most qualified instructors are selected to teach the PSD course. These are individuals with top-notch skills in the technologies coupled with excellent knowledge of how to use them within the Scrum framework. Each instructor brings their individual experiences and areas of expertise to bear, but all students learn from the same core content. This improves a student's ability to pass the Professional Scrum Developer assessment and apply Professional Scrum in their workplace. For more information on the PSD program, training, and certification, visit *www.scrum.org/psd*.

Assess Your Knowledge

Scrum.org also provides tools that you can use to examine and enhance your knowledge of Scrum. The primary aim of these assessments is to provide information about an individual's level of knowledge and thereby to enable improvement.

Each Scrum.org assessment is based on the *Scrum Guide* and developed by Scrum thought leaders with formal input from a wide range of industry experts and then enhanced with input from the larger Scrum community. The assessments are then monitored in an ongoing attempt to ensure their continued integrity and relevance.

- **Professional Scrum Master**   Three levels of Professional Scrum Master assessments (fundamental, advanced, and distinguished) that validate and certify knowledge of Scrum and the ability to apply that knowledge.

- **Professional Scrum Product Owner**   Three levels of Professional Scrum Product Owner assessments (fundamental, advanced, and distinguished) that validate and certify knowledge of Scrum Product Ownership and the ability to apply that knowledge.

- **Professional Scrum Developer**   An assessment that is structured to validate and certify knowledge of the practices and techniques that support building complex products as a Developer on a Scrum Team and the ability to apply that knowledge.

- **Scaled Professional Scrum**   Validates and certifies knowledge of how to scale Scrum and the use of the Nexus Framework and the ability to apply that knowledge. I discuss the Nexus Framework and Scaled Professional Scrum in Chapter 11.

- **Professional Scrum with Kanban**   Validates and certifies knowledge of how Scrum Teams can use Scrum with Kanban to support value creation, measurement, and delivery. I discussed many of these topics in Chapter 9, "Improving Flow."

- **Professional Agile Leadership**   Validates and certifies knowledge of how agility adds value and why leadership's understanding, sponsorship, and support of agile practices are essential.

- **Professional Scrum with User Experience**   Validates a fundamental level of understanding about how to integrate modern UX practices into Scrum to deliver greater value.

Those who achieve a minimum passing assessment score receive certification. All Scrum.org assessments use the most recent version of the English *Scrum Guide* as the source for questions regarding the rules, artifacts, events, and roles of Scrum. Reading the *Scrum Guide* alone will not provide enough preparation for someone to pass an assessment. Questions often ask test-takers to interpret information and apply it to challenging situations, so knowledge gained from personal experience of Scrum, as well as other sources, is typically required. For more information on the various Scrum.org assessments, visit *www.scrum.org/assessments*.

Scrum.org also offers many tools and resources free to the community. Among these tools are several "open" assessments. These assessments have no fee and can be taken anonymously. There is also no certification associated with passing an open assessment. They do allow you to gauge your basic knowledge of Scrum, the Scrum roles, and related practices. The questions on the open assessments do not have the same level of difficulty as the certification assessments. They do, however, provide practice for the actual assessments. For more information on the Scrum.org open assessments, visit *www.scrum.org/open-assessments*.

Become a High-Performance Scrum Team

High-performance Professional Scrum Teams are the best of the best. They have mastered the key pillars of Scrum: self-management, transparency, inspection, and adaptation. They have focus, exhibit courage and openness, believe in and practice commitment, and respect others. They know the rules of Scrum according to the *Scrum Guide*, and they are able to deliver Increments of business value regularly in the form of working product.

It's possible to become a high-performance Scrum Team through continuous improvement, as Figure 10-7 illustrates.

FIGURE 10-7 Teams can progress from not doing Scrum all the way to High-Performance Scrum.

Looking at a distribution curve and contemplating your own situation, you might be wondering, "Is my team doing Scrum?" It turns out that this is a harder question to answer than you might think. One might believe that just reading the *Scrum Guide*, filling all the roles, participating in all the events, and using the artifacts correctly would be enough. Even just the mechanics of doing Scrum is not easy to define.

> **Note** In November 2011, several Professional Scrum Trainers met in Redmond, Washington, prior to Microsoft's Application Lifecycle Management (ALM) Summit. One of the items on the agenda was to create a way to definitively determine whether a team was on the left side of the first dotted line in Figure 10-7. Teams wanted to know if they were "doing Scrum." To determine this, we created a simple, measurable checklist that a team could answer. If all questions were answered in the affirmative, that team was doing Scrum. The determination of "how well" they are doing Scrum and where they are at on their improvement journey will need to be the subject of a future survey.

Here is an unofficial survey, representing the musings of several like-minded Professional Scrum Trainers, to determine if your team is "doing Scrum":

1. Does your Scrum Team maintain an ordered Product Backlog?

2. Does your Scrum Team have a Product Owner, a Scrum Master, and three or more Developers, with an overall size of typically 10 or fewer people?

3. Does your Product Owner actively manage the Product Backlog?

4. Does your Scrum Master actively manage the process?

5. Do you have fixed-length Sprints of one month or less?

6. Do the Developers create a Sprint Backlog during Sprint Planning?

7. Can progress be assessed from the Sprint Backlog?

8. Do the Developers develop a Done Increment of product each Sprint?

9. Does your Scrum Team hold Sprint Reviews and Sprint Retrospectives?

10. Do your stakeholders inspect the Increment and provide feedback?

11. Does the Scrum Team inspect its process and make actionable improvements in follow-up Sprints?

All throughout this book, I have provided patterns to adopt and anti-patterns to avoid when it comes to practicing Professional Scrum—especially in the context of using a tool like Azure DevOps. All of this guidance, in addition to adopting and living the core principles and values of Scrum, will enable you and your fellow team members to become a high-performance Professional Scrum Team.

Chapter Retrospective

Here are the key concepts I covered in this chapter:

- **Professional Scrum Teams know Professional Scrum** Know the rules of Scrum and how to overcome its common challenges. These teams also know that dysfunctions should be identified and removed as they inspect and adapt.

- **High-performance Professional Scrum Teams continually improve** These teams are hungry to do better and take every opportunity to inspect and adapt, remove, or mitigate dysfunction, and continuously improve their game.

- **Remove impediments; don't manage them** Remove your own impediments rather than relying on the Scrum Master to do it for you.

- **Estimate as a team** Help the Product Owner keep the Product Backlog refined, including estimating the size of upcoming PBIs. Estimates will organically improve over time.

- **Assess progress**   Use burndown charts, work item counts, passing tests, flow metrics, or other practices to assess your team's progress toward a goal.

- **Renegotiate scope when needed**   This can happen, and when it does, collaborate with the Product Owner to accommodate the change. Constant changing of scope is a dysfunction.

- **Leave canceling a Sprint up to the Product Owner**   Only the Product Owner can cancel a Sprint. It is a traumatic event and should be avoided if possible.

- **Ensure every PBI is Done before being released**   Every PBI should be completed according to the Definition of Done. Undone work should not be released.

- **Avoid undone work**   Unfinished PBIs cannot be released and shouldn't be inspected during Sprint Review. Instead, they should be put back on the Product Backlog for refinement and consideration to be developed in a future Sprint.

- **Quarantine undone work with feature flags**   Feature flags are a technique where functionality can be selectively excluded, disabled, or enabled from released software.

- **Use spikes to increase knowledge**   Spikes are experiments performed by the Developers to learn and prove feasibility. Consider representing larger spikes in the Product Backlog.

- **Scrum supports fixed-price contracts**   Scrum works as well as any other process when it comes to fixed-price contracts. It works better when a level of trust and sharing of risk exist between the Scrum Team and the customer.

- **Measure performance at a team level**   Measuring performance should be done at a team level, not an individual level.

- **Inspect and adapt**   Take advantage of the built-in Scrum events to ask yourself and your team how are you are doing with the product, as well as the process. Be sure to act on any of any findings that require it.

- **Don't change Scrum**   The framework is already barely sufficient, allowing any number of complementary practices to be implemented. Changing Scrum is usually done to hide underlying dysfunctions.

- **Scrum is like chess**   You either play it as its rules state, or you don't. Scrum and chess do not fail or succeed. They are either played by the rules or not.

CHAPTER 11

Scaled Professional Scrum

As you've learned by now, Scrum is a simple framework for delivering complex products such as software by using an empirical approach in which a team delivers value in small increments, inspects the results, and adapts their approach as needed based on feedback. The Scrum framework consists of a simple set of events, roles, and artifacts, bound together by practices and brought to life by the Scrum Values.

That's fine for a Scrum Team of 10 or fewer members, but what about those larger and more complex products that require more than 10 Developers? More than 50? Any situation where there is a single product, a single Product Backlog, and a single Product Owner but multiple teams of Developers working on the product is known as *scaling*. In other words, scaling occurs whenever there is an opportunity for increased chaos caused by inter-team dependencies as they all work on a common product.

Over the past decade, our industry has seen a flurry of conversations about the "problem" of scaling development. Many organizations have tried to scale and failed. Many books and conference presentations have dived deep. Frameworks of all sorts have been introduced. What is evident is that mindless adoption of agile values and principles won't scale to multiple teams—no matter how hard an organization tries or how much they spend. In other words, flaccid/zombie/mechanical Scrum won't scale. An organization must *nail it* before they *scale it*!

When the signers of the Agile Manifesto got together in 2001, they shared ideas about software development. Their discussion resulted in the Agile Manifesto. Although they may not have foreseen the tumult of scaling agile development—or the lucrative consulting opportunities—they knew that values and principles scale but that practices are context sensitive. Although it is hard to continuously improve, it is the *only way* to scale. Organizations cannot buy their way into successful scaling. They must keep the values and principles that single-team Scrum introduced while continuing to inspect and adapt as they add additional teams to product development.

This chapter covers scaling and how to control the chaos as multiple teams collaborate to build a single product. This chapter also introduces the *Nexus* Scaled Scrum framework as well as how Azure DevOps can support a Nexus to plan and manage work at scale.

> **Note** In this chapter, I am using the customized Professional Scrum process, not the out-of-the-box Scrum process. Please refer to Chapter 3, "Azure Boards," for information on this custom process and how to create it for yourself.

The Nexus Framework

As Scrum has increased in popularity, so has the need for Scrum Teams to work together in delivering products. When organizations want multiple Scrum Teams to work on a product, they often find that the Scrum framework alone is not enough. When multiple teams work on a product, the productivity they enjoy from using single-team Scrum may erode as additional teams share in the work. This is primarily due to the dependencies between the teams. That is because, to scale successfully, teams should not be left as separate units. In fact, the teams themselves must integrate into a connected system and identify as a common group, which is what is referred to as a *Nexus*. A Nexus is a relationship or connection between people or things. Nexus is Scrum at scale.

> **Note** Disclaimer: I am biased toward the Nexus framework. I helped create it along with Ken Schwaber. In addition, we created Scaled Professional Scrum, which is Scrum.org's instantiation of the Nexus framework. You can learn more about the Nexus and Scaled Professional Scrum at *www.scrum.org/nexus*.

The Scrum framework is the foundation for Nexus and, as you can see in Figure 11-1, looks very similar to the Scrum framework. The Nexus framework retains the simplicity of its origins, adding additional roles, events, artifacts, and rules only where necessary to enable successful large product development initiatives. Schwaber describes the Nexus as an "exoskeleton" to Scrum.

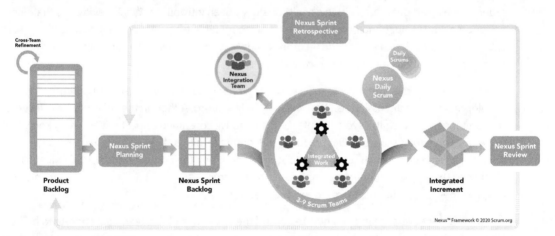

FIGURE 11-1 The Nexus framework.

> **Note** The Nexus contains 3–9 Scrum Teams but only one Product Owner. This is possible because on each of those Scrum Teams, the Product Owner is a *role*, not a separate person. All of those roles are played by one person (for example, Paula).

As you can see, the Scrum framework is still intact within the Nexus framework. The same events, artifacts, and roles exist. You might have also noticed that the Nexus framework adds some additional elements. Refinement becomes a proper event. Sprint Planning, Daily Scrum, Sprint Review, and Sprint Retrospective all have complementary events—identifiable by the word "Nexus" in their name. A Nexus Sprint Backlog, Nexus Sprint Goal, and Integrated Increment have also been added, along with a new role called the Nexus Integration Team. All new Nexus elements are listed in Table 11-1.

TABLE 11-1 The Nexus framework adds additional elements to Scrum.

Additional role	Additional events	Additional artifacts
Nexus Integration Team	Refinement	Nexus Sprint Backlog
	Nexus Sprint Planning	Nexus Sprint Goal
	Nexus Daily Scrum	Integrated Increment
	Nexus Sprint Review	
	Nexus Sprint Retrospective	

The Nexus is about identifying and mitigating dependencies. But what are dependencies? You might imagine a software or component dependency (for example, service A depends on component B). That is just one kind of dependency. Dependencies can also be people, domain, and technology in nature. For example, different Developers on the same or different teams can have different skill sets, different domain experiences and levels, different platforms and tools, and different permissions and access levels across people and teams.

It's especially important to identify cross-team dependencies. These are dependencies where two or more teams within the Nexus must coordinate their work in order to deliver Done functionality (for example, team A handles appointment scheduling and team B handles notifications). Hopefully, over time, each team can become a truly cross-product feature team, alleviating most dependencies.

External dependencies are the riskiest—just the same as for single-team Scrum. For example, a Scrum Team that is required to perform user acceptance testing (UAT) won't stand a chance when trying to scale. External dependencies like these must be solved prior to scaling.

Nexus Process Flow

A Nexus consists of multiple cross-functional Scrum Teams working together to deliver a releasable Integrated Increment by at least the end of each Sprint. Based on dependencies, the teams self-manage and select the most appropriate members to do specific work.

Here is a high-level description of the Nexus process flow:

- **Product Backlog Refinement** The Product Backlog must be decomposed so that dependencies are identified and removed or minimized. PBIs are refined into thinly sliced pieces of functionality and the team likely to do the work is identified.

- **Nexus Sprint Planning** Appropriate representatives from each Scrum Team meet to discuss and review the refined Product Backlog. They select PBIs for each team. Each Scrum Team then plans its own Sprint, interacting with other teams as appropriate. The outcome is a set of Sprint Goals that aligns with the overarching Nexus Sprint Goal, each Scrum Team's Sprint Backlog, and a single Nexus Sprint Backlog. The Nexus Sprint Backlog makes the work of all the Scrum Teams' selected PBIs and any dependencies transparent.

- **Development** All teams frequently integrate their work into a common environment that can be tested to ensure that the integration is done. All teams work from a common Definition of Done.

- **Nexus Daily Scrum** Developers representing each Scrum Team meet daily to identify any current integration issues. This information is transferred back to each Scrum Team's Daily Scrum. Scrum Teams use their Daily Scrum to create a plan for the day, making sure to address any integration issues raised during the Nexus Daily Scrum.

- **Nexus Sprint Review** The Nexus Sprint Review is held at the end of the Sprint for stakeholders to provide feedback on the Integrated Increment that the Nexus has built over the Sprint. All Scrum Teams meet with stakeholders to review the Integrated Increment. Adjustments are made to the Product Backlog.

- **Nexus Sprint Retrospective** Appropriate representatives from each Scrum Team meet to identify shared challenges. Each Scrum Team then holds individual Sprint Retrospectives. Appropriate representatives from each team meet again to discuss any actions needed based on shared challenges. This provides bottom-up intelligence to the entire Nexus.

Nexus Integration Team

The Nexus Integration Team is a new role. It is a Scrum Team within the Nexus that is accountable for ensuring that a Done, Integrated Increment is produced at least once every Sprint. Whereas the individual Scrum Teams are responsible for doing the actual building and delivering of a Done, Integrated Increment, the Nexus Integration Team remains *accountable*. This means that if the individual Scrum Teams can't or won't solve integration issues, the Nexus Integration Team must jump in and do that work. The Nexus Integration Team provides transparent accountability and a focal point for Nexus integration. Integration includes resolving any technical or nontechnical, cross-team constraints that may impede a Nexus's ability to deliver a constantly Integrated Increment.

The Product Owner is part of the Nexus Integration Team. There must also be a Scrum Master and enough Developers to help with the types of dependency and integration issues that the Nexus might face that Sprint. The Scrum Master in the Nexus Integration Team has overall responsibility for ensuring the Nexus framework is understood and enacted. This Scrum Master may also be a Scrum Master in one or more of the Scrum Teams in that Nexus. Nexus Integration Team composition may change between Sprints—ideally during Sprint Retrospective.

The Developers in the Nexus Integration Team typically consist of professionals who are skilled in the use of tools and relevant engineering practices (for example, DevOps). Nexus Integration Team members ensure that the Scrum Teams within the Nexus understand and implement the relevant tools and practices necessary to detect and mitigate dependencies. These teams must also frequently integrate all artifacts according to the Definition of Done.

Nexus Integration Team members are responsible for facilitating, coaching, and guiding the Scrum Teams in a Nexus to acquire, implement, and learn these practices and tools. Only in times of emergency—when the Nexus grinds to a halt and the individual Scrum Teams can't do the work—will the Nexus Integration Team members do that work. Additionally, the Nexus Integration Team coaches the individual Scrum Teams on the necessary development, infrastructural, or architectural practices and standards required by the organization to ensure the development of quality Integrated Increments.

> **Note** It's important to note that the Nexus Integration Team is not a separate team with separate individuals. It is a team of individuals who are already working in the Nexus. The Developers come from other Scrum Teams in the Nexus. They leave that team temporarily when Nexus Integration Team work demands their attention.

Membership in the Nexus Integration Team must take precedence over Scrum Team commitments. For example, a Developer from team #3 who is also on the Nexus Integration Team must drop what they are doing on team #3 and work on the Nexus Integration Team when the Nexus experiences a demanding issue that the teams can't solve for themselves.

Nexus Events

The Nexus framework defines additional events to complement their respective single-team Scrum events. Sprint Planning, Daily Scrum, Sprint Review, and Sprint Retrospectives are augmented at scale to provide opportunities for cross-team collaboration and bottom-up intelligence. The duration of each Nexus event is guided by the length of the corresponding events from single-team Scrum as described in the *Scrum Guide*. The Nexus events are timeboxed in addition to their corresponding Scrum events. The duration of those timeboxes is up to the Nexus to decide.

Refinement

In single-team Scrum, Product Backlog refinement is the act of breaking down and further defining PBIs into smaller more precise items. This is an ongoing activity to add details, such as a description, order, and size. In single-team Scrum, Product Backlog refinement is not an event. It is not even required. The Scrum Team decides if, when, and where this practice will occur. As the Product Backlog goes through seasons of change, more or less refinement will be necessary.

In the Nexus, Product Backlog Refinement is now an event and it is required. The *when* and *where* are up to the Nexus. Refinement of the Product Backlog at scale serves a dual purpose. It identifies dependencies across the teams, and it helps those teams forecast which will deliver which PBIs. This transparency allows the teams to monitor and minimize dependencies. Refinement of PBIs by the Nexus continues until the PBIs are sufficiently independent to be worked on by a single Scrum Team without excessive coordination.

The number, frequency, duration, and attendance of the Refinement event are based on the dependencies and uncertainty inherent in the Product Backlog. PBIs pass through different levels of decomposition from very large and vague requests (for example, epics or features) to more actionable-sized work that a single Scrum Team could deliver inside a Sprint (for example, user stories). Cross-team Refinement should be continuous throughout the Sprint as necessary and appropriate as the Product Owner decides which larger PBIs might be of interest in approaching Sprints and which teams should develop which PBIs. Individual team Product Backlog refinement also occurs continuously in order for each team's PBIs to be ready for selection in Nexus Sprint Planning.

Nexus Sprint Planning

The purpose of Nexus Sprint Planning is to coordinate the high-level activities of all Scrum Teams in a Nexus for a single Sprint. The Product Backlog should be adequately refined with dependencies identified and removed, or at least minimized, before Nexus Sprint Planning. The Product Owner provides domain knowledge and guides selection and priority decisions.

During Nexus Sprint Planning, appropriate representatives from each Scrum Team validate and make adjustments to the ordering of the work as created during prior Refinement events. All members of the Scrum Teams should participate in Nexus Sprint Planning to minimize communication issues.

> **Note** Many of the Nexus events require the "appropriate representatives" to attend. This means that the Nexus should decide which representatives from each Scrum Team should attend. Each team should have at least one representative, but some teams might send more than one. The representatives can rotate each meeting or always be the same. Let the Nexus—and the Scrum Teams within the Nexus—make those decisions.

The Product Owner discusses the Nexus Sprint Goal during Nexus Sprint Planning. The Nexus Sprint Goal describes the purpose or expected outcome that will be achieved by the Scrum Teams during the Sprint. After the overall work for the Nexus is understood, Nexus Sprint Planning continues, with each Scrum Team performing their own individual Sprint Planning. The Scrum Teams should continue to share newly discovered dependencies with other Scrum Teams in the Nexus. Each Scrum Team then develops their own Sprint Backlog—interacting with other teams as appropriate. Nexus Sprint Planning is complete when every Scrum Team has finished their individual Sprint Planning events.

New dependencies may emerge during Nexus Sprint Planning. They should be made transparent and minimized. The sequence of work across teams may also be adjusted. An adequately refined Product Backlog will minimize the emergence of new dependencies during Nexus Sprint Planning. The outcome of Nexus Sprint Planning is a set of Sprint Goals that aligns with the overarching Nexus Sprint Goal, each Scrum Team's Sprint Backlog, and a single Nexus Sprint Backlog. The Nexus Sprint Backlog makes the work of all the Scrum Teams' selected PBIs and any dependencies transparent.

Nexus Daily Scrum

The Nexus Daily Scrum, much like the Daily Scrum, is an event for Developer representatives from the individual Scrum Teams. They will meet to inspect the current state of the Integrated Increment and to identify integration issues or newly discovered cross-team dependencies or cross-team impacts.

During the Nexus Daily Scrum, attendees focus on each team's impact on the Integrated Increment. They discuss whether the previous day's work successfully integrated and what new dependencies or impacts have been identified. The output of this event is a shared understanding of the information that must be shared across teams in the Nexus.

The Developer representatives use the Nexus Daily Scrum to inspect progress toward the Nexus Sprint Goal. At least during every Nexus Daily Scrum, the Nexus Sprint Backlog should be updated to reflect the current understanding of the work of the Scrum Teams within the Nexus. The representatives take these observations, issues, and work that were identified during the Nexus Daily Scrum back to their individual Scrum Teams for planning inside their respective Daily Scrums.

Nexus Sprint Review

The Nexus Sprint Review occurs at the end of the Sprint to provide feedback on the Integrated Increment that the Nexus has built during that Sprint and to adapt the Product Backlog as needed. A Nexus Sprint Review replaces the need for individual Scrum Team Sprint Reviews, because the entire Integrated Increment is typically the focus for capturing feedback from stakeholders. The goal of any Sprint Review is to obtain stakeholder feedback. This feedback will result in a revised Product Backlog.

Because of the amount of Done functionality, it may not be possible to inspect all completed work in detail at the Nexus Sprint Review. Techniques may be necessary to maximize stakeholder feedback—not to mention to keep them awake. One such technique, or practice, is known as a *science fair* (also known as "expo" or "bazaar"). More like an informal gathering or meetup than a formal review, these practices can help improve the Nexus Sprint Review by letting stakeholders see what they want/must see, on demand. For more details on running a science fair Sprint Review, read fellow Professional Scrum Trainer Erik Weber's blog post at *www.scrum.org/resources/blog/sprint-review-technique-science-fair*.

An *offline* Sprint Review may also be an option for a large product with numerous stakeholders—especially if they are primarily remote. This practice enables teams to better engage with stakeholders when those stakeholders cannot attend the Sprint Review. You can create short video demonstrations of the Done functionality and share them with your stakeholders to prompt their feedback or entice them to join the Sprint Review.

Nexus Sprint Retrospective

The Nexus Sprint Retrospective is a formal opportunity for a Nexus to inspect and adapt itself and create a plan for improvement. This is key to ensuring continuous improvement. Identical to single-team Scrum, the Nexus Sprint Retrospective occurs after the Nexus Sprint Review and prior to the next Nexus Sprint Planning. The Nexus Sprint Retrospective is an opportunity for improvement for each team as well as the Nexus as a whole.

A Nexus Sprint Retrospective consists of three parts:

- **Nexus pre-meeting** Appropriate representatives from the Scrum Teams meet and identify issues that have impacted more than a single team. The purpose is to make shared issues transparent to all Scrum Teams.

- **Individual Scrum Team Sprint Retrospectives** Each Scrum Team holds their own, individual Sprint Retrospective—per the *Scrum Guide.* They can use issues raised at the pre-meeting as input to their team discussions. The individual Scrum Teams should plan experiments to address any issues.

- **Nexus post-meeting** Appropriate representatives from the Scrum Teams meet again and agree on how to visualize and track the identified improvements. This allows the Nexus as a whole to adapt.

Because they are common scaling dysfunctions, every Nexus Sprint Retrospective should address the following subjects:

- Was any work left undone? Did the Nexus generate technical debt? If so, why did this happen? How can it be corrected? How can it be prevented?

- Were all artifacts, particularly code, frequently (as often as every day) successfully integrated? If not, why not?

- Was the product successfully built, tested, and deployed often enough to prevent the accumulation of unresolved dependencies? If not, why not?

A retrospective board or other visualization may be helpful to plan and track inspections and adaptations across teams. Using a wiki may not be enough. Having an actionable board may be beneficial. You might want to experiment with the *Retrospectives* extension found in the Azure DevOps Marketplace. You can see an example of it in Figure 11-2 and learn more about it at *https://marketplace.visualstudio.com/items?itemName=ms-devlabs.team-retrospectives.*

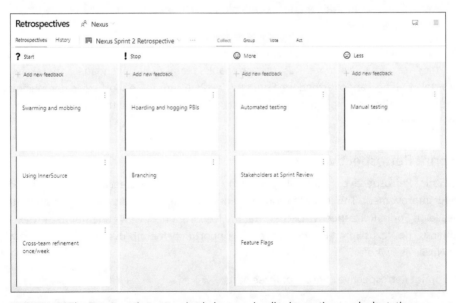

FIGURE 11-2 The Retrospectives extension helps you visualize inspections and adaptations.

Nexus Artifacts

The Nexus framework introduces additional artifacts to complement the existing artifacts described in the *Scrum Guide*. These new artifacts provide additional transparency and opportunities for inspection and adaptation at scale.

It's important to remind you again that there is a single Product Backlog for the entire Nexus. The Product Owner is accountable for the Product Backlog, including its content, availability, and ordering. At scale, the Product Backlog must be understood at a level where dependencies can be detected and minimized. PBIs are often split and refined to a granularity of thinly sliced functionality. PBIs are deemed "ready" for Nexus Sprint Planning when the Scrum Teams can select items to be done with no or minimal coordination with other Scrum Teams.

Nexus Sprint Backlog

In single-team Scrum, the Sprint Backlog contains the Sprint Goal, the forecasted PBIs, and the plan for delivering them. At scale, each team still has their own Sprint Backlog, but the Nexus also keeps and maintains a Nexus Sprint Backlog. The Nexus Sprint Backlog is a composite view of all the forecasted PBIs from the individual Sprint Backlogs of the individual Scrum Teams. It is used to highlight dependencies and the flow of work during the Sprint. It is updated at least daily, often as part of the Nexus Daily Scrum.

The Nexus Sprint Backlog doesn't show the individual teams' *plans*—such as tasks or test cases—only their PBIs. Too much detail would distract from its purpose—to make dependencies transparent so the Nexus can mitigate them. Any visualization of the Nexus Sprint Backlog should show all forecasted PBIs by team and state (for example, to-do, in progress, and done). It should also show dependencies. A manual board is great, but somebody will have to keep it updated.

Unfortunately, Azure Boards doesn't offer a great board or visualization for the Nexus Sprint Backlog; however, with a bit of process customization, an Azure extension, and a little effort, you can come close. Figure 11-3 shows the results of the following efforts:

- Further customize the Professional Scrum process, adding a *WIP* state to the PBI work item type. The new state maps to the *In-Progress* state category.

- Install the Query Based Boards extension. You can learn more about this extension here: *https://marketplace.visualstudio.com/items?itemName=realdolmen.EdTro-AzureDevOps-Extensions-QueryBasedBoards-Public*.

- Adopt a PBI naming convention prefixing the title with the team name or initials.

- Create a "Nexus Sprint Backlog" query that returns all PBIs for the current Sprint.

- Set predecessor/successor links accordingly to represent cross-PBI dependencies. These dependencies are typically only cross-team. Cross-PBI dependencies, within the same team, can be managed through that team's natural execution sequencing—the same as with single-team Scrum.

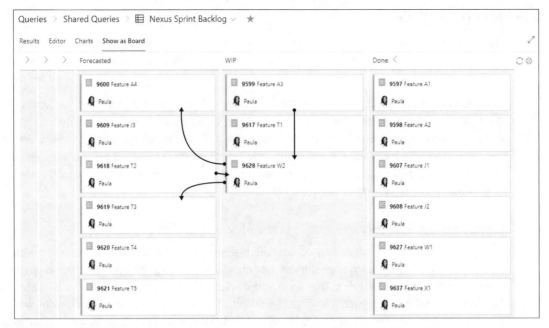

FIGURE 11-3 The Query Based Boards extension can display an actionable Nexus Sprint Backlog.

Nexus Sprint Goal

The Nexus Sprint Goal is an objective for the Sprint. It captures the essence of all the work that the Scrum Teams within the Nexus will perform. Stakeholders will inspect the functionality that the Nexus developed to achieve the Nexus Sprint Goal at the Nexus Sprint Review in order to provide feedback. At the Nexus Daily Scrum, the representatives will inspect progress toward the Nexus Sprint Goal, adjusting their Sprint Backlog and the Nexus Sprint Backlog to reflect their current understanding of the work of the Scrum Teams within the Nexus.

Much like the Sprint Goal for single-team Scrum, the Nexus Sprint Goal gives the Nexus flexibility in developing the scope of what's been forecasted. That said, the Nexus Sprint Goal is different from a Sprint Goal, because it crosses teams. In that regard, it is probably closer to the single-team Scrum concept of a release goal or a Product Goal.

> **Tip** If it is hard to create a Nexus Sprint Goal, then maybe all that is shared amongst the teams is a codebase? Maybe the teams are not all building a cohesive product. Maybe they shouldn't be in a Nexus.

The Nexus Sprint Goal is met through the implementation of PBIs by multiple teams. As the Nexus works, it keeps the Nexus Sprint Goal in mind. It gives an individual within a Nexus meaning, purpose, and focus for their specific tasks. When a Nexus has more than 80 Developers, it's easy for each of them to get lost in the noise. They should keep focused on the big picture. The Nexus Sprint Goal is that big picture.

Integrated Increment

In single-team Scrum, an Increment is a concrete stepping-stone toward the Product Goal. Each Increment is additive to all prior Increments and thoroughly verified, ensuring that all Increments work together. In order to provide value, the Increment must be usable. Work cannot be considered part of an Increment unless it meets the Definition of Done.

At scale, there is still the concept of an Increment, but it contains the sum of all integrated work completed by a Nexus. An Integrated Increment must be usable and releasable, which means it must meet the Nexus Definition of Done. The Integrated Increment may itself be integrated into another, larger, working Increment.

> **Note** The Nexus Integration Team is accountable for ensuring that a Done Integrated Increment (the combined work completed by a Nexus) is produced at least once every Sprint.

Definition of Done

A Nexus is based on transparency. All Scrum Teams as well as the Nexus Integration Team work within a Nexus and the organization to ensure that transparency is apparent across all artifacts. They also make the state (for example, integrated or not integrated) of the Integrated Increment transparent. All Scrum Teams in the Nexus should be made aware of integration problems.

Decisions made based on the state of Nexus artifacts are only as effective as the level of their transparency. Incomplete or partial information will lead to incorrect or flawed decisions. The impact of those decisions can be magnified at the scale of Nexus. The product must be developed in a way that dependencies are detected and resolved before technical debt becomes unacceptable. A lack of complete transparency will make it impossible to guide a Nexus effectively to minimize risk and maximize value.

The Definition of Done is applied to the Integrated Increment developed each Sprint. It defines what must be true for the Increment to be integrated, usable, and releasable by the Product Owner. All Scrum Teams of a Nexus adhere to this Definition of Done. Individual Scrum Teams may choose to apply a more stringent Definition of Done within their own teams but cannot apply less rigorous criteria. The Integrated Increment can be considered Done only when the work from each team correctly integrates with the work from all other teams.

> **Note** The Nexus Integration Team is responsible for a Definition of Done that can be applied to the Integrated Increment developed each Sprint. This Definition may be based on existing Definitions of Done, product constraints, and other conventions of the development organization.

As the Nexus improves, it is expected that its Definition of Done will expand to include more stringent criteria for higher quality. New definitions, as used, may uncover work to be done in previously Done Increments. The Nexus should inspect that work and adapt a plan for making it Done per the current definition.

Nexus Support in Azure DevOps

The bad news is that there is no first-class support for Nexus in Azure DevOps. The good news is that Nexus is just Scrum, so everything I've presented in the book thus far still applies. Only a couple of tweaks are required in order to support multiple teams working on a single Product Backlog. I will call them out in this section.

Configuring Additional Teams

When you create a new Azure DevOps project, a single, default team is created. It has the same name as the project. For example, when I created the Fabrikam project, a "Fabrikam Team" was created. For single-team Scrum, just add the Product Owner, Scrum Master, and Developers to that team and you are good to go. Backlogs, boards, and dashboards are automatically configured for this default team so that the team members can start managing, planning, and executing work.

> **Tip** Azure DevOps allows you to rename the default team. I recommend renaming it to the *Nexus Integration Team*. Since the Product Owner is a permanent member of the Nexus Integration Team, they will have visibility to all PBIs in the Product Backlog.

Because the Nexus requires a minimum of three Scrum Teams, you'll be adding teams. Select a short, unique name for each team. You'll also must select the team members and identify team administrators. I'm a fan of making team members Project Administrators as well—so that Azure DevOps doesn't block the team. The default behavior is to create an area with the same name as the team. (More on that in a moment.) After creating the teams, you can further socialize the teams by selecting a team logo or avatar. Figure 11-4 shows the Scrum Teams for a medium-sized Nexus along with their respective team images.

Each team needs at least one team administrator to be able to configure, customize, and manage the team-related activities. These activities include adding team members and other team administrators, as well as configuring agile tools and team assets. By default, the person who created the team is a team administrator. Naturally, I'm a fan of making *all* team members a team administrator so that nobody gets blocked by the tool.

FIGURE 11-4 The Teams page lists all the teams in the Nexus.

> **Note** Don't confuse *team administrator* with *project administrator*. Although both act as a permission container, the team administrator is more focused on managing the agile tools and assets for a team, rather than the project as a whole. Ideally, everyone on the Scrum Team is both a project administrator and a team administrator. To learn more about the team administrator, visit *https://aka.ms/add-team-administrator*.

Each team gets access to its own suite of agile tools and team assets. These tools, which are listed in Table 11-2, provide teams with the ability to work autonomously while still collaborating with other teams across the Nexus. Each team can configure and customize their own tools to support how they work. These tools reference the team's *default area*, *iteration*, and *selected Sprints* to filter automatically the set of work items they display. I'll discuss those settings in a moment.

TABLE 11-2 Team-specific tools and assets.

Planning and tracking	Collaborating	Monitoring and learning
Product Backlog	Team alerts	Analytics by team
Forecasting	Team favorites	Velocity
Epic and feature backlog	Team group filter	Burndown/burnup
Kanban board	Dashboards	Cumulative flow diagram
Sprint Backlog		
Taskboard		

After creating a new team, the following settings should be configured:

- **Membership** Add the Product Owner, Scrum Master, Developers, and any stakeholders to the team. Be sure to give the Scrum Team members project administrator permissions and also make them team administrators. Consider reader (read-only) permission for any stakeholders, unless you trust them making changes to the Product Backlog.

- **Team image** Select an awesome avatar or representation for the team. This will up the social experience within Azure DevOps.

- **Notifications** Configure any team-level notifications for work items, code (Azure Repos), builds/releases (Azure Pipelines), and artifacts (Azure Artifacts).

- **Dashboards** Create a team dashboard with widgets that radiate the information that the team desires.

- **Backlogs** Select the backlogs that the team will use (Epics, Features, etc.).

- **Working days** Configure which days of the week (for example, Monday through Friday) that the team will be working.

- **Default iteration** This iteration is automatically assigned whenever a work item is created from the team context.

- **Backlog iteration** This iteration determines which items appear on the team's backlogs and boards.

- **Selected iterations** The iterations selected will appear in the Planning pane in the Backlogs page as well as on the Sprints page (Sprint Backlog and Taskboard).

- **Default area** The area is automatically assigned whenever a work item is created. The default area must be one of the areas selected for the team.

- **Selected areas** The product areas that this team will develop/look after. The selected areas will determine which items show up on a team's backlog.

Configuring Areas

Most of the tools in Azure Boards operate using system queries that reference the team's area settings. For example, a team's *default area* filters the work items that appear on that team's backlog. Also, work items that are created using Azure Boards are auto-assigned the areas and iterations based on team defaults. Teams can be associated with one or more areas and include the sub-areas as well.

When introducing the Nexus, the area hierarchy must be updated, to introduce the team areas into the hierarchy. By default, these team areas are created automatically when the teams were created. You can also create new areas manually to represent teams. After the team areas are created, you will want to move existing areas under those teams to establish a hierarchical structure. Drag and drop is your friend. You can see an area hierarchy before (pre-Nexus) and after introducing additional teams in Figure 11-5.

BEFORE
Areas
Fabrikam
Customers
Dashboard
Alerts
Messages
Tips
Employees
Mobile
Security
Profile
Reports
Social
Tickets
Assignment
Escalation
AFTER
Areas
Teams
Fabrikam
Avengers
Avengers
Dashboard
Alerts
Messages
Tips
Justice League
Justice League
Customers
Employees
Titans
Titans
Mobile
Security
Watchmen
Watchmen
Profile
Reports
Social
X-Men
X-Men
Tickets
Assignment
Escalation

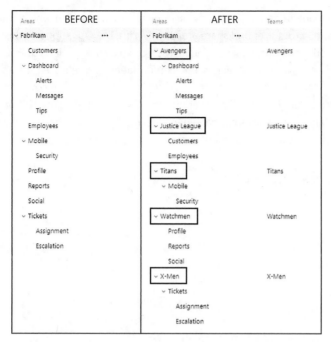

FIGURE 11-5 Adding teams to the area hierarchy to represent ownership of various product areas.

When I created the five teams, I let Azure DevOps create these placeholder areas. They are not required and may even be considered added noise and confusion for some users. For example, business stakeholders, with access to Azure Boards, may wonder about the "Justice League" part of the product. If you want, you can remove those team placeholder areas and just indicate which areas are selected by which team(s). As you can see in Figure 11-6, the list of areas is simpler and less confusing. You can also see that multiple teams are covering the Employees and Tickets areas. I'll talk about how to do that in a moment.

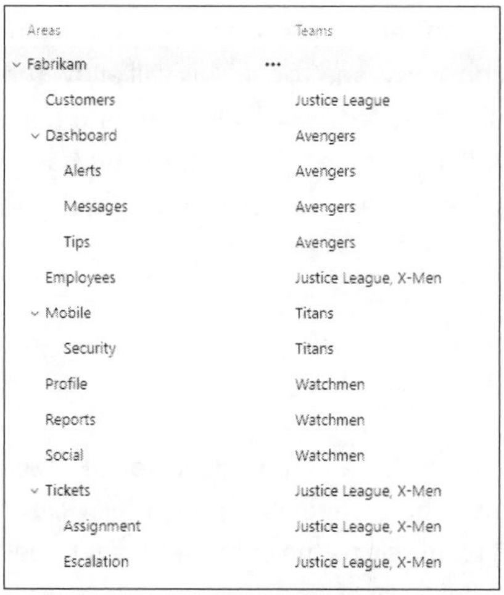

Areas
Teams
Fabrikam
Customers
Justice League
Dashboard
Avengers
Alerts
Avengers
Messages
Avengers
Tips
Avengers
Employees
Justice League, X-Men
Mobile
Titans
Security
Titans
Profile
Watchmen
Reports
Watchmen
Social
Watchmen
Tickets
Justice League, X-Men
Assignment
Justice League, X-Men
Escalation
Justice League, X-Men

FIGURE 11-6 Removing placeholder areas results in a cleaner hierarchy.

Whether or not you use pseudo team areas, it's pretty obvious which team looks after which areas. In practice, however, teams can and do cross over and share areas and Azure DevOps supports this as well. After the areas are configured at the project level, you just select which areas will be used by which team. This is done by going to each individual team's area configuration and selecting the applicable areas. Multiple areas may be selected, including sub-areas. Also, multiple teams can cover the same area—as was the case with the Employees and Tickets area in our last example. Figure 11-7 shows these two areas selected for the X-Men team. Sub-areas have been included for each area as well. The default area will be assigned to work items that are created from the team context. It must be one of the related areas.

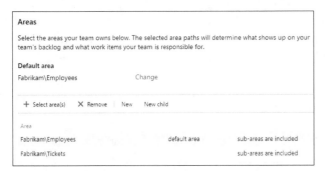

FIGURE 11-7 A team can look after multiple product areas.

> **Tip** If there's a chance that you'll add new areas under existing areas in the future, be sure to *include sub-areas* in order to future-proof the functionality of Azure Boards. If you don't and a new child area gets added with PBIs, they won't be visible on your Product Backlog!

> **Note** Initially, individual Scrum Teams may "specialize" in certain product areas. Over time, those teams will improve, learning new domains, technology, and tools. This will allow them to work in increasingly more product areas. Eventually those teams will blossom into feature teams—allowing them to deliver value across many/all areas of the product. As this happens, cross-team dependencies will drop. Someday a team may even be able to get rid of all the team areas and just set their default to the root!

Configuring Iterations

As previously mentioned, iterations support teams who want to plan and track their work. For Scrum Teams, they allow PBI, Task, Test Case, and other work items to be grouped by Sprint. Much like areas, a Nexus defines iterations at the project level and then each team selects the ones they want to be active.

Configuring Sprints with their start and end dates is the same for a Nexus as it is with a single team. I covered this in Chapter 4, "The Pre-game." With Azure DevOps and Nexus, it is possible to create different Sprint schedules (cadences) for different teams. For example, some Scrum Teams in a Nexus may have a 4-week Sprint whereas others have a 2-week Sprint. If this is the case, you simply define a node under the top project node for each team or cadence (for example, *Fabrikam/Titans* or *Fabrikam/2-Week Sprints*), and then define the Sprints under those nodes.

> **Tip** Even though Azure DevOps and Nexus allow teams to have different cadences—different Sprint start dates and lengths—you shouldn't do this. It will be confusing to know which team is on which Sprint. More importantly, how can the teams plan, execute, and deliver as a cohesive Nexus if the events don't align? Please weigh all factors before doing this!

After the iterations (Sprints) have been configured at the project level, the individual teams can set a default iteration, set a backlog iteration, and select their active iterations. The iterations selected will appear in the Planning pane in the Backlogs page as well as on the Sprints page (Sprint Backlog and Taskboard). Rather than seeing all Sprints, a team can keep the list of selected Sprints small, possibly only including the current Sprint and a few future Sprints to support their planning horizon.

The *default iteration* is automatically assigned whenever a work item is created from the team context. For example, if Sprint 3 is selected, then any work items created by this team will have its Iteration set to Sprint 3. The default is @CurrentIteration, which is a macro that always equates to a team's current Sprint. This macro is not supported on versions of Team Foundation Server older than 2015, or when used in a version of Visual Studio/Team Explorer older than 2015, or when used by Microsoft Excel.

> **Note** As of this writing, when the default iteration is set to the @CurrentIteration macro, PBIs added to the Product Backlog will be assigned to the current Sprint—which is confusing. Though helpful when adding Task work items to the Sprint Backlog or Taskboard, or Test Case work items to the Sprint's test plan, PBIs added to the Product Backlog should *always* be set to the iteration root (no Sprint) so that they can be refined and forecasted in a future Sprint Planning. Hopefully, Microsoft will correct this dysfunctional behavior.

Managing the Product Backlog

When viewing the Product Backlog, only PBIs in the areas selected for the team will be displayed. The default team (for example, the Nexus Integration Team) should be able to see all items in the Product Backlog. This is possible because that team's only selected area is the root and includes sub-areas. It's important that the Product Owner be able to have a view of the entire Product Backlog.

The other Scrum Teams will only see their PBIs in the Product Backlog. This filtered set depends on which areas are selected by that team. Scrum Teams new to the Nexus may be fairly specialized in their domain and technology capabilities and only be able to work in a few product areas. For example, as you can see in Figure 11-8, the Justice League team's backlog shows only a subset of the full Product Backlog—just those PBIs with an area of Customers, Employees, or Tickets. Scrum Teams who have improved to the point of being true feature teams may eventually see *all* PBIs in their backlog— because they are capable of working in any part of the product.

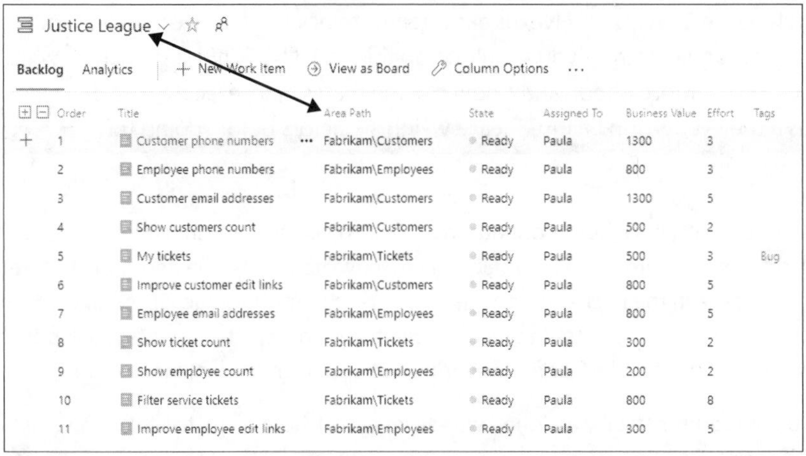

Order	Title	Area Path	State	Assigned To	Business Value	Effort	Tags
1	Customer phone numbers	Fabrikam\Customers	Ready	Paula	1300	3	
2	Employee phone numbers	Fabrikam\Employees	Ready	Paula	800	3	
3	Customer email addresses	Fabrikam\Customers	Ready	Paula	1300	5	
4	Show customers count	Fabrikam\Customers	Ready	Paula	500	2	
5	My tickets	Fabrikam\Tickets	Ready	Paula	500	3	Bug
6	Improve customer edit links	Fabrikam\Customers	Ready	Paula	800	5	
7	Employee email addresses	Fabrikam\Employees	Ready	Paula	800	5	
8	Show ticket count	Fabrikam\Tickets	Ready	Paula	300	2	
9	Show employee count	Fabrikam\Employees	Ready	Paula	200	2	
10	Filter service tickets	Fabrikam\Tickets	Ready	Paula	800	8	
11	Improve employee edit links	Fabrikam\Employees	Ready	Paula	300	5	

(Backlog | Analytics | + New Work Item | View as Board | Column Options | ... — Justice League)

FIGURE 11-8 Each team's view of the Product Backlog is filtered based on their selected areas.

Tip Can a user be a member of more than one Azure DevOps team? Yes. Should they be? Probably not, unless they are the Product Owner or the Scrum Master of those teams. This is not because it's difficult or confusing to switch back and forth between backlogs in Azure Boards, but more about the drop in focus when a Developer jumps from team to team. As I've said before, growing long-lived teams of full-time, dedicated team members is preferable.

Chapter Retrospective

Here are the key concepts I covered in this chapter:

- **Nexus** A relationship or connection between people or things.

- **Nexus framework** A framework, similar to Scrum, consisting of roles, events, artifacts, and rules that bind and weave together the work of approximately three to nine Scrum Teams working on a single Product Backlog to build an Integrated Increment that meets a goal. The Nexus framework can be considered an "exoskeleton" to the Scrum framework, allowing it to scale.

- **Scaled Professional Scrum** Scrum.org's instantiation of the Nexus framework. Scaled Professional Scrum includes more than 40 complementary practices for scaling Scrum. You can see that list here: *www.scrum.org/scaled-professional-scrum-nexus-practices*.

- **Nexus Integration Team** A new role that exists to coordinate, coach, and supervise the application of Nexus and the operation of Scrum so that the best outcomes are derived. The Nexus Integration Team is a Scrum Team and, thus, consists of the Product Owner, a Scrum Master, and three or more Developers.

- **Refinement** Similar to single-team Scrum but even more important at scale so that dependencies, especially those involving more than a single team, are identified and removed or minimized. Refinement is a required event in Nexus.

- **Nexus Sprint Planning** Similar to single-team Scrum but even more important at scale to coordinate the high-level activities of all Scrum Teams. Sprint Planning occurs as a whole as well as for each team within the Nexus Sprint Planning event.

- **Nexus Daily Scrum** A Daily Scrum pre-meeting in which appropriate Developer representatives from each Scrum Team inspect the current state of the Integrated Increment and identify integration issues or newly discovered cross-team dependencies or cross-team impacts. The Daily Scrum occurs for each team after the Nexus Daily Scrum.

- **Nexus Sprint Review** Similar to single-team Scrum, a review held at the end of the Sprint to allow stakeholders to provide feedback on the Integrated Increment and to adapt the Product Backlog if needed. The Nexus Sprint Review replaces the need for individual Scrum Team Sprint Reviews.

- **Science fair** A Sprint Review technique that enables many stakeholders to inspect just the Done functionality that they are interested in. The science fair is an informal gathering, much like a meetup, and is also known as an "expo" or a "bazaar."

- **Nexus Sprint Retrospective** A formal opportunity for a Nexus to inspect and adapt itself and create a plan for improvement to be enacted during the next Sprint to ensure continuous improvement. The Sprint Retrospective occurs for each team within the Nexus Sprint Retrospective.

- **Nexus Sprint Backlog** A composite view of the PBIs from the Sprint Backlogs of the individual Scrum Teams. The Nexus Sprint Backlog is used to highlight dependencies and the flow of work during the Sprint. It should be updated at least daily.

- **Nexus Sprint Goal** An objective set for the Sprint. It is the sum, aggregate, or theme of all the work and Sprint Goals of the Scrum Teams within the Nexus. The Nexus Sprint Goal should be shared with all members of all teams to be used as a guiding light during development.

- **Integrated Increment** Represents the current sum of all integrated work completed by a Nexus. The Integrated Increment must be usable and releasable, which means it must meet the Definition of Done. The Integrated Increment is inspected at the Nexus Sprint Review.

- **Definition of Done**   Similar to single-team Scrum but even more important at scale because it is applied to the Integrated Increment developed each Sprint. Individual Scrum Teams may choose to apply a more stringent Definition of Done within their own teams, but they cannot apply less rigorous criteria. All Scrum Teams of a Nexus adhere to this common Definition of Done.

- **Team Areas**   Determine which PBIs appear on the team's backlog and Kanban board. Teams can select one or more areas and an area can be selected by more than one team. The default team is where the Product Owner will primarily hang out and should have the root area selected, including sub-areas to see all PBIs in the Product Backlog. Some Nexus rename this default team to the Nexus Integration Team.

- **Feature teams**   A long-lived, cross-functional, cross-component team that can complete many/all end-to-end customer features. Feature teams will have many team areas selected—indicating that they can work in all of those product areas. High-performance feature teams may have one single area selected—the root, including all sub-areas.

Index

Q-R

Plug into learning at

MicrosoftPressStore.com

The Microsoft Press Store by Pearson offers:

- Free U.S. shipping

- Buy an eBook, get three formats – Includes PDF, EPUB, and MOBI to use with your computer, tablet, and mobile devices

- Print & eBook Best Value Packs

- eBook Deal of the Week – Save up to 50% on featured title

- Newsletter – Be the first to hear about new releases, announcements, special offers, and more

- Register your book – Find companion files, errata, and product updates, plus receive a special coupon* to save on your next purchase